# Lecture Notes in Computer Science          8943

Commenced Publication in 1973
Founding and Former Series Editors:
Gerhard Goos, Juris Hartmanis, and Jan van Leeuwen

More information about this series at http://www.springer.com/series/7409

Thepchai Supnithi · Takahira Yamaguchi · Jeff Z. Pan
Vilas Wuwongse · Marut Buranarach (Eds.)

# Semantic Technology

4th Joint International Conference, JIST 2014
Chiang Mai, Thailand, November 9–11, 2014
Revised Selected Papers

 Springer

*Editors*

Thepchai Supnithi
National Electronics and Computer
    Technology Center
Pathum Thani
Thailand

Takahira Yamaguchi
Keio University Fac. of Science &
    Technology
Yokohama
Japan

Jeff Z. Pan
Department of Computing Science
University of Aberdeen
Aberdeen
United Kingdom

Vilas Wuwongse
Asian University
Chonburi
Thailand

Marut Buranarach
National Electronics Computer Tech. Ctr.,
    Sci. Park
Pathum Thani
Thailand

ISSN 0302-9743
Lecture Notes in Computer Science
ISBN 978-3-319-15614-9
DOI 10.1007/978-3-319-15615-6

ISSN 1611-3349    (electronic)

ISBN 978-3-319-15615-6    (eBook)

Library of Congress Control Number: 2015932667

LNCS Sublibrary: – SL 3 Information Systems and Applications, incl. Internet/Web and HCI

Springer Cham Heidelberg New York Dordrecht London

Printed on acid-free paper

Springer International Publishing AG Switzerland is part of Springer Science+Business Media
(www.springer.com)

# Preface

This volume contains the papers presented at JIST 2014: The 4th Joint International Semantic Technology Conference held during November 9–11, 2014 in Chiang Mai, Thailand. JIST 2014 was co-hosted by National Electronics and Computer Technology Center (NECTEC), Chiang Mai University, Sirindhorn International Institute of Technology (SIIT), Thailand, and Korea Institute of Science and Technology Information (KISTI), Korea. JIST is one of the leading conferences in its field in the Asia region with high participation from Japan, Korea, and China. It often attracts many participants from Europe and USA. The main topics of the conference include Semantic Technologies such as Semantic Web, Social Web, Linked Data, and so on.

The theme of the JIST 2014 conference was "Open Data and Semantic Technology." JIST 2014 conference consisted of main technical tracks including regular paper track (full and short papers), in-use track and special track, poster and demo session, two workshops, and four tutorials. There were a total of 71 submissions for the regular paper track from 19 countries, which included Asian, European, African, North American, and South American countries.

Most papers were reviewed by three reviewers and the results were rigorously discussed by the program chairs. Twenty full papers (29%) and 12 short, in-use track and special track papers (17%) were accepted. The paper topics are divided into eight categories: Ontology and Reasoning (7 papers), Linked Data (7 papers), Learning and Discovery (3 papers), RDF and SPARQL (3 papers), Ontological Engineering (3 papers), Semantic Social Web (3 papers), Search and Querying (3 papers), and Applications of Semantic Technology (3 papers).

We would like to thank the JIST Steering Committee, Organizing Committee, and Program Committee for their significant contributions. We also would like to specially thank the co-hosts and sponsors for their support in making JIST 2014 a successful and memorable event. Finally, we would like to express our appreciation to all speakers and participants of JIST 2014. This book is an outcome of their contributions.

December 2014

Thepchai Supnithi
Takahira Yamaguchi
Jeff Z. Pan
Vilas Wuwongse
Marut Buranarach

# Organization

JIST 2014 was organized by National Electronics and Computer Technology Center (NECTEC), Sirindhorn International Institute of Technology (SIIT), Chiang Mai University, Thailand, and Korea Institute of Science and Technology Information (KISTI), Korea.

## General Chairs

Jeff Z. Pan                 University of Aberdeen, UK
Vilas Wuwongse              Asian University, Thailand

## Program Chairs

Thepchai Supnithi           NECTEC, Thailand
Takahira Yamaguchi          Keio University, Japan

## In-use Track Chairs

Chutiporn Anutariya         Asian University, Thailand
Alan Wu                     Oracle, USA

## Special Track Chairs

Pasi Gabriella              University of Milan, Italy
Kouji Kozaki                Osaka University, Japan

## Poster and Demo Chairs

Hanmin Jung                 KISTI, Korea
Ekawit Nantajeewarawat      SIIT, Thailand
Kultida Tuamsuk             Khon Kaen University, Thailand

## Workshop Chairs

Thatsanee Charoenporn       Burapha University, Thailand
Sasiporn Usanavasin         SIIT, Thailand

## Tutorial Chair

Boontawee Suntisrivaraporn  SIIT, Thailand

## Industrial Chairs

Key-Sun Choi                          KAIST, Korea
Takahiro Kawamura                     Toshiba, Japan
Virach Sornlertlamvanich              TPA, Thailand

## Publicity Chairs

Jeerayut Chaijaruwanich               Chiang Mai University, Thailand
Krich Intratip                        Royal Thai Army, Thailand

## Local Organizing Chairs

Marut Buranarach                      NECTEC, Thailand
Rattasit Sukhahuta                    Chiang Mai University, Thailand
Thanaruk Theeramunkong               SIIT, Thailand

## Local Organizing Committee

Prachya Boonkwan                      NECTEC, Thailand
Nopphadol Chalortham                  Chiang Mai University, Thailand
Varin Chouvatut                       Chiang Mai University, Thailand
Wattana Jindaluang                    Chiang Mai University, Thailand
Ubolwan Khamkaew                      NECTEC, Thailand
Krit Kosawat                          NECTEC, Thailand
Kanyanut Kriengket                    NECTEC, Thailand
Nattapol Kritsuthikul                 NECTEC, Thailand
Wasan Na Chai                         NECTEC, Thailand
Jeeravan Nuananun                     NECTEC, Thailand
Taneth Ruangrajitpakorn               NECTEC, Thailand
Kanchana Saengthongpattana            NECTEC, Thailand
Vipas Sutantayawalee                  NECTEC, Thailand
Ratsameetip Wita                      Chiang Mai University, Thailand

# Sponsors

### Platinum Sponsor

 Electronic Transactions Development Agency (Public Organization)

### Gold Sponsor

 Siam Commercial Bank Public Company Limited

### Bronze Sponsor

 Electronic Government Agency (Public Organization)

# Program Committee

| | |
|---|---|
| Chutiporn Anutariya | Asian University, Thailand |
| Paolo Bouquet | University of Trento, Italy |
| Marut Buranarach | NECTEC, Thailand |
| Nopphadol Chalortham | Chiang Mai University, Thailand |
| Chantana Chantrapornchai | Kasetsart University, Thailand |
| Huajun Chen | Zhejiang University, China |
| Gong Cheng | Nanjing University, China |
| Paola Di Maio | ISTCS.org and IIT Mandi, India |
| Stefan Dietze | L3S Research Center, Germany |
| Dejing Dou | University of Oregon, USA |
| Jae-Hong Eom | Seoul National University, Korea |
| Zhiqiang Gao | Southeast University, China |
| Volker Haarslev | Concordia University, Canada |
| Masahiro Hamasaki | AIST, Japan |
| Sungkook Han | Wonkwang University, Korea |
| Koiti Hasida | University of Tokyo, Japan |
| Itaru Hosomi | NEC Service Platforms Research Laboratories, Japan |
| Wei Hu | Nanjing University, China |

| Eero Hyvönen | Aalto University and University of Helsinki, Finland |
| Ryutaro Ichise | National Institute of Informatics, Japan |
| Krich Intratip | Royal Thai Army, Thailand |
| Vahid Jalali | Samsung Research America, USA |
| Hanmin Jung | KISTI, Korea |
| Jason Jung | Chung-Ang University, Korea |
| Takahiro Kawamura | Toshiba Corp., Japan |
| Pyung Kim | Jeonju National University of Education, Korea |
| Yoshinobu Kitamura | Osaka University, Japan |
| Seiji Koide | National Institute of Informatics, Japan |
| Kouji Kozaki | Osaka University, Japan |
| Myungjin Lee | LIST, Korea |
| Seungwoo Lee | KISTI, Korea |
| Tony Lee | Saltlux, Inc., Korea |
| Alain Leger | Orange Labs, France |
| Yuan-Fang Li | Monash University, Australia |
| Mihhail Matskin | Royal Institute of Technology, Sweden |
| Yutaka Matsuo | University of Tokyo, Japan |
| Riichiro Mizoguchi | JAIST, Japan |
| Takeshi Morita | Aoyama Gakuin University, Japan |
| Ralf Möller | Hamburg University of Technology, Germany |
| Shinichi Nagano | Toshiba Corp., Japan |
| Ekawit Nantajeewarawat | SIIT, Thailand |
| Leo Obrst | MITRE, USA |
| Ikki Ohmukai | National Institute of Informatics, Japan |
| Jeff Pan | University of Aberdeen, UK |
| Ratchata Peachavanish | Thammasat Univeristy, Thailand |
| Guilin Qi | Southeast University, China |
| Yuzhong Qu | Nanjing University, China |
| Ulrich Reimer | University of Applied Sciences St. Gallen, Switzerland |
| Marco Ronchetti | University of Trento, Italy |
| Twittie Senivongse | Chulalongkorn University, Thailand |
| Siraya Sitthisarn | Thaksin University, Thailand |
| Giorgos Stoilos | National Technical University of Athens, Greece |
| Umberto Straccia | ISTI-CNR, Italy |
| Boontawee Suntisrivaraporn | SIIT, Thailand |
| Thepchai Supnithi | NECTEC, Thailand |
| Hideaki Takeda | National Institute of Informatics, Japan |
| Kerry Taylor | CSIRO and Australian National University, Australia |
| Anni-Yasmin Turhan | Dresden University of Technology, Germany |
| Sasiporn Usanavasin | SIIT, Thailand |
| Tomas Vitvar | Oracle, Czech Republic |

| | |
|---|---|
| Holger Wache | University of Applied Sciences Northwestern Switzerland, Switzerland |
| Haofen Wang | East China University of Science and Technology, China |
| Krzysztof Wecel | Poznan University of Economics, Poland |
| Gang Wu | Northeastern University, China |
| Honghan Wu | University of Aberdeen, UK |
| Vilas Wuwongse | Asian University, Thailand |
| Bin Xu | Tsinghua University, China |
| Takahira Yamaguchi | Keio University, Japan |
| Jun Zhao | Lancaster University, UK |
| Amal Zouaq | Royal Military College of Canada, Canada |

## Additional Reviewers

Bazzanella, Barbara
Bortoli, Stefano
Cheng, Jingwei
Dojchinovski, Milan
Hwang, Myunggwon
Kafle, Sabin
Kim, Taehong
Kuchař, Jaroslav
Liu, Haishan

Morita, Takeshi
Ratanajaipan, Photchanan
Ruan, Tong
Seon, Choong-Nyoung
Vlasenko, Jelena
Xu, Hang
Yang, Mengdong
Zhang, Xiaowang
Zhou, Zhangquan

# Contents

**Learning and Discovery**

**RDF and SPARQL**

**Ontological Engineering**

**Semantic Social Web**

**Search and Querying**

**Applications of Semantic Technology**

# Ontology and Reasoning

# Revisiting Default Description Logics – and Their Role in Aligning Ontologies

Kunal Sengupta[1], Pascal Hitzler[1]([⊠]), and Krzysztof Janowicz[2]

[1] Wright State University, Dayton, OH 45435, USA
pascal.hitzler@wright.edu
[2] University of California, Santa Barbara, USA

**Abstract.** We present a new approach to extend the Web Ontology Language (OWL) with the capabilities to reason with defaults. This work improves upon the previously established results on integrating defaults with description logics (DLs), which were shown to be decidable only when the application of defaults is restricted to named individuals in the knowledge base. We demonstrate that the application of defaults (integrated with DLs) does not have to be restricted to named individuals to retain decidability and elaborate on the application of defaults in the context of ontology alignment and ontology-based systems.

## 1 Introduction and Motivation

The wide adoption of linked data principles has led to an enormous corpus of semantically enriched data being shared on the web. Researchers have been building (semi-)automatic matching systems [1,24] to build links (correspondences) between various conceptual entities as well as instances in the linked data. These systems are commonly known as ontology matching/alignment systems. The correspondences generated by these systems are represented using some standard knowledge representation language such as the web ontology language (OWL). However, due to the amount of heterogeneity present in the linked data and the web, OWL does not seem to be a completely suitable language for this purpose as we discuss in the following.

One key aspect of the web (or the world) is variety. There are subtle differences in how a conceptual entity and its relation to other entities is perceived depending on the geographical location, culture, political influence, etc. [15]. To give a simple example consider the concept of marriage, in some conservative parts of the world, marriage stands for a relationship between two individuals of opposite genders whereas in other more liberal places the individuals involved may have the same gender. Consider the axioms in Figure 1, let axioms (1) to (8) represent a part of ontology A (the conservative perspective) and axioms (9) to (13) represent a part of ontology B (the liberal perspective). It would be safe to assume that an ontology matching system would output the axioms (14) to (16) as the correspondences between these two ontologies. This however, leads to a logical inconsistency under OWL semantics when the two ontologies are

© Springer International Publishing Switzerland 2015
T. Supnithi et al.(Eds.): JIST 2014, LNCS 8943, pp. 3–18, 2015.
DOI: 10.1007/978-3-319-15615-6_1

$$a{:}has\,Wife \sqsubseteq a{:}hasSpouse \quad (1)$$
$$symmetric(a{:}hasSpouse) \quad (2)$$
$$\exists a{:}hasSpouse.a{:}Female \sqsubseteq a{:}Male \quad (3)$$
$$\exists a{:}hasSpouse.a{:}Male \sqsubseteq a{:}Female \quad (4)$$
$$a{:}has\,Wife(a{:}john, a{:}mary) \quad (5)$$
$$a{:}Male(a{:}john) \quad (6)$$
$$a{:}Female(a{:}mary) \quad (7)$$
$$a{:}Male \sqcap a{:}Female \sqsubseteq \bot \quad (8)$$

$$symmetric(b{:}hasSpouse) \quad (9)$$
$$b{:}hasSpouse(b{:}mike, b{:}david) \quad (10)$$
$$b{:}Male(b{:}david) \quad (11)$$
$$b{:}Male(b{:}mike) \quad (12)$$
$$b{:}Female(b{:}anna) \quad (13)$$

$$a{:}hasSpouse \equiv b{:}hasSpouse \quad (14)$$
$$a{:}Male \equiv b{:}Male \quad (15)$$
$$a{:}Female \equiv b{:}Female \quad (16)$$

**Fig. 1.** Running example with selected axioms

merged based on the given correspondences: From axioms (10), (11), (14), and (3), we derive *a:Female(b:mike)* which together with axioms (12), (15), and (8) results in an inconsistency as we derive *a:Male(b:mike)* and *a:Female(b:mike)* while axiom (8) states *a:Male $\sqcap$ a:Female $\sqsubseteq \bot$*.

This drives the need for an alignment language which could handle such subtle differences in perspectives. We propose an extension of description logics based on defaults to be used as an ontology alignment language. Using the notion of defaults we could re-state axiom (14) to an axiom which would semantically mean: every pair of individuals in *b:hasSpouse* is also in *a:hasSpouse* (and vice versa) unless it leads to a logically inconsistency. And those pairs which lead to an inconsistency are treated as exceptions to this axiom. In such a setting the pair *(b:mike, b:David)* would be treated as an exception to the re-stated axiom and would not cause an inconsistency any more.

A *default* is a kind of an inference rule that enables us to model some type of stereotypical knowledge such as "birds usually fly," or "humans usually have their heart on the left side of their chest." Default logic, which formalizes this intuition, was introduced by Ray Reiter [21] in the 80s, and it is one of the main approaches towards non-monotonic reasoning. In fact, it was the starting point for one of the primary approaches to logic programming and non-monotonic reasoning today, namely the stable model semantics [8] and answer set programming [20].

Reiter's approach is so powerful because exceptions to the default rules are implicitly handled by the logic, so that it is not left to the knowledge modeler to know all relevant exceptions and to take care of them explicitly, as is required in OWL-based ontology modeling. In fact, defaults in the general sense of Reiter still appear to be one of the most intuitive ways of formally modeling this type of stereotypical reasoning [11].

Alas, a paper by Baader and Hollunder [2], published almost 20 years ago, seemed to put an early nail into the coffin of default-extended description logics.

Therein, the authors show that a certain extension of the description logic $\mathcal{ALC}$[1] becomes undecidable if further extended with Reiter defaults. Since decidability was (and still is) a key design goal for description logics, this result clearly was a showstopper for further development of default-extended description logics. Of course, Baader and Hollunder also provided a quick fix: If we impose that the default rules only apply to known individuals (i.e., those explicitly present in the knowledge base), then decidability can be retained. However, this semantics for defaults is rather counter-intuitive, as it implies that default rules never apply to unknown individuals. In other words: to unknown individuals the defaults do, by default, not apply. Arguably, this is not a very intuitive semantics for defaults.

In this paper, we show that there is still a path of development for default-extended description logics, and that they may yet attain a useful standing in ontology modeling. In fact, we will present a way to extend decidable description logics with defaults which transcends the approach by Baader and Hollunder while retaining decidability: in our approach, defaults rules do apply to unknown individuals. We refer to the type of default semantics which we introduce as *free defaults*. Indeed, the contributions of this paper are (1) A new semantics for defaults (free defaults) in description logics and thereby OWL, such that the application of defaults are not limited to named individuals in the knowledge base, (2) We show that reasoning under this new semantics is decidable, which improves upon the results shown in [2], (3) Adding default role inclusion axioms also yields a decidable logic, and (4) We introduce the use of free defaults as a basis for a new language for ontology alignment and show some application scenarios in where defaults could play a major role and show how our approach covers these scenarios.

Let us briefly look at some of the related work published in recent years. There is in fact a plethora of publications on integrating description logics and non-monotonic formalisms, and it is not feasible to give all relevant proposals sufficient credit. We thus refer the interested reader to [18] for pointers to some of the most important approaches to date, including their relationships.

A series of work has recently been published which is related to typicality reasoning in description logics [9], in which the authors provide a minimal model semantics to achieve typicality. While our work is also based on preferred models, our goal is to follow some of the central ideas of default logic and to adapt it to provide a model-theoretic approach to defaults in DLs. Exact formal relationships between different proposals remain to be investigated. Other approaches that could be used to simulate defaults include circumscription [4,23], while again the exact relationship between the approaches remains to be investigated. Also [3] talks about defeasible inclusions in the tractable fragments of DL, which again follows a similar intuition. We understand our proposal and results as a contribution to the ongoing discussion about the best ways to capture non-monotonic reasoning for Semantic Web purposes.

---

[1] $\mathcal{ALC}$ is a very basic description logic which, among other things, constitutes the core of OWL 2 DL [13,14].

The remainder of this paper is organized as follows. Section 2 introduces preliminaries which form the basis required for the understanding of our work. In section 3, we discuss the semantics of free defaults for description logics, in the case of subclass defaults. In section 4 we give decidability results. In section 5 we show that adding free subrole defaults also retains decidability. In section 6 we discuss examples which illustrate the potential and relevance of free defaults to ontology modeling. We conclude in section 7.

## 2    Preliminaries

### 2.1    Default Logic

Default logic [21] is a form of non-monotonic logic which allows us to add inference rules called default rules on top of the conceptual knowledge. A *default rule* (or simply *default*) is of the form

$$\frac{\alpha : \beta_1, \ldots, \beta_n}{\gamma},$$

where $\alpha, \beta_i, \gamma$ are first order formulae. $\alpha$ is called the *pre-requisite* of the rule, $\beta_1, \ldots, \beta_n$ are its *justifications* and $\gamma$ its *consequent*. A default rule is *closed* if all the formulae that occur in the default are closed first order formulae, otherwise the default rule is called *open*. A *default theory* is further defined as a pair $(\mathcal{D}, \mathcal{W})$, where $\mathcal{D}$ is a set of defaults and $\mathcal{W}$ is a set of closed first order formulae. A default theory is *closed* if all the default rules in the set $\mathcal{D}$ are closed, otherwise it is called an *open* default theory.

The intuitive meaning of a default rule is that if $\alpha$ is true, and if furthermore assuming $\beta_1, \ldots, \beta_n$ to be true does not result in an inconsistency, then $\gamma$ is entailed. The formal semantics of a default theory is defined in terms of a notion of extension. An extension of a default theory is a completion (i.e., closure under entailment) of a possibly incomplete theory. The following describes formally the notion of an extension, directly taken from [21].

**Definition 1.**    *Let $\Delta = (\mathcal{D}, \mathcal{W})$ be a closed default theory, so that every default of $\mathcal{D}$ has the form*

$$\frac{\alpha : \beta_1, \ldots, \beta_n}{\gamma},$$

*where $\alpha, \beta_1, \ldots, \beta_n, \gamma$ are all closed formulae of L (a first order language). For any set of closed formulae $S \subseteq L$, let $\Gamma(S)$ be the smallest set satisfying the following three properties:*
- $\mathcal{W} \subseteq \Gamma(S)$
- *$\Gamma(S)$ is closed under entailment.*
- *If $\frac{\alpha : \beta_1, \ldots \beta_n}{\gamma} \in \mathcal{D}$, $\alpha \in \Gamma(S)$, and $\neg\beta_1, \ldots, \neg\beta_n \notin \Gamma(S)$, then $\gamma \in \Gamma(S)$.*

*A set of closed formulae $E \subseteq L$ is an extension of $\Delta$ if $\Gamma(E) = E$, i.e. if E is a fixed point of the operator $\Gamma$.*

The complexity of reasoning with (variants of) default logic is generally very high [10], and the same holds for most other non-monotonic logics, unless severe restrictions are put in place.[2]

In this paper we deal with a special type of defaults called *normal defaults*, and give it our own semantics which satisfies the intuition we intend to serve. We do this, rather than attempt to build on the more general approach by Reiter, because we strive for a simple but useful approach [23]. Normal defaults are those in which the justification and consequent are the same. We observe that these kinds of defaults have a variety of applications as we will see in section 6.

## 2.2 Description Logics

We briefly introduce the basic description logic (DL) $\mathcal{ALC}$, although our approach also works for more expressive description logics.

Let $\mathsf{N}_C, \mathsf{N}_R$ and $\mathsf{N}_I$ be countably infinite sets of concept names, role names and individual names, respectively. The set of $\mathcal{ALC}$ *concepts* is the smallest set that is created using the following grammar where $A \in \mathsf{N}_C$ denotes an atomic concept, $R \in \mathsf{N}_R$ is a role name and $C, D$ are concepts.

$$C ::= \top \mid \bot \mid A \mid \neg C \mid C \sqcap D \mid C \sqcup D \mid \exists R.C \mid \forall R.C$$

An $\mathcal{ALC}$ *TBox* is a finite set of axioms of the form $C \sqsubseteq D$, called *general concept inclusion* (GCI) *axioms*, where $C$ and $D$ are concepts. An $\mathcal{ALC}$ *ABox* is a finite set of axioms of the form $C(a)$ and $R(a, b)$, which are called *concept* and *role assertion axioms*, where $C$ is a concept, $R$ is a role and $a, b$ are individual names. An $\mathcal{ALC}$ *knowledge base* is a union of an $\mathcal{ALC}$ ABox and an $\mathcal{ALC}$ TBox

The semantics is defined in terms of interpretations $\mathcal{I} = (\Delta^{\mathcal{I}}, \cdot^{\mathcal{I}})$, where $\Delta^{\mathcal{I}}$ is a non-empty set called the *domain* of interpretation and $\cdot^{\mathcal{I}}$ is an interpretation function which maps each individual name to an element of the domain $\Delta^{\mathcal{I}}$ and interprets concepts and roles as follows:

$$\top^{\mathcal{I}} = \Delta^{\mathcal{I}}, \quad \bot^{\mathcal{I}} = \emptyset, \quad A^{\mathcal{I}} \subseteq \Delta^{\mathcal{I}}, \quad R^{\mathcal{I}} \subseteq \Delta^{\mathcal{I}} \times \Delta^{\mathcal{I}},$$
$$(\neg C)^{\mathcal{I}} = \Delta^{\mathcal{I}} \setminus C^{\mathcal{I}}, \quad (C_1 \sqcap C_2)^{\mathcal{I}} = C_1^{\mathcal{I}} \cap C_2^{\mathcal{I}}, \quad (C_1 \sqcup C_2)^{\mathcal{I}} = C_1^{\mathcal{I}} \cup C_2^{\mathcal{I}},$$
$$(\forall r.C)^{\mathcal{I}} = \{x \in \Delta^{\mathcal{I}} \mid (x, y) \in r^{\mathcal{I}} \text{ implies } y \in C^{\mathcal{I}}\},$$
$$(\exists r.C)^{\mathcal{I}} = \{x \in \Delta^{\mathcal{I}} \mid \text{ there is some } y \text{ with } (x, y) \in r^{\mathcal{I}} \text{ and } y \in C^{\mathcal{I}}\}$$

An interpretation $\mathcal{I}$ *satisfies* (is a *model* of) a GCI $C \sqsubseteq D$ if $C^{\mathcal{I}} \subseteq D^{\mathcal{I}}$, a concept assertion $C(a)$ if $a^{\mathcal{I}} \in C^{\mathcal{I}}$, a role assertion $R(a, b)$ if $(a^{\mathcal{I}}, b^{\mathcal{I}}) \in R^{\mathcal{I}}$. We say $\mathcal{I}$ *satisfies* (is a *model* of) a knowledge base $K$ if it satisfies every axiom in $K$. $K$ is *satisfiable* if such a model $\mathcal{I}$ exists.

The negation normal form of a concept $C$, denoted by $\mathsf{NNF}(C)$, is obtained by pushing the negation symbols inward, as usual, such that negation appears only in front of atomic concepts, e.g., $\mathsf{NNF}(\neg(C \sqcup D)) = \neg C \sqcap \neg D$.

We will occasionally refer to additional description logic constructs which are not contained in $\mathcal{ALC}$. Please refer to [14] for further background on description logics, and how they relate to the Web Ontology Language OWL [13].

---

[2] An exception is [17] for tractable description logics, but the practical usefulness of that approach for default modeling still needs to be shown.

# 3  Semantics of Free Defaults

In this section, we introduce the semantics of free defaults. We restrict our attention to normal defaults and show that reasoning in this setting is decidable in general when the underlying DL is also decidable. Normal defaults are very intuitive and we observe that there are many applications in practice where normal defaults can be very useful—see section 6. We also provide a DL syntax to encode default rules in the knowledge bases. For our purposes, a normal default rule is of the form $\dfrac{A : B}{B}$, where $A$ and $B$ are class names,[3] i.e., the justification and conclusion of the default rule are the same. For a description logic $\mathcal{L}$ we are going to represent the same rule in the form of an axiom $A \sqsubseteq_d B$, where $A$ and $B$ are $\mathcal{L}$-concepts and $\sqsubseteq_d$ represents (*free*) *default subsumption*. We refer to statements of the form $A \sqsubseteq_d B$ as (*free*) *default rules* or *default axioms*.

**Definition 2.** *Let KB be a description logic knowledge base, and let $\delta$ be a set of default axioms of the form $C \sqsubseteq_d D$, where $C$ and $D$ are concepts appearing in KB. Then we call the pair $(KB, \delta)$ a default-knowledge-base.*

The semantics of the default subsumption can be informally stated as follows: if $C \sqsubseteq_d D$, then every named individual in $C$ can also be assumed to be in $D$, unless it results in a logical inconsistency. Also, if $C \sqsubseteq_d D$, then every unnamed individual in $C$ is also in $D$, i.e., for unnamed individuals $\sqsubseteq_d$ behaves exactly the same as $\sqsubseteq$. Furthermore, we say a named individual $a$ satisfies a default axiom $C \sqsubseteq_d D$ if (1) $a^{\mathcal{I}} \in C^{\mathcal{I}}, D^{\mathcal{I}}$ or (2) $a^{\mathcal{I}} \in (\neg C)^{\mathcal{I}}$. The intuition behind the semantics of free defaults is to maximize the sets of the named individuals that satisfy the default axioms while maintaining the consistency of the knowledge base.

The following notations will be needed to formalize this intuition for the semantics of free defaults.

**Definition 3.** *For a default-knowledge-base $(KB, \delta)$, we define the following.*
- $\mathsf{Ind}_{KB}$ *is the set of all the named individuals occurring in KB.*
- $\mathsf{P}(\mathsf{Ind}_{KB})$ *is the power set of $\mathsf{Ind}_{KB}$.*
- $\mathsf{P}^n(\mathsf{Ind}_{KB})$ *is the set of n-tuples obtained from the Cartesian product: $\mathsf{P}(\mathsf{Ind}_{KB})$ $\times \cdots$ n times $\times \mathsf{P}(\mathsf{Ind}_{KB})$, where n is the cardinality of $\delta$.*

The notion of interpretation for the default-knowledge-bases $(KB, \delta)$ remains the same as that of the underlying DL of the knowledge base $KB$.[4] Additionally, given an interpretation $\mathcal{I}$, we define $\delta^{\mathcal{I}}$ to be the tuple $(X_1^{\mathcal{I}}, \ldots, X_n^{\mathcal{I}})$, where each $X_i^{\mathcal{I}}$ is the set of interpreted named individuals that satisfy the $i^{th}$ default $C_i \sqsubseteq_d D_i$ in the sense that $X_i^{\mathcal{I}} = (C_i^{\mathcal{I}} \cap D_i^{\mathcal{I}} \cap \Delta_{Ind}^{\mathcal{I}}) \cup ((\neg C_i)^{\mathcal{I}} \cap \Delta_{Ind}^{\mathcal{I}})$ with $\Delta_{Ind}^{\mathcal{I}} = \{a^{\mathcal{I}} \mid a \in \mathsf{Ind}_{KB}\} \subseteq \Delta^{\mathcal{I}}$ being the set of interpreted individuals occurring in the knowledge base. We now need to define a preference relation over the interpretations such that we can compare them on the basis of the sets of named individuals satisfying each default.

---

[3] We will lift this to roles in section 5.

[4] See section 2.

**Definition 4.** *(Preference relation* $>_{KB,\delta}$*) Given a knowledge base KB and a set of default axioms* $\delta$*. Let* $\mathcal{I}$ *and* $\mathcal{J}$ *be two interpretations of the pair* $(KB, \delta)$*, then we say that* $\mathcal{I}$ *is preferred over* $\mathcal{J}$ *or* $\mathcal{I} >_{KB,\delta} \mathcal{J}$ *if all of the following hold.*

1. $a^{\mathcal{I}} = a^{\mathcal{J}}$ *for all* $a \in \mathsf{N}_I$
2. $X_i^{\mathcal{I}} \supseteq X_i^{\mathcal{J}}$ *for all* $1 \leq i \leq |\delta|$*, where* $X_i^{\mathcal{I}} \in \delta^{\mathcal{I}}$ *and* $X_i^{\mathcal{J}} \in \delta^{\mathcal{J}}$*.*
3. $X_i^{\mathcal{I}} \supset X_i^{\mathcal{J}}$ *for some* $1 \leq i \leq |\delta|$*, where* $X_i^{\mathcal{I}} \in \delta^{\mathcal{I}}$ *and* $X_i^{\mathcal{J}} \in \delta^{\mathcal{J}}$*.*

The concept of a model under the semantics of free defaults would be the one which is maximal with respect to the above relation.

**Definition 5.** *(d-model) Given* $(KB, \delta)$*, we call* $\mathcal{I}$ *a d-model of KB with respect to a set of defaults* $\delta$*, written* $\mathcal{I} \models_d (KB, \delta)$*, if all of the following hold.*

1. $\mathcal{I}$ *satisfies all axioms in KB.*
2. $C_i^{\mathcal{I}} \setminus \Delta_{Ind}^{\mathcal{I}} \subseteq D_i^{\mathcal{I}}$*, for each* $(C_i \sqsubseteq_d D_i) \in \delta$*.*
3. *There is no interpretation* $\mathcal{J} >_{KB,\delta} \mathcal{I}$ *satisfying conditions 1 and 2 above.*

*Furthermore, if* $(KB, \delta)$ *has at least one model, then the default knowledge base is said to be d-satisfiable.*

The following proposition is obvious from the definition of d-model.

**Proposition 1.** *If* $\mathcal{I}$ *is a d-model of the default-knowledge-base* $(KB, \delta)$*, then* $\mathcal{I}$ *is a classical model of KB.*

For default theories two types of entailments are usually considered: credulous and skeptical [21]. A logical formula is a credulous entailment if it is true in at least one of the extensions of the default theory. Skeptical entailment requires the logical formula to be true in all the extensions. We follow the skeptical entailment approach as it fits better to the description logic semantics.[5]

**Definition 6.** *(d-entailment) Given a default-knowledge-base* $(KB, \delta)$ *and DL axiom* $\alpha$*,* $\alpha$ *is d-entailed by* $(KB, \delta)$ *if it holds in all the d-models of* $(KB, \delta)$*.*

## 4  Decidability

In this section we show that the tasks of checking for d-satisfiability and d-entailment for default-knowledge-bases are decidable in general. Let $(KB, \delta)$ be a default-knowledge-base where KB is encoded in a decidable DL $\mathcal{L}$ which supports nominal concept expressions. We show that finding a d-model for $(KB, \delta)$ is also decidable. For some $\mathcal{P} = (X_1, \ldots, X_n) \in \mathsf{P}^n(\mathsf{Ind}_{KB})$, let $KB_{\mathcal{P}}$ be the knowledge base that is obtained by adding the following axioms to KB, for each $C_i \sqsubseteq_d D_i \in \delta$:

1. $\overline{X_i} \equiv (C_i \sqcap D_i \sqcap \{a_1, \ldots, a_k\}) \sqcup (\neg C_i \sqcap \{a_1, \ldots, a_k\})$, where $\overline{X_i}$ is the nominal expression $\{x_1, \ldots, x_m\}$ containing exactly all the named individuals in $X_i$, and $\{a_1, \ldots, a_k\} = \mathsf{Ind}_{KB}$.
2. $C_i \sqcap \neg\{a_1, \ldots, a_k\} \sqsubseteq D_i$, where $\{a_1, \ldots, a_k\} = \mathsf{Ind}_{KB}$.

---

[5] Whether credulous entailment is useful in a Semantic Web context is to be determined.

The first step in the above construction is useful to identify the sets of default-satisfying individuals. The extensions of the $\overline{X}_i$s represent those sets. The second step ensures all the unnamed individuals satisfy the default axioms. Notice that $KB_{\mathcal{P}}$ as constructed using the above rewriting steps makes it fall under the expressivity of the DL $\mathcal{L}$, and construction of $KB_{\mathcal{P}}$ requires only a finite number of steps since $\delta$ is a finite set. Furthermore, we can compute an order $\succ$ on the set $\mathsf{P}^n(\mathsf{Ind}_{KB})$ based on the $\supseteq$-relation, defined as follows: Let $\mathcal{P}_1, \mathcal{P}_2 \in \mathsf{P}^n(\mathsf{Ind}_{KB})$, then $\mathcal{P}_1 \succ \mathcal{P}_2$ iff
1. $X_{1i} \supseteq X_{2i}$ for each $X_{1i} \in \mathcal{P}_1$ and $X_{2i} \in \mathcal{P}_2$ and
2. $X_{1i} \supset X_{2i}$ for some $X_{1i} \in \mathcal{P}_1$ and $X_{2i} \in \mathcal{P}_2$.

**Lemma 1.** *Given a default-knowledge-base $(KB, \delta)$, if $KB_{\mathcal{P}}$ is classically satisfiable for some $\mathcal{P} \in \mathsf{P}^n(\mathsf{Ind}_{KB})$, then $(KB, \delta)$ has a d-model.*

*Proof.* Let $\mathcal{P}_1 \in \mathsf{P}^n(\mathsf{Ind}_{KB})$ such that $KB_{\mathcal{P}_1}$ has a classical model $\mathcal{I}$. Then there are two possible cases.

In the first case there is no $\mathcal{P}_x \in \mathsf{P}^n(\mathsf{Ind}_{KB})$ such that $\mathcal{P}_x \succ \mathcal{P}_1$ and $KB_{\mathcal{P}_x}$ has a classical model. In this case $\mathcal{I}$ satisfies all the conditions of a d-model: (1) $\mathcal{I}$ satisfies all axioms of $KB$ since $KB \subseteq KB_{\mathcal{P}}$. (2) $\mathcal{I}$ satisfies condition 2 of the definition of d-model, this follows from the second step of the construction of $KB_{\mathcal{P}}$. (3) This follows directly from the assumption for this case. So $\mathcal{I}$ is a d-model for $(KB, \delta)$ in this case.

The second case is when there is some $\mathcal{P}_x \in \mathsf{P}^n(\mathsf{Ind}_{KB})$ for which there is a classical model $\mathcal{I}$ for $KB_{\mathcal{P}_x}$ and $\mathcal{P}_x \succ \mathcal{P}_1$. Again, there are two possibilities as in case of $\mathcal{P}_1$. Either the first case above holds for $\mathcal{P}_x$, or there is some $\mathcal{P}_y \succ \mathcal{P}_x \in \mathsf{P}^n(\mathsf{Ind}_{KB})$ for which the second case holds. In the latter situation, the argument repeats, eventually giving rise to an ascending chain with respect to the order $\succ$ on $\mathsf{P}^n(\mathsf{Ind}_{KB})$. However, since $\mathsf{P}^n(\mathsf{Ind}_{KB})$ is finite this chain has a maximal element and thus the first case applies. Therefore, there is a d-model for $(KB, \delta)$. $\square$

The following theorem is a direct consequence of Lemma 1 and the finiteness of $\delta$.

**Theorem 1.** *The task of determining d-satisfiability of default-knowledge-bases is decidable.*

It should be noted that in case of Reiter's defaults it is known that for normal default theories an extension always exists, but in the case of free defaults it can be easily seen that there might be some default-knowledge-bases which do not have a d-model. This is not completely satisfactory of course. However, at this stage it is unknown whether a stronger result can be obtained without giving up decidability. Though the notion of d-satisfiability is important for checking that the default-knowledge-base modelled is consistent and can be used for reasoning-based query services, the more interesting problem in the case of default-knowledge-bases is to handle d-entailment inference services. As it can be observed that d-entailment checking is not directly reducible to satisfiability checking of the default-knowledge-base,[6] we define a mechanism of checking d-entailments and show that this is also decidable.

**Proposition 2.** *Let $(KB, \delta)$ be a default-knowledge-base. If $\mathcal{I}$ is a d-model of $(KB, \delta)$, then there exists $\mathcal{P} \in \mathsf{P}^n(\mathsf{Ind}_{KB})$ such that $\mathcal{I}$ is a classical model of $KB_{\mathcal{P}}$ and all classical models of $KB_{\mathcal{P}}$ are d-models of $(KB, \delta)$.*

*Proof.* Given $\mathcal{I}$ we construct a $\mathcal{P} \in \mathsf{P}^n(\mathsf{Ind}_{KB})$ as follows: Given a default $C_i \sqsubseteq_d D_i \in \delta$, let $X_i$ be the maximal subset of $\mathsf{Ind}_{KB}$ such that $X_i^{\mathcal{I}} \subseteq (C_i \sqcap D_i)^{\mathcal{I}} \cup (\neg C_i)^{\mathcal{I}}$. Given all these $X_i's$, let $\mathcal{P} = \{X_1, \ldots, X_n\}$.

Clearly, $\mathcal{I}$ is then a classical model of $KB_{\mathcal{P}}$.

Furthermore, since $\mathcal{I}$ is a d-model of $(KB, \delta)$, there is no $\mathcal{P}_x \in \mathsf{P}^n(\mathsf{Ind}_{KB})$ such that $KB_{\mathcal{P}_x}$ has a classical model and $P_x \succ \mathcal{P}$. By construction of $KB_{\mathcal{P}}$ all classical models of $KB_{\mathcal{P}}$ satisfy the three d-model conditions for $(KB, \delta)$ because (1) $KB \subseteq KB_{\mathcal{P}}$, all axioms of $KB$ are satisfied, (2) the second step of the construction of $KB_{\mathcal{P}}$ ensures the second condition of d-model is satisfied. (3) Since $\mathcal{P}$ satisfies the maximality condition, all classical models of $KB_{\mathcal{P}}$ also satisfy the maximality condition of being a d-model ensured by step one of the construction of $KB_{\mathcal{P}}$. $\square$

Consider the two sets

$$\mathcal{P}_{KB\text{-}d} = \{\mathcal{P} \in \mathsf{P}^n(\mathsf{Ind}_{KB}) \mid KB_{\mathcal{P}} \text{ is classically satisfiable}\}$$

$$\mathcal{P}_{KB\text{-}d\text{-}model} = \{\mathcal{P} \in \mathcal{P}_{KB\text{-}d} \mid \mathcal{P} \text{ is maximal w.r.t. } \succ\}$$

and note that they are both computable in finite time. We refer to all the $KB_{\mathcal{P}}$'s generated from all $\mathcal{P} \in \mathcal{P}_{KB\text{-}d\text{-}model}$ as *d-model generating knowledge bases*.

**Lemma 2.** *A DL axiom $\alpha$ is d-entailed by a default-knowledge-base $(KB, \delta)$ iff it is classically entailed by every $KB_{\mathcal{P}}$ obtained from KB, $\delta$, and all $\mathcal{P} \in \mathcal{P}_{KB\text{-}d\text{-}model}$.*

*Proof.* This is a consequence of Proposition 2, since all classical models of each $\{KB_{\mathcal{P}} \mid \mathcal{P} \in \mathcal{P}_{KB\text{-}d\text{-}model}\}$ are also the d-models of the knowledge base. $\square$

We assume $KB$ is in a decidable description logic $\mathcal{L}$ that supports nominals and full negation. It is a well known result that all common inference tasks are reducible to a satisfiability check in DLs that support full negation [5]. Furthermore, $KB_{\mathcal{P}}$ is constructed by adding GCIs involving concept expressions using nominals and conjunctions, so we can safely assume that $KB_{\mathcal{P}}$ also falls under the DL $\mathcal{L}$. Hence, all the d-model generating knowledge bases are in $\mathcal{L}$.

**Theorem 2.** *(Decidability of d-entailment) Let $\mathcal{L}$ be a decidable DL with full negation and nominal support. Then the tasks of subsumption checking, instance checking, and class satisfiability are decidable for default-knowledge-bases with KB in $\mathcal{L}$.*

---

[6] This is due to non-monotonicity of the logic.

*Proof.* Given a default knowledge base $(KB, \delta)$, then by Lemma 2 the inference tasks can be reduced as follows:

- Subsumption: $C \sqsubseteq D$ is d-entailed by $(KB, \delta)$ iff $KB_{\mathcal{P}} \cup \{C \sqcap \neg D\}$ is classically unsatisfiable for all $\mathcal{P} \in \mathcal{P}_{\text{KB-d-model}}$.
- Instance checking: $C(a)$ is d-entailed by $(KB, \delta)$ iff $KB_{\mathcal{P}} \cup \{\neg C(a)\}$ is classically unsatisfiable for all $\mathcal{P} \in \mathcal{P}_{\text{KB-d-model}}$.
- Class satisfiability: a class $C$ is satisfiable iff $C \sqsubseteq \bot$ is not d-entailed by $(KB, \delta)$.

Consider the task of checking $(KB, \delta) \models_d C \sqsubseteq D$. Then $(KB, \delta) \models_d C \sqsubseteq D$ iff $KB_{\mathcal{P}} \cup \{C \sqcap \neg D\}$ is classically unsatisfiable for all $\mathcal{P} \in \mathcal{P}_{\text{KB-d-model}}$. Since $\mathcal{P}_{\text{KB-d-model}}$ is finitely computable and checking classical satisfiability is decidable in $\mathcal{L}$, checking the satisfiability for each $KB_{\mathcal{P}}$ is decidable. Hence, checking $(KB, \delta) \models_d C \sqsubseteq D$ is decidable. Similar arguments hold for the other tasks. $\square$

# 5   Default Role Inclusion Axioms

So far we have restricted our attention to default concept inclusions. We made this restriction for the purpose of obtaining a clearer presentation of our approach. However, as may be clear by now, we can also carry over our approach to cover default role inclusions, and we discuss this briefly in the following.

We use the notation $R \sqsubseteq_d S$ for free (normal) role defaults. As in the case of default concept inclusion axioms for role defaults, we restrict the exceptions to these defaults to be pairs of named individuals only. The intuitive semantics of $R \sqsubseteq_d S$ is that for every pair $(a, b)$ of named individuals in the knowledge base, if $R$ holds then assume $S$ also holds unless it leads to an inconsistency. For all other pairs of individuals (with at least one unnamed individual), if $R$ holds then $S$ also holds. We extend the definition of default-knowledge-bases and adjust the other definitions in the following.

**Definition 7.** *Let $KB$ be a knowledge base in a decidable DL and let $\delta$ be a set of default axioms of the form $C \sqsubseteq_d D$ or $R \sqsubseteq_d S$, where $C, D$ and $R, S$ are respectively concepts and roles appearing in $KB$. Then we call $(KB, \delta)$ a default-knowledge-base. Furthermore:*

- *The definition of $\mathsf{Ind}_{KB}, \mathsf{P}(\mathsf{Ind}_{KB}), \mathsf{P}^n(\mathsf{Ind}_{KB})$ carry over from Definition 3, where $n$ is the number of axioms of the form $C \sqsubseteq_d D$ in $\delta$.*
- *$\mathsf{P}(\mathsf{Ind}_{KB} \times \mathsf{Ind}_{KB})$ denotes the power set of $\mathsf{Ind}_{KB} \times \mathsf{Ind}_{KB}$.*
- *$\mathsf{P}^m(\mathsf{Ind}_{KB} \times \mathsf{Ind}_{KB})$ is the set of m-tuples obtained from the Cartesian product: $\mathsf{P}(\mathsf{Ind}_{KB} \times \mathsf{Ind}_{KB}) \times \ldots_{m\,times} \times \mathsf{P}(\mathsf{Ind}_{KB} \times \mathsf{Ind}_{KB})$, where $m$ is the number of default role axioms in $\delta$.*

For simplicity of presentation we assume that $\delta$ is arranged such that all default concept inclusion axioms appear before all default role inclusion axioms. Now, consider the set $\mathcal{D}_{KB} = \mathsf{P}^n(\mathsf{Ind}_{KB}) \times \mathsf{P}^m(\mathsf{Ind}_{KB} \times \mathsf{Ind}_{KB})$ which is a set of tuples, where each tuple is of the form $((X_1, \ldots, X_n), (Y_1, \ldots, Y_m))$ such that $(X_1, \ldots, X_n) \in \mathsf{P}^n(\mathsf{Ind}_{KB})$ and $(Y_1, \ldots, Y_m) \in \mathsf{P}^m(\mathsf{Ind}_{KB} \times \mathsf{Ind}_{KB})$. An interpretation for default-knowledge-bases with default role inclusion axioms should now map $\delta$ to a tuple as follows. $\delta^{\mathcal{I}} = (\mathcal{X}^{\mathcal{I}}, \mathcal{Y}^{\mathcal{I}}) \in \mathcal{D}_{KB}$, where $\mathcal{X}^{\mathcal{I}} = (X_1^{\mathcal{I}}, \ldots, X_n^{\mathcal{I}})$ and $\mathcal{Y}^{\mathcal{I}} = (Y_1^{\mathcal{I}}, \ldots, Y_m^{\mathcal{I}})$ such that $X_i^{\mathcal{I}} = (C_i^{\mathcal{I}} \cap D_i^{\mathcal{I}} \cap \Delta_{Ind}^{\mathcal{I}}) \cup ((\neg C)^{\mathcal{I}} \cap \Delta_{Ind}^{\mathcal{I}})$ and $Y_j^{\mathcal{I}} = (R_j^{\mathcal{I}} \cap S_j^{\mathcal{I}} \cap (\Delta_{Ind}^{\mathcal{I}} \times \Delta_{Ind}^{\mathcal{I}})) \cup ((\neg R_j)^{\mathcal{I}} \cap \Delta_{Ind}^{\mathcal{I}} \times \Delta_{Ind}^{\mathcal{I}})$, for all $C_i \sqsubseteq_d D_i \in \delta$ and $R_j \sqsubseteq_d S_j \in \delta$. In other words $X_i^{\mathcal{I}}$ denotes the extension of the named individuals that satisfy the $i^{th}$ default concept axioms and $Y_j^{\mathcal{I}}$ denotes the extension of pairs of named individuals that satisfy the $j^{th}$ default role axiom.

To ensure the maximal application of the default axioms we need the preference relation to be adapted to this setting.

**Definition 8.** *(Preference relation $>_{KB,\delta}$) Given a knowledge base KB, a set of default axioms $\delta$, and $\mathcal{I}$ and $\mathcal{J}$ be two interpretations of $(KB, \delta)$. We say that $\mathcal{I}$ is preferred over $\mathcal{J}$, written $\mathcal{I} >_{KB,\delta} \mathcal{J}$, if*
1. *conditions 1-4 of Definition 4 hold,*
2. $Y_i^{\mathcal{I}} \supseteq Y_i^{\mathcal{J}}$ *for all $1 \leq i \leq m$, where $Y_i^{\mathcal{I}} \in \mathcal{Y}^{\mathcal{I}}$ and $Y_i^{\mathcal{J}} \in \mathcal{Y}^{\mathcal{J}}$,*
3. $Y_i^{\mathcal{I}} \supset Y_i^{\mathcal{J}}$ *for some $1 \leq j \leq m$, where $Y_i^{\mathcal{I}} \in \mathcal{Y}^{\mathcal{I}}$ and $Y_i^{\mathcal{J}} \in \mathcal{Y}^{\mathcal{J}}$.*
*where m is the number of role inclusion axioms in $\delta$.*

The definition of d-model now carries over from Definition 5, the only difference being that the new definition $>_{KB,\delta}$ of the preference relation is used when default role axioms are also included.

To show the decidability of reasoning with any default-knowledge-base $(KB, \delta)$ with role defaults, we assume that $KB$ is in a decidable DL $\mathcal{L}$ which supports nominal concept expression, boolean role constructors, concept products, and the universal role $\mathcal{U}$. In [22], it was shown that expressive DLs can be extended with boolean role constructors for simple roles without compromising on complexity and decidability. For some tuple $\mathcal{P} \equiv ((X_1, \ldots, X_n), (Y_1, \ldots, Y_m)) \in \mathcal{D}_{KB}$, let $KB_{\mathcal{P}}$ be the knowledge base that is obtained by adding the following axioms to $KB$. For each $C_i \sqsubseteq_d D_i \in \delta$ add the following.
1. $\overline{X_i} \equiv (C_i \sqcap D_i \sqcap \{a_1, \ldots, a_k\}) \sqcup (\neg C_i \sqcap \{a_1, \ldots, a_k\})$, where $\overline{X_i}$ is the nominal expression $\{x_1, \ldots, x_m\}$ containing exactly the named individuals in $X_i$, and $\{a_1, \ldots, a_k\} = \mathsf{Ind}_{KB}$.
2. $C_i \sqcap \neg\{a_1, \ldots, a_k\} \sqsubseteq D_i$, where $\{a_1, \ldots, a_k\} = \mathsf{Ind}_{KB}$.
And for each $R_j \sqsubseteq_d S_j \in \delta$, add the following.

1. For each $(a, b) \in Y_j$, add the ABox axiom $R_{a,b}(a, b)$ and the axiom

$$\{x\} \sqcap \exists R_{a,b}.\{y\} \sqsubseteq \{a\} \sqcap \exists \mathcal{U}.(\{y\} \sqcap \{b\}),$$

where $R_{a,b}$ is a fresh role name, and $\{x\}$ and $\{y\}$ are so-called *nominal schemas* as introduced in [19]: They are a kind of nominal variables, which can stand for any nominal. In fact, this axiom can easily be cast into a set of axioms not containing nominal schemas, as shown in [19]. The axiom just given enforces that $R_{a,b}^{\mathcal{I}} \cap (\Delta_{Ind}^{\mathcal{I}} \times \Delta_{Ind}^{\mathcal{I}}) = \{(a, b)\}$.
2. $\bigsqcup_{(a,b) \in Y_j} R_{a,b} \equiv (R_j \sqcap D_j \sqcap \mathcal{U}_g) \sqcup R_j \sqcap \neg \mathcal{U}_g$, where $\mathcal{U}_g \equiv \mathsf{Ind}_{KB} \times \mathsf{Ind}_{KB}$.
3. $R_j \sqcap \neg \mathcal{U}_g \equiv S_j$, where $\mathcal{U}_g = \mathsf{Ind}_{KB} \times \mathsf{Ind}_{KB}$.

The construction just given for role defaults is analogous to the one for class inclusion defaults, with the exception that we do not have a nominal constructor for roles. However, for the specific setting we have here, we can obtain the same result by using the axioms from points 1 and 2 just given.

It should also be noted that the above outlined construction of $KB_{\mathcal{P}}$ can be computed in a finite number of steps.

The remainder of the decidability argument for d-entailment now carries over easily from section 4, and we omit the details. It should be noted that the availability of boolean role constructors is required for our argument, and that corresponding simplicity restrictions may apply, depending on the concrete case.

## 6   Application of Defaults in Ontology Alignment

Variety and semantic heterogeneity are at the very core of many fields like the Semantic Web, Big Data etc. To give a concrete example, many interesting scientific and societal questions cannot be answered from within one domain alone but span across disciplines. Studying the impact of climate change, for instance, requires to consider data and models from climatology, economics, biology, ecology, geography, and the medical science. While all these disciplines share an overlapping set of terms, the meanings of these terms clearly differ between them. A street, for instance, is a *connection* between A and B from the view point of transportation science, and, at the same time, a disruptive *separation* which cuts through habitats from the view point of ecology. Even within single domains, the used terminology differs over time and space [16]. The idea that this variety should be 'resolved' is naive at best. The defaults extension proposed in this work can thus support more robust ontology alignments that respect variety and still allow us to share and integrate the heterogeneous data. In the following we give some concrete examples that cannot be sufficiently addressed with existing ontology alignment frameworks, but would benefit from the proposed extension.

Consider the axioms in Figure 2. The ontology fragment consisting of (17) to (21) reflects a certain perspective on canals valid in a transportation application. In contrast, the axioms (22) to (24) reflect a different but equally valid perspective from an agricultural perspective. Typically, ontology alignment systems would default to a syntactic matching of shared primitives such as *AgriculturalField* or *IrrigationCanal*. However, applied to these two ontology fragments this would yield a logical inconsistency in which some waterbodies would have to be land masses at the same time. Using our proposed free defaults, only certain canals from $a$ would qualify as canals in $b$, avoiding the inconsistencies.

While in the above example the inconsistency was largely caused by the cardinality restrictions, other cases involve concrete domains. For instance, each US state (and the same argument can be made between counties as well) has its own legally binding definition of what distinguishes a town from a city. Thus, to query the Linked Data cloud for towns it is required to take these local definitions into account. Otherwise one would, among hundreds of thousands of small municipalities, also retrieve Los Angeles, CA or Stuttgart, Germany.[7]

---

[7] In case of DBpedia via `dbpedia:Stuttgart rdf:type dbpedia-owl:Town`.

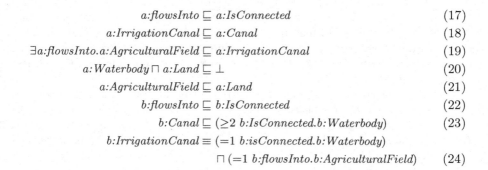

$$a{:}flowsInto \sqsubseteq a{:}IsConnected \qquad (17)$$

$$a{:}IrrigationCanal \sqsubseteq a{:}Canal \qquad (18)$$

$$\exists a{:}flowsInto.a{:}AgriculturalField \sqsubseteq a{:}IrrigationCanal \qquad (19)$$

$$a{:}Waterbody \sqcap a{:}Land \sqsubseteq \bot \qquad (20)$$

$$a{:}AgriculturalField \sqsubseteq a{:}Land \qquad (21)$$

$$b{:}flowsInto \sqsubseteq b{:}IsConnected \qquad (22)$$

$$b{:}Canal \sqsubseteq (\geq 2 \; b{:}IsConnected.b{:}Waterbody) \qquad (23)$$

$$b{:}IrrigationCanal \equiv (=1 \; b{:}isConnected.b{:}Waterbody)$$
$$\sqcap (=1 \; b{:}flowsInto.b{:}AgriculturalField) \qquad (24)$$

**Fig. 2.** Fragments of two ontologies, (17)-(21), respectively (22)-(24), to be aligned

In several cases these state-specific distinctions solely depend on the population count and, thus, could be handled using existing alignment systems. However, in other cases they are driven by administrative divisions of geographic space, are based on historical reasons, or other properties. As argued before, our free defaults can handle these cases.

Let us now return to the example from section 1 and discuss it in more technical depth. We showed that an alignment using axiom (14) leads to inconsistency. Now consider instead using the approach of free defaults, by replacing axiom (14) with $b{:}hasSpouse \sqsubseteq_d a{:}hasSpouse$. As per our semantics the pair $(b{:}mike, b{:}david)$ will act as an exception to the default role inclusion that we just added and $a{:}hasSpouse(b{:}mike, b{:}david)$ will not hold anymore. On the other hand if we also add the axiom $a{:}hasSpouse \sqsubseteq_d b{:}hasSpouse$ then $b{:}hasSpouse(a{:}john, a{:}mary)$ will also hold.

To see this formally, consider all the axioms of figure 1 except (14) to be $KB$ and let $\delta \equiv \{(a{:}hasSpouse \sqsubseteq_d b{:}hasSpouse), (b{:}hasSpouse \sqsubseteq_d a{:}hasSpouse)\}$ and consider an interpretation $\mathcal{I}$ such that $(a{:}hasSpouse)^{\mathcal{I}} = \{(a{:}john^{\mathcal{I}}, a{:}mary^{\mathcal{I}})\}$ and $(b{:}hasSpouse)^{\mathcal{I}} = \{(b{:}mike^{\mathcal{I}}, b{:}david^{\mathcal{I}}), (a{:}john^{\mathcal{I}}, a{:}mary^{\mathcal{I}})\}$. Note that forcing $(b{:}mike^{\mathcal{I}}, b{:}david^{\mathcal{I}})$ in the extension of $a{:}hasSpouse$ will result in an inconsistency because of the reasons mentioned in section 1. On the other hand, if we consider an interpretation $\mathcal{J}$ such that $(a{:}hasSpouse)^{\mathcal{J}} = \{(a{:}john^{\mathcal{J}}, a{:}mary^{\mathcal{J}})\}$ and $(b{:}hasSpouse)^{\mathcal{J}} = \{(b{:}mike^{\mathcal{J}}, b{:}david^{\mathcal{J}})\}$, then clearly $\mathcal{I} >_{KB,\delta} \mathcal{J}$, because the extension of b:hasSpouse in $\mathcal{I}$ is greater than that of $\mathcal{J}$. So, $\mathcal{I}$ is preferred over $\mathcal{J}$. In fact $\mathcal{I}$ is also a d-model of this default-knowledge-base.

The above example shows that in case of ontology alignments, using default constructs to map terms could help us avoid potentially inconsistent merged knowledge bases due to subtle semantic differences in the different ontologies.

As final example, we want to discuss the implications our approach has with respect to the use and abuse of owl:sameAs in linked data [12],[8] where this Web ontology language (OWL) construct is used extensively to provide links between

---

[8] Note owl:sameAs is OWL representation of individual equality in DLs

different datasets. However, owl:sameAs is, semantically, a strong equality which equates entities, and this strong (specification-compliant) interpretation easily leads to complications. For instance, consider two linked datasets $a$ and $b$ where $a$ contains the axioms: $a{:}airport(a{:}kennedy)$ and $a{:}airport \sqsubseteq a{:}place$, and $b$ contains axioms $b{:}president(b{:}kennedy)$ and $b{:}president \sqsubseteq b{:}person$ plus the disjointness axiom $b{:}person \sqcap a{:}place \sqsubseteq \bot$.

Now, if some text-based co-reference resolution system identifies $a{:}kennedy$ and $b{:}kennedy$ as the same object, then it will result in a link such as $owl{:}sameAs$ ($a{:}kennedy, b{:}kennedy$). Obviously, this yields to an inconsistency because of the disjointness axiom. However, if we use defaults this could be expressed as $\{a{:}kennedy\} \sqsubseteq_d \{b{:}kennedy\}$, which essentially is another way of saying that $a{:}kennedy$ is identical to $b{:}kennedy$ *unless* it causes an inconsistency. While [12] argues for a set of variants of equality with differing semantic strength, automatic identification of the exact variant of equality to be used is yet another matter, and presumably rather difficult to accomplish. So for automated co-reference resolution, we would argue that the use of free defaults, which semantically recover from erroneous guesses by the alignment system, are a much more suitable solution.

## 7   Conclusion

In this paper, we have provided a new semantics for embedding defaults into description logics, and have shown that reasoning tasks are decidable in our setting. Both the decidable logic from [2] and our work are variants of Reiter's defaults [21]. But the approach in [2] is very restricted and arguably violates some key intuitions. Our proposal provides an improvement, mainly because with our free defaults, the application of defaults is not limited to named individuals. However, we impose that exceptions to the default rules only occur in the named individuals of the knowledge base. Also, our approach to the semantics is model-theoretic whereas most of the previous work on defaults has been mainly based on fixed point semantics [6,7]. We have furthermore given a thorough motivation of the usefulness of free defaults in the context of ontology alignments. Through the examples in section 6, it is shown that the new semantics that we have introduced in this paper is useful when dealing with the integration of heterogeneous ontologies. We believe that our work provides a foundation for a new and more powerful ontology alignment language.

Whether defaults over DLs are decidable when we allow exceptions to also occur over unnamed individuals, is still an open question and we intend to investigate this in the future. Future work also includes smart algorithmization and implementation of d-entailment tasks mentioned in this paper. A naive algorithm can easily be developed by searching for maximal d-model generating tuples from $\mathsf{P}^n(\mathsf{Ind}_{KB})$, i.e. by searching for all maximal $\mathcal{P}$s in $\mathsf{P}^n(\mathsf{Ind}_{KB})$ for which $KB_{\mathcal{P}}$ has a classical model and then using the process outlined in Theorem 2. Although this reasoning procedure appears to be decidable it is very expensive and thus not feasible for practical use. However, the algorithmization could be made smarter by using some optimization techniques. For instance, $P^n(\mathsf{Ind}_{KB})$ could be represented as an ordered set of tuples where each tuple is a collection of comparable $P$s sorted by the $\succ$ relation. The algorithm would then look for maximally satisfying $P$s for each tuple by performing a binary search on every tuple. This should significantly improve the performance of the naive approach since the number of steps to find all suitable $P$s has been reduced by a large factor. These and other optimizations will be central to our investigation of algorithmizations.

**Acknowledgements.** This work was supported by the National Science Foundation under award 1017225 "III: Small: TROn—Tractable Reasoning with Ontologies" and award 1440202 "EarthCube Building Blocks: Collaborative Proposal: GeoLink—Leveraging Semantics and Linked Data for Data Sharing and Discovery in the Geosciences."

# References

1. Aguirre, J.L., Eckert, K., Euzenat, J., Ferrara, A., van Hage, W.R., Hollink, L., Meilicke, C., Nikolov, A., Ritze, D., Scharffe, F., Shvaiko, P., Sváb-Zamazal, O., dos Santos, C.T., Jiménez-Ruiz, E., Grau, B.C., Zapilko, B.: Results of the ontology alignment evaluation initiative 2012. In: Shvaiko, P., Euzenat, J., Kementsietsidis, A., Mao, M., Noy, N.F., Stuckenschmidt, H. (eds.) Proceedings of the 7th International Workshop on Ontology Matching, Boston, MA, USA, 11 November 2012, CEUR Workshop Proceedings, vol. 946. CEUR-WS.org (2012)
2. Baader, F., Hollunder, B.: Embedding defaults into terminological knowledge representation formalisms. J. Autom. Reasoning **14**(1), 149–180 (1995)
3. Bonatti, P.A., Faella, M., Sauro, L.: Defeasible inclusions in low-complexity DLs. Artificial Intelligence (JAIR) **42**, 719–764 (2011)
4. Bonatti, P.A., Lutz, C., Wolter, F.: The Complexity of Circumscription in Description Logic. Journal of Artificial Intelligence Research **35**, 717–773 (2009)
5. Buchheit, M., Donini, F.M., Schaerf, A.: Decidable reasoning in terminological knowledge representation systems. In: Proceedings of the 13th International Joint Conference on Artifical Intelligence, IJCAI 1993, vol. 1, pp. 704–709. Morgan Kaufmann Publishers Inc., San Francisco (1993)
6. Dao-Tran, M., Eiter, T., Krennwallner, T.: Realizing default logic over description logic knowledge bases. In: Sossai, C., Chemello, G. (eds.) ECSQARU 2009. LNCS, vol. 5590, pp. 602–613. Springer, Heidelberg (2009)
7. Eiter, T., Lukasiewicz, T., Schindlauer, R., Tompits, H.: Combining answer set-programming with description logics for the semantic web. In: Proc. of the 9th Int. Conf. on the Principles of Knowledge Representation and Reasoning (KR 2004). AAAI Press (2004)
8. Gelfond, M., Lifschitz, V.: The stable model semantics for logic programming. In: Kowalski, R.A., Bowen, K.A. (eds.) Proceedings of the 5th International Conference and Symposium on Logic Programming, pp. 1070–1080. MIT Press (1988)
9. Giordano, L., Gliozzi, V., Olivetti, N., Pozzato, G.L.: A non-monotonic description logic for reasoning about typicality. Artificial Intelligence **195**, 165–202 (2013)
10. Gottlob, G.: Complexity results for nonmonotonic logics. J. Log. Comput. **2**(3), 397–425 (1992)

11. Grimm, S., Hitzler, P.: Semantic Matchmaking of Web Resources with Local Closed-World Reasoning. International Journal of Electronic Commerce **12**(2), 89–126 (2007)

12. Halpin, H., Hayes, P.J., McCusker, J.P., McGuinness, D.L., Thompson, H.S.: When owl:sameAs isn't the same: an analysis of identity in linked data. In: Patel-Schneider, P.F., Pan, Y., Hitzler, P., Mika, P., Zhang, L., Pan, J.Z., Horrocks, I., Glimm, B. (eds.) ISWC 2010, Part I. LNCS, vol. 6496, pp. 305–320. Springer, Heidelberg (2010)

13. Hitzler, P., Krötzsch, M., Parsia, B., Patel-Schneider, P.F., Rudolph, S. (eds.): OWL 2 Web Ontology Language: Primer. W3C Recommendation 27 October 2009 (2009). http://www.w3.org/TR/owl2-primer/

14. Hitzler, P., Krötzsch, M., Rudolph, S.: Foundations of Semantic Web Technologies. Chapman & Hall/CRC, Boca Raton (2009)

15. Janowicz, K.: The role of space and time for knowledge organization on the Semantic Web. Semantic Web **1**(1–2), 25–32 (2010)

16. Janowicz, K., Hitzler, P.: The digital earth as knowledge engine. Semantic Web **3**(3), 213–221 (2012)

17. Knorr, M., Alferes, J., Hitzler, P.: Local Closed-World Reasoning with Description Logics under the Well-founded Semantics. Artificial Intelligence **175**(9–10), 1528–1554 (2011)

18. Knorr, M., Hitzler, P., Maier, F.: Reconciling OWL and non-monotonic rules for the Semantic Web. In: Raedt, L.D., et al. (eds.) ECAI 2012–20th European Conference on Artificial Intelligence, Montpellier, France, 27–31 August 2012, pp. 474–479. IOS Press, Amsterdam (2012)

19. Krötzsch, M., Maier, F., Krisnadhi, A.A., Hitzler, P.: A better uncle for OWL: nominal schemas for integrating rules and ontologies. In: Sadagopan, S., et al. (eds.) Proceedings of the 20th International World Wide Web Conference, WWW 2011, Hyderabad, India, 2011, pp. 645–654. ACM, New York (2011)

20. Leone, N., Faber, W.: The DLV project: a tour from theory and research to applications and market. In: Garcia de la Banda, M., Pontelli, E. (eds.) ICLP 2008. LNCS, vol. 5366, pp. 53–68. Springer, Heidelberg (2008)

21. Reiter, R.: A logic for default reasoning. Artif. Intell. **13**(1–2), 81–132 (1980)

22. Rudolph, S., Krötzsch, M., Hitzler, P.: Cheap Boolean role constructors for description logics. In: Hölldobler, S., Lutz, C., Wansing, H. (eds.) JELIA 2008. LNCS (LNAI), vol. 5293, pp. 362–374. Springer, Heidelberg (2008)

23. Sengupta, K., Krisnadhi, A.A., Hitzler, P.: Local closed world semantics: grounded circumscription for OWL. In: Aroyo, L., Welty, C., Alani, H., Taylor, J., Bernstein, A., Kagal, L., Noy, N., Blomqvist, E. (eds.) ISWC 2011, Part I. LNCS, vol. 7031, pp. 617–632. Springer, Heidelberg (2011)

24. Shvaiko, P., Euzenat, J.: Ontology matching: state of the art and future challenges. IEEE Trans. Knowl. Data Eng. **25**(1), 158–176 (2013)

# On Desirable Properties of the Structural Subsumption-Based Similarity Measure

Suwan Tongphu and Boontawee Suntisrivaraporn[✉]

School of Information, Computer and Communication Technology,
Sirindhorn International Institute of Technology, Thammasat University,
Phra Nakhon Si Ayutthaya, Pathumthani, Thailand
stongphu@gmail.com, sun@siit.tu.ac.th

**Abstract.** Checking for subsumption relation is the main reasoning service readily available in classical DL reasoners. With their binary response stating whether two given concepts are in the subsumption relation, it is adequate for many applications relied on the service. However, in several specific applications, there often exists the case that requires an investigation for concepts that are not directly in a subclass-superclass relation but shared some commonality. In this case, providing merely a crisp response is apparently insufficient. To achieve this, the similarity measure for DL $\mathcal{ELH}$, which is inspired by the homomorphism-based structural subsumption characterization, has been introduced. To ensure that the proposed method reaches the performance, in this work, desirable properties for concept similarity measure are checked and compared with those previously reported in other classical works.

**Keywords:** Concept similarity · Non-standard reasoner · Description logic · Structural subsumption

## 1 Introduction

Knowledge representation is one such major research area that has a long range of development and mainly focuses on an investigation for well-founded ways to model, share, and interpret the knowledge. One modeling formalism is an exploitation of the family of Description Logics (DLs) which allows various types of reasoning services. Among those readily available in classical DL reasoners, concept subsumption (i.e. identification of subclass-superclass relationships) is one of the most prominent services. Despite its usefulness, classical subsumption reasoners merely response with a binary result (i.e. whether two given concepts are in a subclass-superclass relation). This capability seems adequate for many applications. However, in some situations, there may be the case that the two concepts do not align that way but share some commonality. This special case, on the other hand, turns into account in many specific applications. For example, in hospitals, once a doctor has diagnosed medical conditions of a patient and identified what the illness is, he/she may need to investigate further for other possible illnesses of similar but not exactly the same conditions.

© Springer International Publishing Switzerland 2015
T. Supnithi et al.(Eds.): JIST 2014, LNCS 8943, pp. 19–32, 2015.
DOI: 10.1007/978-3-319-15615-6_2

**Table 1.** Syntax and semantics of the Description Logic $\mathcal{ELH}$

| Name | Syntax | Semantics |
|------|--------|-----------|
| top | $\top$ | $\Delta^{\mathcal{I}}$ |
| concept name | $A$ | $A^{\mathcal{I}} \subseteq \Delta^{\mathcal{I}}$ |
| conjunction | $C \sqcap D$ | $C^{\mathcal{I}} \cap D^{\mathcal{I}}$ |
| existential restriction | $\exists r.C$ | $\{x \in \Delta^{\mathcal{I}} \mid \exists y \in \Delta^{\mathcal{I}} : (x,y) \in r^{\mathcal{I}} \wedge y \in C^{\mathcal{I}}\}$ |
| concept inclusion | $A \sqsubseteq D$ | $A^{\mathcal{I}} \subseteq D^{\mathcal{I}}$ |
| concept equivalence | $A \equiv C$ | $A^{\mathcal{I}} = C^{\mathcal{I}}$ |
| role inclusion | $r \sqsubseteq s$ | $r^{\mathcal{I}} \subseteq s^{\mathcal{I}}$ |

Our first introduction to $\mathcal{EL}$ concept similarity measure [13] and its sample application [8] have shown its usability specifically in but not limited to one of the most popular medical-domain ontologies, SNOMED CT [12]. In this work, we extend the algorithm to a more expressive DL $\mathcal{ELH}$. Therefore, role names of the same hierarchy are taken into account. Moreover, to ensure that the proposed method reaches the performance and holds satisfactory features, in this work, desirable properties for concept similarity measure are proofed and compared with those previously reported in other works.

In the next section, notions of the DL and necessary backgrounds are introduced. Section 3 and Section 4 provide details on the proposed method and its properties, respectively. Concluding remarks are given in the last section.

## 2    Preliminaries

Let CN and RN be a set of concept names and a set of role names. In Description Logics (DLs), complex $\mathcal{ELH}$ concept descriptions can be built using a set of constructors shown in the upper part of Table 1. The background knowledge about the domain called *terminology box* or *TBox* can then be devised using a set of ontological axioms shown in the second part of Table 1. A TBox is *unfoldable* if, for each concept name, there is only one concept definition and there is neither direct nor indirect concept definition that refers to the concept itself. Figure 1 shows an example of the unfoldable Tbox $\mathcal{O}_{\mathsf{med}}$.

An interpretation $\mathcal{I} = (\Delta^{\mathcal{I}}, \cdot^{\mathcal{I}})$ comprises of *interpretation domain* $\Delta^{\mathcal{I}}$ and *interpretation function* $\cdot^{\mathcal{I}}$. The interpretation function maps every concept name $A \in$ CN to a subset $A^{\mathcal{I}} \subseteq \Delta^{\mathcal{I}}$, every role name $r \in$ RN to a binary relation $r^{\mathcal{I}} \subseteq \Delta^{\mathcal{I}} \times \Delta^{\mathcal{I}}$, and every individual $x \in$ Ind to an element $x^{\mathcal{I}} \in \Delta^{\mathcal{I}}$. The last column of Table 1 depicts the semantics for $\mathcal{ELH}$ constructors and terminological axioms. An interpretation $\mathcal{I}$ is a model of a TBox $\mathcal{O}$ if it satisfies every axiom defined in $\mathcal{O}$. Let $C$, $D$ and $E$ be concept descriptions, $C$ is subsumed by $D$

(written $C \sqsubseteq D$) iff $C^{\mathcal{I}} \subseteq D^{\mathcal{I}}$ in every model $\mathcal{I}$ and if $C \sqsubseteq D \sqsubseteq E$, then $C \sqsubseteq E$. Moreover, $C, D$ are equivalent (written $C \equiv D$) iff $C \sqsubseteq D$ and $D \sqsubseteq C$, i.e. $C^{\mathcal{I}} = D^{\mathcal{I}}$ for all interpretations $\mathcal{I}$.

By introducing a set of fresh concept names [13], a concept inclusion can be transformed to an equivalent form. Without losing of generality, we assume that all concept names can be expanded (i.e. they can be replaced by the definition) and has the following form:

$$\prod_{i \leq m} P_i \sqcap \prod_{j \leq n} \exists.r_j C_j$$

where $1 \leq i \leq m$ and $1 \leq j \leq n$. This is true both for defined concepts (i.e. concepts that appear on the left-hand side of a definition) and primitive concepts (i.e. concepts that appear only on the right-hand side of a definition). For any primitive concept $P$, $P \sqsubseteq \top$ therefore $P \equiv X \sqcap \top \equiv X$ where $X$ is a fresh concept name. For convenience, we denote by $\mathcal{P}_C = \{P_1, \ldots, P_m\}$ and $\mathcal{E}_C = \{\exists r_1 C_1, \ldots, \exists r_n C_n\}$ the set of top-level primitive concepts and the set of top-level existential restrictions. Also, we denote by $\mathcal{R}_r = \{s | r \sqsubseteq^* s\}$ the set of all super roles where $r$ and $s$ are role names, $*$ is the transitive closure, and $r \sqsubseteq^* s$ if $r = s$ or $r_i \sqsubseteq r_{i+1} \in \mathcal{O}$ where $r_1 = r$ and $r_n = s$. The following demonstrates the expanded form of the concept AspirationOfMucus defined in $\mathcal{O}_{\mathsf{med}}$ (see Figure 1).

$$X \sqcap \mathsf{RespiratoryDisorder} \sqcap \exists \mathsf{agent.Mucus}$$

where $X$ is a fresh concept name.

Let $\mathcal{T} = (V, E, rt, \ell, \rho)$ be the $\mathcal{ELH}$ description tree [13] w.r.t. an unfoldable TBox, where $V$ is a set of nodes, $E \subseteq V \times V$ is a set of edges, $rt$ is the root, $\ell : V \to 2^{\mathsf{CN^{pri}}}$ is a node labeling function, and $\rho : E \to 2^{\mathsf{RN}}$ is an edge labeling function. Definition 1 defines a homomorphism mapping. Let $\mathcal{T}_C$ and $\mathcal{T}_D$ be $\mathcal{ELH}$ description trees w.r.t. the concept $C$ and $D$, Theorem 1 depicts a characterization of $C \sqsubseteq D$ based on a homomorphism that maps the root of $\mathcal{T}_D$ to the root of $\mathcal{T}_C$.

**Definition 1 (Homomorphism).** *Let $\mathcal{T}$ and $\mathcal{T}'$ be two $\mathcal{ELH}$ description trees as previously defined. There exists a homomorphism $h$ from $\mathcal{T}$ to $\mathcal{T}'$ written $h : \mathcal{T} \to \mathcal{T}'$ iff the following conditions are satisfied:*

    *(i) $\ell(v) \subseteq \ell'(h(v))$*
    *(ii) For each successor $w$ of $v$ in $\mathcal{T}$, $h(w)$ is a successor of $h(v)$ with $\rho(v, w) \subseteq \rho'(h(v), h(w))$*

**Theorem 1 ([9]).** *Let $C, D$ be $\mathcal{ELH}$ concept descriptions and $\mathcal{T}_C, \mathcal{T}_D$ be $\mathcal{ELH}$ concept description trees w.r.t. $C$ and $D$. Then, $C \sqsubseteq D$ iff there exists a homomorphism $h : \mathcal{T}_D \to \mathcal{T}_C$ which maps the root of $\mathcal{T}_D$ to the root of $\mathcal{T}_C$.*

By using Theorem 1 together with properties of homomorphism mapping defined in Definition 1, Corollary 1 and Corollary 2 hold due to an associativity and commutativity of concept conjunction.

**Corollary 1.** *Let $C$ and $D$ be concept names. Then $C \sqsubseteq D$ iff $\mathcal{P}_D \subseteq \mathcal{P}_C$ and for each $\exists r.D' \in \mathcal{E}_D$ there exists $\exists s.C'$ such that $s \sqsubseteq^* r$ and $C' \sqsubseteq D'$.*

**Corollary 2.** *Let $C$ and $D$ be concept names, then $\mathcal{E}_D \cong \mathcal{E}_C$ iff for each $\exists r.D' \in \mathcal{E}_D$ there exists $\exists s.C'$ such that $s \sqsubseteq^* r$, $r \sqsubseteq^* s$, $C' \sqsubseteq D'$, and $D' \sqsubseteq C'$. Moreover, $C \equiv D$ iff $\mathcal{P}_D = \mathcal{P}_C$ (i.e. $\mathcal{P}_D \subseteq \mathcal{P}_C$ and $\mathcal{P}_C \subseteq \mathcal{P}_D$) and $\mathcal{E}_D \cong \mathcal{E}_C$.*

| | | |
|---:|:---:|:---|
| AspirationOfMucus | $\equiv$ | AspirationSyndromes $\sqcap$ $\exists$agent.Mucus |
| AspirationOfMilk | $\equiv$ | AspirationSyndromes $\sqcap$ InhalationOfLiquid $\sqcap$ $\exists$agent.Milk $\sqcap$ $\exists$assocWith.Milk |
| Hypoxia | $\equiv$ | RespiratoryDisorder $\sqcap$ BloodGasDisorder $\sqcap$ $\exists$interprets.OxygenDelivery |
| Hypoxemia | $\equiv$ | RespiratoryDisorder $\sqcap$ BloodGasDisorder $\sqcap$ $\exists$interprets.OxygenDelivery $\sqcap$ $\exists$site.ArterialSystem |
| AspirationSyndromes | $\sqsubseteq$ | RespiratoryDisorder |
| agent | $\sqsubseteq$ | assocWith |

**Fig. 1.** Examples of $\mathcal{ELH}$ concept descriptions defined in $\mathcal{O}_{\mathsf{med}}$

Based on the property of concept subsumption and homomorphism mapping, in the next section, we introduce the notion of homomorphism degree hd and concept similarity sim.

## 3   Homomorphism Degree

Let $\mathcal{T} = (V, E, rt, \ell, \rho)$ be the $\mathcal{ELH}$ description tree as previously defined. Then, the degree of having a homomorphism from $\mathcal{T}_D$ to $\mathcal{T}_C$ is defined by Definition 2.

**Definition 2 (Homomorphism degree)**
*Let $\mathbf{T}^{\mathcal{ELH}}$ be the set of all $\mathcal{ELH}$ description trees. The homomorphism degree function hd : $\mathbf{T}^{\mathcal{ELH}} \times \mathbf{T}^{\mathcal{ELH}} \to [0,1]$ is inductively defined as follows:*

$$\mathsf{hd}(\mathcal{T}_D, \mathcal{T}_C) := \mu \cdot \mathsf{p\text{-}hd}(\mathcal{P}_D, \mathcal{P}_C) + (1 - \mu) \cdot \mathsf{e\text{-}set\text{-}hd}(\mathcal{E}_D, \mathcal{E}_C), \qquad (1)$$

*where $|\cdot|$ represents the set cardinality, $\mu = \frac{|\mathcal{P}_D|}{|\mathcal{P}_D \cup \mathcal{E}_D|}$ and $0 \leq \mu \leq 1$;*

$$\mathsf{p\text{-}hd}(\mathcal{P}_D, \mathcal{P}_C) := \begin{cases} 1 & \text{if } \mathcal{P}_D = \emptyset \\ \frac{|\mathcal{P}_D \cap \mathcal{P}_C|}{|\mathcal{P}_D|} & \text{otherwise,} \end{cases} \qquad (2)$$

$$\mathsf{e\text{-}set\text{-}hd}(\mathcal{E}_D, \mathcal{E}_C) := \begin{cases} 1 & \text{if } \mathcal{E}_D = \emptyset \\ \sum\limits_{\epsilon_i \in \mathcal{E}_D} \frac{\max\{\mathsf{e\text{-}hd}(\epsilon_i, \epsilon_j) : \epsilon_j \in \mathcal{E}_C\}}{|\mathcal{E}_D|} & \text{otherwise,} \end{cases} \qquad (3)$$

*where $\epsilon_i, \epsilon_j$ are existential restrictions; and*

$$\text{e-hd}(\exists r.X, \exists s.Y) := \gamma(\nu + (1 - \nu) \cdot \text{hd}(\mathcal{T}_X, \mathcal{T}_Y)) \tag{4}$$

*where* $\gamma = \frac{|\mathcal{R}_r \cap \mathcal{R}_s|}{|\mathcal{R}_r|}$ *and* $0 \leq \nu < 1$.

The meaning of $\mu$ and $\nu$ are similar to those defined in our previous work [13] and are set to $\frac{|\mathcal{P}_C|}{|\mathcal{P}_C \cup \mathcal{E}_C|}$ and 0.4, respectively. However, in this work, we introduce the notion of $\gamma$ which is the proportion of common roles between $r$ and $s$ against all those respect to $r$. For a special case where $\gamma = 0$, this means that there is no role commonality, therefore, further computation for all successors should be omitted. For the case that $0 < \gamma < 1$, this reveals that there exists some commonality. Moreover, if $\gamma = 1$, both $r$ and $s$ are totally similar and thus considered logically equivalent.

**Proposition 1.** *Let* $C, D$ *be* $\mathcal{ELH}$ *concept descriptions, and* $\mathcal{O}$ *an* $\mathcal{ELH}$ *unfoldable TBox. Then, the following are equivalent:*

1. $C \sqsubseteq_{\mathcal{O}} D$
2. $\text{hd}(\mathcal{T}_D, \mathcal{T}_C) = 1$,

*where* $X$ *is the equivalent expanded concept description w.r.t.* $\mathcal{O}$, *and* $\mathcal{T}_X$ *is its corresponding* $\mathcal{ELH}$ *description tree, with* $X \in \{C, D\}$.

*Proof.* $(1 \implies 2)$ To prove this, we need to show that for each $v \in V_D$, there exists $h(v) \in V_C$ such that $\text{p-hd}(\cdot, \cdot) = 1$ and $\text{e-set-hd}(\cdot, \cdot) = 1$ (only for those non-leaf nodes). Let $d$ be the depth of $\mathcal{T}_D$. Since $C \sqsubseteq_{\mathcal{O}} D$, by Theorem 1 there exists a homomorphism from $\mathcal{T}_D$ to $\mathcal{T}_C$. For the induction base case where $d = 0$ and $C = P_1 \sqcap \ldots \sqcap P_m$, there exists a mapping from $rt_D$ to $rt_C$ such that $\ell_D(v) \subseteq \ell_C(h(v))$ (i.e. $\text{hd} = \text{p-hd} = 1$). For the induction step where $C = P_1 \sqcap \ldots \sqcap P_m \sqcap \exists r_1 C_1 \sqcap \ldots \sqcap \exists r_n C_n$ there exists a mapping from each $v$ to $h(v)$ such that $\ell_D(v) \subseteq \ell_C(h(v))$ (i.e. $\text{p-hd}(\cdot, \cdot) = 1$) and $\rho_D(v, w) \subseteq \rho_C(h(v), h(w))$ (i.e. $\text{e-set-hd}(\cdot, \cdot) = 1$) where $w$ and $h(w)$ are successors of $v$ and $h(v)$, respectively. For the case where $v$ is a leaf, this is similar to the base case (i.e. $\text{p-hd}(\cdot, \cdot) = 1$).

$(2 \implies 1)$ By Definition 2, $\text{hd}(\mathcal{T}_D, \mathcal{T}_C) = 1$ means $\text{p-hd}(\mathcal{P}_D, \mathcal{P}_C) = 1$ and $\text{e-set-hd}(\mathcal{E}_D, \mathcal{E}_C) = 1$ (in case that the tree has child nodes), therefore for each $P \in \mathcal{P}_D$ there exists $P \in \mathcal{P}_C$ (i.e. $\mathcal{P}_D \subseteq \mathcal{P}_C$) and for each $\exists r.D' \in \mathcal{E}_D$ there exists $\exists s.C' \in \mathcal{E}_C$ such that $s \sqsubseteq^* r$ and $C' \sqsubseteq D'$. By Corollary 1, this implies that $C \sqsubseteq D$.

The homomorphism degree function provides a numerical value that represents structural similarity of one concept description when compared to another concept description. Since both directions constitute the degree of the two concepts being equivalent, our similarity measure for $\mathcal{ELH}$ concept descriptions is defined by means of these values. Definition 3 defines a similarity between concepts.

**Definition 3.** *Let $C, D$ be $\mathcal{ELH}$ concept descriptions, and $\mathcal{O}$ an $\mathcal{ELH}$ unfoldable TBox. The* degree of similarity *between $C$ and $D$, in symbols $\mathsf{sim}(C, D)$, is defined as:*

$$\mathsf{sim}(C, D) := \frac{\mathsf{hd}(\mathcal{T}_C, \mathcal{T}_D) + \mathsf{hd}(\mathcal{T}_D, \mathcal{T}_C)}{2}, \tag{5}$$

*where $X$ is the equivalent expanded concept description w.r.t. $\mathcal{O}$, and $\mathcal{T}_X$ is its corresponding $\mathcal{ELH}$ description tree, with $X \in \{C, D\}$.*

*Example 1.* To be more illustrative, consider concepts defined in $\mathcal{O}_{\mathsf{med}}$ (see Figure 1). From a classical DL reasoner's point of view, it is clear that the concept AspirationOfMilk (AMK) and AspirationOfMucus (AMC) are not in the subsumption relation, i.e. there is no relationship between the two concepts, despite the fact that they are both disorders in a group of AspirationSymdromes. Moreover, it is intuitive to argue that AspirationOfMilk is more similar to AspirationOfMucus than to Hypoxemia or to Hypoxia. Consider the expanded form of AMK and AMC.

$$\text{AMK} \equiv X \sqcap \text{RespiratoryDisorder} \sqcap \text{InhalationOfLiquid}$$
$$\sqcap \exists\text{agent.Milk} \sqcap \exists\text{assocWith.Milk}$$

$$\text{AMC} \equiv X \sqcap \text{RespiratoryDisorder} \sqcap \exists\text{agent.Mucus}$$

where $X$ is a fresh concept. The following shows sample computation steps for $\mathsf{hd}(\mathcal{T}_{\mathsf{AMK}}, \mathcal{T}_{\mathsf{AMC}})$:

$$\mathsf{hd}(\mathcal{T}_{\mathsf{AMK}}, \mathcal{T}_{\mathsf{AMC}}) := \tfrac{3}{5}\mathsf{p\text{-}hd}(\mathcal{P}_{\mathsf{AMK}}, \mathcal{P}_{\mathsf{AMC}}) + \tfrac{2}{5}\mathsf{e\text{-}set\text{-}hd}(\mathcal{E}_{\mathsf{AMK}}, \mathcal{E}_{\mathsf{AMC}})$$

$$:= \tfrac{3}{5}(\tfrac{2}{3}) + \tfrac{2}{5}\sum_{\epsilon_i \in \mathcal{E}_{\mathsf{AMK}}} \frac{max\{\mathsf{e\text{-}hd}(\epsilon_i, \epsilon_j) : \epsilon_j \in \mathcal{E}_{\mathsf{AMC}}\}}{|\mathcal{E}_{\mathsf{AMK}}|}$$

$$:= \tfrac{3}{5}(\tfrac{2}{3}) + \tfrac{2}{5}(\tfrac{1}{2})\sum_{\epsilon_i \in \mathcal{E}_{\mathsf{AMK}}} max\{\mathsf{e\text{-}hd}(\epsilon_i, \epsilon_j) : \epsilon_j \in \mathcal{E}_{\mathsf{AMC}}\}$$

$$:= \tfrac{3}{5}(\tfrac{2}{3}) + \tfrac{2}{5}(\tfrac{1}{2})(\tfrac{2}{5} + \tfrac{2}{5})$$

$$//\text{Where} \sum_{\epsilon_i \in \mathcal{E}_{\mathsf{AMK}}} max\{\mathsf{e\text{-}hd}(\epsilon_i, \epsilon_j) : \epsilon_j \in \mathcal{E}_{\mathsf{AMC}}\} = \tfrac{2}{5} + \tfrac{2}{5}; \text{ see belows}$$

$$:= 0.56$$

The computation for the sub-description corresponding to $\epsilon_i = \exists\text{agent.Milk}$ and $\epsilon_j = \exists\text{agent.Mucus}$ is as follows:

$$\mathsf{e\text{-}hd}(\epsilon_i, \epsilon_j) := \gamma(\nu + (1 - \nu) \cdot \mathsf{hd}(\mathcal{T}_{\mathsf{Milk}}, \mathcal{T}_{\mathsf{Mucus}}))$$
$$:= \tfrac{2}{2}(\tfrac{2}{5} + 0) := \tfrac{2}{5}$$

With the sub-description $\epsilon_i = \exists\text{assocWith.Milk}$ and $\epsilon_j = \exists\text{agent.Mucus}$, we have

$$\mathsf{e\text{-}hd}(\epsilon_i, \epsilon_j) := \gamma(\nu + (1 - \nu) \cdot \mathsf{hd}(\mathcal{T}_{\mathsf{Milk}}, \mathcal{T}_{\mathsf{Mucus}}))$$
$$:= \tfrac{1}{1}(\tfrac{2}{5} + 0) := \tfrac{2}{5}.$$

**Table 2.** Homomorphism degrees among concepts defined in $\mathcal{O}_{med}$ where HPX, HPM and ASD stand for Hypoxia, Hypoxemia and AspirationSyndromes, respectively

| hd($\downarrow, \rightarrow$) | AMC | AMK | HPX | HPM | ASD |
|---|---|---|---|---|---|
| AspirationOfMucus | 1.0 | 0.56 | 0.333 | 0.25 | 1.0 |
| AspirationOfMilk | 0.8 | 1.0 | 0.333 | 0.25 | 1.0 |
| Hypoxia | 0.333 | 0.2 | 1.0 | 0.75 | 0.5 |
| Hypoxemia | 0.333 | 0.2 | 1.0 | 1.0 | 0.5 |
| AspirationSyndromes | 0.667 | 0.4 | 0.333 | 0.25 | 1.0 |

The reverse direction can be computed by:

$$\text{hd}(\mathcal{T}_{AMC}, \mathcal{T}_{AMK}) := \tfrac{2}{3}\text{p-hd}(\mathcal{P}_{AMC}, \mathcal{P}_{AMK}) + \tfrac{1}{3}\text{e-set-hd}(\mathcal{E}_{AMC}, \mathcal{E}_{AMK})$$

$$:= \tfrac{2}{3}(\tfrac{2}{2}) + \tfrac{1}{3}\sum_{\epsilon_i \in \mathcal{E}_{AMC}} \frac{max\{\text{e-hd}(\epsilon_i, \epsilon_j):\epsilon_j \in \mathcal{E}_{AMK}\}}{|\mathcal{E}_{AMC}|}$$

$$:= \tfrac{2}{3}(\tfrac{2}{2}) + \tfrac{1}{3}(\tfrac{1}{1})\sum_{\epsilon_i \in \mathcal{E}_{AMC}} max\{\text{e-hd}(\epsilon_i, \epsilon_j) : \epsilon_j \in \mathcal{E}_{AMK}\}$$

$$:= \tfrac{2}{3}(\tfrac{2}{2}) + \tfrac{1}{3}(\tfrac{1}{1})(\tfrac{2}{5})$$

//Where $max\{\text{e-hd}(\epsilon_i, \epsilon_j) : \epsilon_j \in \mathcal{E}_{AMC}\} = \tfrac{2}{5}$; see belows

$$:= 0.8$$

The computation for the sub-description corresponding to $\epsilon_i = \exists\text{agent.Mucus}$ and $\epsilon_j = \exists\text{agent.Milk}$ is as follows:

$$\text{e-hd}(\epsilon_i, \epsilon_j) := \gamma(\nu + (1 - \nu) \cdot \text{hd}(\mathcal{T}_{Mucus}, \mathcal{T}_{Milk}))$$
$$:= \tfrac{2}{2}(\tfrac{2}{5} + 0) := \tfrac{2}{5}.$$

With $\epsilon_i = \exists\text{agent.Mucus}$ and $\epsilon_j = \exists\text{assocWith.Milk}$, the computation for the sub-description is as follows;

$$\text{e-hd}(\epsilon_i, \epsilon_j) := \gamma(\nu + (1 - \nu) \cdot \text{hd}(\mathcal{T}_{Mucus}, \mathcal{T}_{Milk}))$$
$$:= \tfrac{1}{2}(\tfrac{2}{5} + 0) := \tfrac{1}{5}.$$

Table 2 and Table 3 show homomorphism degrees and similarity degrees among all concepts defined in $\mathcal{O}_{med}$. It is obvious that the results we obtained are as expected.

It is to be mentioned that the similarity measure sim first introduced in [13] is quite similar to *simi* proposed by [10] since they are both recursive-based method. In fact, the meaning of the weighting parameter $\nu$ used in sim and $\omega$ in *simi* are identical and similarly defined. Likewise, the operators that represent the t-conorm, and fuzzy connector are relatively used but differently defined. However, unlike the work proposed by [10], the use of $\mu$ and the way

**Table 3.** Similarity degrees among concepts defined in $\mathcal{O}_{med}$

| sim($\downarrow$, $\rightarrow$) | AMC | AMK | HPX | HPM | ASD |
|---|---|---|---|---|---|
| AspirationOfMucus | 1.0 | 0.68 | 0.333 | 0.292 | 0.833 |
| AspirationOfMilk | - | 1.0 | 0.267 | 0.225 | 0.7 |
| Hypoxia | - | - | 1.0 | 0.875 | 0.417 |
| Hypoxemia | - | - | - | 1.0 | 0.375 |
| AspirationSyndromes | - | - | - | - | 1.0 |

it is weighted, which determines how important the primitive concepts are to be considered, is defined. The other is obviously the distinction of their inspirations. While *simi* is inspired by the Jaccard Index [5], sim is, on the other hand, motivated by the homomorphism-based structural subsumption characterization. In sim, as a pre-process, concept names are to be transformed into an $\mathcal{ELH}$ concept description tree. Taking this as an advantage, a bottom-up approach, which allows rejection of unnecessary recursive calls and reuses of solutions to subproblems, can be alternatively devised.

## 4   Desirable Properties for Concept Similarity Measure

This section describes desirable properties for concept similarity measure and provides corresponding mathematical proofs. At the end of the section, a comparison of satisfactory properties between sim and those significantly reported in other classical works is made available.

Definition 4 determines important properties for concept similarity measure introduced by [10]. These are believed to be desirable features and thus checked for satisfaction. Theorem 2 states the characteristics of sim.

**Definition 4.** *Let $C$, $D$ and $E$ be $\mathcal{ELH}$ concept, the similarity measure is*

- *i. symmetric iff* $\text{sim}(C, D) = \text{sim}(D, C)$,
- *ii. equivalence closed iff* $\text{sim}(C, D) = 1 \Longleftrightarrow C \equiv D$,
- *iii. equivalence invariant if* $C \equiv D$ *then* $\text{sim}(C, E) = \text{sim}(D, E)$,
- *iv. subsumption preserving if* $C \sqsubseteq D \sqsubseteq E$ *then* $\text{sim}(C, D) \geq \text{sim}(C, E)$,
- *v. reverse subsumption preserving if* $C \sqsubseteq D \sqsubseteq E$ *then* $\text{sim}(C, E) \leq \text{sim}(D, E)$,
- *vi. structurally dependent Let $C_i$ and $C_j$ be atoms in $C$ where $C_i \not\sqsubseteq C_j$, the concept $D' := \bigcap_{i \leq n} C_i \sqcap D$ and $E' := \bigcap_{i \leq n} C_i \sqcap E$ satisfies the condition* $\lim_{n \to \infty} \text{sim}(D', E') = 1$,
- *vii. satisfying triangle inequality iff* $1 + \text{sim}(D, E) \geq \text{sim}(D, C) + \text{sim}(C, E)$.

**Theorem 2.** *The similarity-measure sim is:*

- *i. symmetric,*
- *ii. equivalence closed,*

*iii. equivalence invariant,*
*iv. subsumption preserving,*
*v. structurally dependent,*
*vi. not reverse subsumption preserving, and*
*vii. not satisfying triangle inequality.*

*Proof.*   i. By Definition 3, it is obvious that $\mathsf{sim}(C, D) = \mathsf{sim}(D, C)$.

ii. ($\Longrightarrow$) By Definition 2, $\mathsf{sim}(C, D) = 1$ iff $\mathsf{hd}(\mathcal{T}_C, \mathcal{T}_D) = 1$ and $\mathsf{hd}(\mathcal{T}_D, \mathcal{T}_C) = 1$. By Proposition 1, these imply that $C \sqsubseteq D$ and $D \sqsubseteq C$. Therefore, $C \equiv D$. ($\Longleftarrow$) Assume $C \equiv D$, then $C \sqsubseteq D$ and $D \sqsubseteq C$. Using the same proposition, this ensures that $\mathsf{hd}(\mathcal{T}_C, \mathcal{T}_D) = 1$, and $\mathsf{hd}(\mathcal{T}_D, \mathcal{T}_C) = 1$, which means $\mathsf{sim}(C, D) = 1$.

iii. $C \equiv D$ iff $C \sqsubseteq D$ and $D \sqsubseteq C$. By using Corollary 2, we have $\mathcal{P}_C = \mathcal{P}_D$ and $\mathcal{E}_C \cong \mathcal{E}_D$. Therefore, $\mathcal{T}_C = \mathcal{T}_D$ and this implies $\mathsf{hd}(\mathcal{T}_C, \mathcal{T}_E) = \mathsf{hd}(\mathcal{T}_D, \mathcal{T}_E)$ and $\mathsf{hd}(\mathcal{T}_E, \mathcal{T}_C) = \mathsf{hd}(\mathcal{T}_E, \mathcal{T}_D)$. Such that $\mathsf{sim}(C, E) = \mathsf{sim}(D, E)$.

iv. We need to show that

$$\frac{\mathsf{hd}(\mathcal{T}_C, \mathcal{T}_D) + \mathsf{hd}(\mathcal{T}_D, \mathcal{T}_C)}{2} \geq \frac{\mathsf{hd}(\mathcal{T}_C, \mathcal{T}_E) + \mathsf{hd}(\mathcal{T}_E, \mathcal{T}_C)}{2}$$

Since $C \sqsubseteq D$ and $D \sqsubseteq E$, then $C \sqsubseteq E$. By Proposition 1, $\mathsf{hd}(\mathcal{T}_E, \mathcal{T}_C) = 1$ and $\mathsf{hd}(\mathcal{T}_D, \mathcal{T}_C) = 1$. Therefore, it suffices to show that

$$\mathsf{hd}(\mathcal{T}_C, \mathcal{T}_D) \geq \mathsf{hd}(\mathcal{T}_C, \mathcal{T}_E)$$

If expanded, on both sizes of the upper equation, we have $\mu = \frac{|\mathcal{P}_C|}{|\mathcal{P}_C \cup \mathcal{E}_C|}$. Hence, it is adequate to show that $\mathsf{p\text{-}hd}(\mathcal{P}_C, \mathcal{P}_D) \geq \mathsf{p\text{-}hd}(\mathcal{P}_C, \mathcal{P}_E)$ and $\mathsf{e\text{-}set\text{-}hd}(\mathcal{E}_C, \mathcal{E}_D) \geq \mathsf{e\text{-}set\text{-}hd}(\mathcal{E}_C, \mathcal{E}_E)$. For the first part, we show that

$$\frac{|\mathcal{P}_C \cap \mathcal{P}_D|}{|\mathcal{P}_C|} \geq \frac{|\mathcal{P}_C \cap \mathcal{P}_E|}{|\mathcal{P}_C|} \tag{6}$$

$$|\mathcal{P}_C \cap \mathcal{P}_D| \geq |\mathcal{P}_C \cap \mathcal{P}_E|$$

By Corollary 1, $C \sqsubseteq D \sqsubseteq E$ ensures that $\mathcal{P}_E \subseteq \mathcal{P}_D \subseteq \mathcal{P}_C$. Therefore

$$|\mathcal{P}_D| \geq |\mathcal{P}_E|$$

and Equation 6 is true. For the second part, we show that

$$\sum_{\epsilon_i \in \mathcal{E}_C} \frac{max\{\mathsf{e\text{-}hd}(\epsilon_i, \epsilon_j) : \epsilon_j \in \mathcal{E}_D\}}{|\mathcal{E}_C|} \geq \sum_{\epsilon_i \in \mathcal{E}_C} \frac{max\{\mathsf{e\text{-}hd}(\epsilon_i, \epsilon_j) : \epsilon_j \in \mathcal{E}_E\}}{|\mathcal{E}_C|} \tag{7}$$

$$\sum_{\epsilon_i \in \mathcal{E}_C} max\{\mathsf{e\text{-}hd}(\epsilon_i, \epsilon_j) : \epsilon_j \in \mathcal{E}_D\} \geq \sum_{\epsilon_i \in \mathcal{E}_C} max\{\mathsf{e\text{-}hd}(\epsilon_i, \epsilon_j) : \epsilon_j \in \mathcal{E}_E\}.$$

Let $\hat{\epsilon}_i \in \mathcal{E}_E$ such that $\text{e-hd}(\epsilon_i, \hat{\epsilon}_i) = max\{\text{e-hd}(\epsilon_i, \epsilon_j) : \epsilon_j \in \mathcal{E}_E\}$, but since $\hat{\epsilon}_i \in \mathcal{E}_E \subseteq \mathcal{E}_D$, then $max\{\text{e-hd}(\epsilon_i, \epsilon_j) : \epsilon_j \in \mathcal{E}_D\} \geq \text{e-hd}(\epsilon_i, \hat{\epsilon}_i)$. Therefore, Equation 7 is true.

v. Let $D' := \prod_{i \leq n} C_i \sqcap D$, $E' := \prod_{i \leq n} C_i \sqcap E$, and $n = n_\mathcal{P} + n_\mathcal{E}$ be the number of all atom sequences in $C$ where $n_\mathcal{P}$ and $n_\mathcal{E}$ be the number of primitive concepts and the number existential restrictions, respectively. To prove this, we consider the following case distinction.

(a) if $n_\mathcal{P} \to \infty$ and $n_\mathcal{E}$ is finite, it suffices to show that $\lim\limits_{n_\mathcal{P} \to \infty} \mu = 1$ and $\lim\limits_{n_\mathcal{P} \to \infty} \text{p-hd}(\mathcal{P}_{D'}, \mathcal{P}_{E'}) = 1$. Therefore, $\text{hd}(D', E') = \text{hd}(E', D') = 1$ and these imply that $\text{sim}(D', E') = 1$. From Equation 2, we have

$$
\begin{aligned}
\mu &= \frac{|\mathcal{P}_{D'}|}{|\mathcal{P}_{D'} \cup \mathcal{E}_{D'}|} \\
&= \frac{|\mathcal{P}_{D'}|}{|\mathcal{P}_{D'}| + |\mathcal{E}_{D'}|} \\
&= \frac{|\mathcal{P}_C| + |\mathcal{P}_D|}{|\mathcal{P}_C| + |\mathcal{P}_D| + |\mathcal{E}_{D'}|} \\
&= \frac{n_\mathcal{P} + |\mathcal{P}_D|}{n_\mathcal{P} + |\mathcal{P}_D| + |\mathcal{E}_{D'}|}
\end{aligned}
\tag{8}
$$

Since $|\mathcal{P}_D|$ and $|\mathcal{E}_{D'}|$ are constant, $\lim\limits_{n_\mathcal{P} \to \infty} \mu = \lim\limits_{n_\mathcal{P} \to \infty} \frac{n_\mathcal{P} + |\mathcal{P}_D|}{n_\mathcal{P} + |\mathcal{P}_D| + |\mathcal{E}_{D'}|} = 1$. For the second part, we have

$$
\begin{aligned}
\text{p-hd}(\mathcal{P}_{D'}, \mathcal{P}_{E'}) &= \frac{|\mathcal{P}_{D'} \cap \mathcal{P}_{E'}|}{|\mathcal{P}_{D'}|} \\
&= \frac{|\mathcal{P}_C| + |\mathcal{P}_D \cap \mathcal{P}_E|}{|\mathcal{P}_C| + |\mathcal{P}_D|} \\
&= \frac{n_\mathcal{P} + |\mathcal{P}_D \cap \mathcal{P}_E|}{n_\mathcal{P} + |\mathcal{P}_D|}
\end{aligned}
$$

where $|\mathcal{P}_D \cap \mathcal{P}_E|$ and $|\mathcal{P}_D|$ are constant. Thus,

$$
\lim\limits_{n_\mathcal{P} \to \infty} \text{p-hd}(\mathcal{P}_{D'}, \mathcal{P}_{E'}) = \lim\limits_{n_\mathcal{P} \to \infty} \frac{n_\mathcal{P} + |\mathcal{P}_D \cap \mathcal{P}_E|}{n_\mathcal{P} + |\mathcal{P}_D|} = 1. \tag{9}
$$

(b) if $n_\mathcal{E} \to \infty$ and $n_\mathcal{P}$ is finite, it suffices to show that $\lim\limits_{n_\mathcal{E} \to \infty} \mu = 0$ and $\lim\limits_{n_\mathcal{E} \to \infty} \text{e-set-hd}(\mathcal{E}_{D'}, \mathcal{E}_{E'}) = 1$ which implies $\text{hd}(D', E') = \text{hd}(E', D') = 1$, and $\text{sim}(D', E') = 1$. From Equation 8, the value of $\mu$ is as follows:

$$
\begin{aligned}
\mu &= \frac{|\mathcal{P}_C| + |\mathcal{P}_D|}{|\mathcal{P}_C| + |\mathcal{P}_D| + |\mathcal{E}_{D'}|} \\
&= \frac{|\mathcal{P}_C| + |\mathcal{P}_D|}{|\mathcal{P}_C| + |\mathcal{P}_D| + n_\mathcal{E} + |\mathcal{E}_D|}
\end{aligned}
$$

Since $|\mathcal{P}_C|$, $|\mathcal{P}_D|$ and $|\mathcal{E}_D|$ are constant, by taking limit, we have

$$\lim_{n_\mathcal{E} \to \infty} \mu = \lim_{n_\mathcal{E} \to \infty} \frac{|\mathcal{P}_C| + |\mathcal{P}_D|}{|\mathcal{P}_C| + |\mathcal{P}_D| + n_\mathcal{E} + |\mathcal{E}_D|} = 0.$$

To show that $\lim_{n_\mathcal{E} \to \infty} \text{e-set-hd}(\mathcal{E}_{D'}, \mathcal{E}_{E'}) = 1$, we have

$$\text{e-set-hd}(\mathcal{E}_{D'}, \mathcal{E}_{E'}) = \sum_{e_i \in \mathcal{E}_{D'}} \frac{\max\{\text{e-hd}(e_i, e_j) : e_j \in \mathcal{E}_{E'}\}}{|\mathcal{E}_{D'}|}$$

$$= \frac{\sum_{e_i \in \mathcal{E}_{D'}} \max\{\text{e-hd}(e_i, e_j) : e_j \in \mathcal{E}_{E'}\}}{|\mathcal{E}_{D'}|}$$

$$= \frac{\sum_{e_i \in \mathcal{E}_C} \max\{\text{e-hd}(e_i, e_j) : e_j \in \mathcal{E}_{E'}\} + \sum_{e_i \in \mathcal{E}_D} \max\{\text{e-hd}(e_i, e_j) : e_j \in \mathcal{E}_{E'}\}}{|\mathcal{E}_C \cup \mathcal{E}_D|}$$

Since $\mathcal{E}_C \subseteq \mathcal{E}_{E'}$, for each $\epsilon_i \in \mathcal{E}_C$ there exists $\epsilon_j \in \mathcal{E}_{E'}$ such that $\epsilon_i = \epsilon_j$. Thus,

$$\text{e-set-hd}(\mathcal{E}_{D'}, \mathcal{E}_{E'}) = \frac{n_\mathcal{E} + p}{|\mathcal{E}_C| + |\mathcal{E}_D|}$$

$$= \frac{n_\mathcal{E} + p}{n_\mathcal{E} + |\mathcal{E}_D|}$$

where $p = \sum_{e_i \in \mathcal{E}_D} \max\{\text{e-hd}(e_i, e_j) : e_j \in \mathcal{E}_{E'}\}$, and $p \leq |\mathcal{E}_D|$. Therefore, the following is true.

$$\lim_{n_\mathcal{E} \to \infty} \text{e-set-hd}(\mathcal{E}_{D'}, \mathcal{E}_{E'}) = \lim_{n_\mathcal{E} \to \infty} \frac{n_\mathcal{E} + p}{n_\mathcal{E} + |\mathcal{E}_D|} = 1. \quad (10)$$

(c) if $n_\mathcal{P} \to \infty$ and $n_\mathcal{E} \to \infty$, it suffices to show that $\lim_{n_\mathcal{P} \to \infty} \text{p-hd}(\mathcal{P}_{D'}, \mathcal{P}_{E'}) = 1$ and $\lim_{n_\mathcal{E} \to \infty} \text{e-set-hd}(\mathcal{E}_{D'}, \mathcal{E}_{E'}) = 1$. But these follow from Equation 9 and Equation 10.

vi. Consider a counter example defined in Figure 2. It is obvious that $C \sqsubseteq D \sqsubseteq E$. By definition,

$$\text{sim}(C, E) := \frac{\text{hd}(\mathcal{T}_C, \mathcal{T}_E) + \text{hd}(\mathcal{T}_E, \mathcal{T}_C)}{2}$$

$$:= \frac{0.4250 + 1}{2}$$

$$:= 0.7125$$

and

$$\text{sim}(D, E) := \frac{\text{hd}(\mathcal{T}_D, \mathcal{T}_E) + \text{hd}(\mathcal{T}_E, \mathcal{T}_D)}{2}$$

$$:= \frac{0.3333 + 1}{2}$$

$$:= 0.6667.$$

Apparently, there exists the case $\text{sim}(C, E) \not\sqsubseteq \text{sim}(D, E)$.

$$
\begin{aligned}
E &\equiv \exists r.(F \sqcap G) \\
D &\equiv \exists r.(F \sqcap G) \sqcap \exists s.F \sqcap \exists s.G \\
C &\equiv \exists r.(F \sqcap G) \sqcap \exists s.F \sqcap \exists s.G \sqcap \exists r.(F \sqcap H)
\end{aligned}
$$

**Fig. 2.** Examples of $\mathcal{ELH}$ concept descriptions

vii. Providing the concept description $C$, $D$, and $E$ defined in Figure 2, the following demonstrates the case $1 + \mathsf{sim}(D,E) \ngeq \mathsf{sim}(D,C) + \mathsf{sim}(C,E)$. Here, we have

$$
\mathsf{sim}(D,E) := \frac{\mathsf{hd}(\mathcal{T}_D,\mathcal{T}_E)+\mathsf{hd}(\mathcal{T}_E,\mathcal{T}_D)}{2} := \frac{0.3333+1}{2}
$$
$$
:= 0.6667
$$

and

$$
\mathsf{sim}(D,C) := \frac{\mathsf{hd}(\mathcal{T}_D,\mathcal{T}_C)+\mathsf{hd}(\mathcal{T}_C,\mathcal{T}_D)}{2} := \frac{1+0.9250}{2}
$$
$$
:= 0.9625
$$

and

$$
\mathsf{sim}(C,E) := \frac{\mathsf{hd}(\mathcal{T}_C,\mathcal{T}_E)+\mathsf{hd}(\mathcal{T}_E,\mathcal{T}_C)}{2} := \frac{0.4250+1}{2}
$$
$$
:= 0.7125.
$$

By applying a summation, it is obvious that $1.6667 \ngeq 1.675$ .

**Table 4.** A comparison on concept-similarity properties

| Similarity Measure | DL | symmetric | equi. closed | equi invariant | sub. preserving | struc. dependent | rev. sub. preserving | triangle inequality |
|---|---|---|---|---|---|---|---|---|
| sim | $\mathcal{ELH}$ | ✓ | ✓ | ✓ | ✓ | ✓ | | |
| Lehmann and Turhan [10] | $\mathcal{ELH}$ | ✓ | ✓ | ✓ | ✓ | ✓ | | |
| Jaccard [5] | $\mathcal{L}_0$ | ✓ | ✓ | ✓ | ✓ | ✓ | ✓ | ✓ |
| Janowicz and Wilkes [7] | $\mathcal{SHI}$ | ✓ | | | | | | ✓ |
| Janowicz [6] | $\mathcal{ALCHQ}$ | ✓ | | | | | | ✓ |
| d'Amato et al. [2] | $\mathcal{ALC}$ | | | | | | | |
| Fanizzi and d'Amato [4] | $\mathcal{ALN}$ | ✓ | | ✓ | | ✓ | ✓ | |
| d'Amato et al. [1] | $\mathcal{ALC}$ | ✓ | | ✓ | | ✓ | ✓ | |
| d'Amato et al. [3] | $\mathcal{ALE}$ | ✓ | | ✓ | | ✓ | ✓ | |

To ensure that our proposed method reaches the performance, Table 4 compares desirable properties of sim and those previously reported in other classical works. Except than the work proposed by [5], which allows only concept conjunction, our approach and that proposed by [10] apparently hold significant features.

# 5   Conclusion

To this end, we have expanded a concept similarity measure for $\mathcal{EL}$ to take into account also role hierarchy. Comparing to other related works, the measure has been proved that it is outperforming and indeed identical to $simi$ in terms of satisfaction of desirable properties.

Particularly, the proposed algorithm is inspired by the homomorphism-based structural subsumption characterization. With the top-down approach, a similarity degree is recursively computed, and as a nature of recursion, there is a chance that the number of unnecessary recursive calls will be greatly increase. Fortunately, as being computed based on description trees, an optimized version of the algorithm that allows rejection of needless computation can be alternatively devised in a reversed direction and this is regarded as one target in our future works. The other directions of possible future works are an extension of the algorithm to a general TBox (i.e. a handling to concepts with cyclic definition) and to a more expressive DL. Lastly, we also aim at setting up experiments on comprehensive terminologies (e.g. SNOMED CT [12] and Gene Ontology [11]) and making a comparison among results obtained from different methods.

**Acknowledgments.** This work is partially supported by the National Research University (NRU) project of Thailand Office for Higher Education Commission; and by Center of Excellence in Intelligent Informatics, Speech and Language Technology, and Service Innovation (CILS), Thammasat University.

# References

1. d'Amato, C., Fanizzi, N., Esposito, F.: A semantic similarity measure for expressive description logics. In: Proceedings of Convegno Italiano di Logica Computazionale, CILC 2005 (2005)
2. d'Amato, C., Fanizzi, N., Esposito, F.: A dissimilarity measure for $\mathcal{ALC}$ concept descriptions. In: Proceedings of the 2006 ACM Symposium on Applied Computing, SAC 2006, pp. 1695–1699. ACM, New York (2006). http://doi.acm.org/10.1145/1141277.1141677
3. d'Amato, C., Staab, S., Fanizzi, N.: On the influence of description logics ontologies on conceptual similarity. In: Gangemi, A., Euzenat, J. (eds.) EKAW 2008. LNCS (LNAI), vol. 5268, pp. 48–63. Springer, Heidelberg (2008). http://dx.doi.org/10.1007/978-3-540-87696-0_7
4. Fanizzi, N., d'Amato, C.: A similarity measure for the aln description logic. Proceedings of Convegno Italiano di Logica Computazionale, CILC 2006, pp. 26–27 (2006)

5. Jaccard, P.: Étude comparative de la distribution florale dans une portion des Alpes et des Jura. Bulletin del la Société Vaudoise des Sciences Naturelles **37**, 547–579 (1901)

6. Janowicz, K.: Sim-DL: towards a semantic similarity measurement theory for the description logic $\mathcal{ALCNR}$ in geographic information retrieval. In: Meersman, R., Tari, Z., Herrero, P. (eds.) OTM 2006 Workshops. LNCS, vol. 4278, pp. 1681–1692. Springer, Heidelberg (2006). http://dx.doi.org/10.1007/11915072_74

7. Janowicz, K., Wilkes, M.: SIM-DL$_A$: a novel semantic similarity measure for description logics reducing inter-concept to inter-instance similarity. In: Aroyo, L., Traverso, P., Ciravegna, F., Cimiano, P., Heath, T., Hyvönen, E., Mizoguchi, R., Oren, E., Sabou, M., Simperl, E. (eds.) ESWC 2009. LNCS, vol. 5554, pp. 353–367. Springer, Heidelberg (2009). http://dx.doi.org/10.1007/978-3-642-02121-3_28

8. Jirathitikul, P., Nithisansawadikul, S., Tongphu, S., Suntisrivaraporn, B.: A similarity measuring service for snomed ct: structural analysis of concepts in ontology. In: 2014 11th International Conference on Electrical Engineering/Electronics, Computer, Telecommunications and Information Technology (ECTI-CON), pp. 1–6, May 2014

9. Lehmann, J., Haase, C.: Ideal downward refinement in the $\mathcal{EL}$ description logic. Technical report, University of Leipzig (2009). http://jens-lehmann.org/files/2009_ideal_operator_el_tr.pdf

10. Lehmann, K., Turhan, A.-Y.: A framework for semantic-based similarity measures for $\mathcal{ELH}$-concepts. In: del Cerro, L.F., Herzig, A., Mengin, J. (eds.) JELIA 2012. LNCS, vol. 7519, pp. 307–319. Springer, Heidelberg (2012)

11. Michael Ashburner, C.A.: Creating the Gene Ontology Resource: Design and Implementation. Genome Research **11**(8), 1425–1433 (2001). http://dx.doi.org/10.1101/gr.180801

12. Schulz, S., Suntisrivaraporn, B., Baader, F., Boeker, M.: SNOMED reaching its adolescence: Ontologists' and logicians' health check. International Journal of Medical Informatics **78**(Supplement 1), S86–S94 (2009)

13. Suntisrivaraporn, B.: A similarity measure for the description logic $\mathcal{EL}$ with unfoldable terminologies. In: International Conference on Intelligent Networking and Collaborative Systems (INCoS-13). pp. 408–413 (2013)

# A Graph-Based Approach to Ontology Debugging in DL-Lite

Xuefeng Fu$^{(\boxtimes)}$, Yong Zhang, and Guilin Qi

School of Computer Science and Engineering, Southeast University, Nanjing, China
{fxf,zhangyong,gqi}@seu.edu.cn

**Abstract.** Ontology debugging is an important nonstandard reasoning task in ontology engineering which provides the explanations of the causes of incoherence in an ontology. In this paper, we propose a graph-based algorithm to calculate minimal incoherence-preserving subterminology (MIPS) of an ontology in a light-weight ontology language, DL-Lite. We first encode a DL-Lite ontology to a graph, then calculate all the MIPS of an ontology by backtracking some pairs of nodes in the graph. We implement the algorithm and conduct experiments over some real ontologies. The experimental results show that our debugging system is efficient and outperforms the existing systems.

**Keywords:** *DL-Lite* · Ontology · Debugging · MIPS · Graph

## 1 Introduction

Ontologies play an important role in the semantic web, as it allows information to be shared in a semantically unambiguous way. However, the development and maintenance of an ontology are complex and error-prone. Thus, an ontology can easily become logically inconsistent. Ontology debugging [14], which aims to pinpoint the causes of logical inconsistencies, has become one of key issues in ontology engineering.

To debug an ontology, one can compute minimal unsatisfiability preserving subterminology (MUPS) of an ontology w.r.t. an unsatisfiable concept [9,12,14] or compute minimal incoherence preserving subterminology (MIPS) of an ontology [12,18]. MUPSs are useful for relating sets of axioms to the unsatisfiability of specific concepts and a MIPS of TBox is the minimal sub-TBox of $\mathcal{T}$ which is incoherent. It is argued that the notion of MIPS is more useful than that of MUPS to repair an ontology as every MIPS is a MUPS and removing one axiom from each MIPS can resolve the incoherence of an ontology. However, existing methods to compute all MIPS rely on the computation of all MUPS of an ontology w.r.t. all unsatisfiable concepts, thus are not efficient in some cases.

Recently, there is an increasing interest in inconsistency handling in DL-Lite, which is a lightweight ontology language supporting tractable reasoning. In [4], an inconsistent tolerant semantics is proposed for DL-Lite ontologies. This method does not consider debugging and repair of an ontology. In [21], the

© Springer International Publishing Switzerland 2015
T. Supnithi et al.(Eds.): JIST 2014, LNCS 8943, pp. 33–46, 2015.
DOI: 10.1007/978-3-319-15615-6_3

authors propose a debugging algorithm by backtracking techniques. This algorithm is further optimized by a module extraction method given in [8]. However, the algorithm presented in this paper only considers the problem of computing MUPS.

In this paper, we provide a graph-based algorithm to find all MIPS of an ontology in DL-Lite. In our approach, we encode the ontology into a graph, and derive subsumption relations between two concepts (or roles) by reachability of two nodes. The computation of all MIPS of an ontology can be achieved by backtracking pairs of nodes in the graph, thus avoiding the computation of all MUPS of the ontology w.r.t. all unsatisfiable concepts.

We implement our algorithm and develop an ontology debugging system. We then evaluate the performance of the algorithm by conducting experiments on some (adapted) real ontologies. The experimental results show the efficiency and scalability of our algorithm, and show that our system outperforms existing systems.

The rest of the paper is organized as follows. We first introduce some basics of debugging in DL-Lite and graph construction in Section 2. We then present our graph-based debugging algorithm in Section 3. After that, we present the evaluation results in Section 4. Section 5 discusses related work. Finally, we conclude this paper in Section 6.

## 2   Preliminaries

### 2.1   Ontology Debugging in DL-Lite

In our work, we consider DL-Lite$_{FR}$, which is an important language in DL-Lite that stands out for its tractable reasoning and efficient query answering [5]. We start with the introduction of DL-Lite$_{core}$, which is the core language for the DL-Lite family [1]. The complex concepts and roles of DL-Lite$_{core}$ are defined as follows: (1) $B \rightarrow A \mid \exists R$, (2) $R \rightarrow P \mid P^-$, (3) $C \rightarrow B \mid \neg B$, (4) $E \rightarrow R \mid \neg R$, where $A$ denotes an atomic concept, $P$ an atomic role, $B$ a basic concept, and $C$ a general concept. A basic concept which can be either an atomic concept or a concept of the form $\exists R$, where $R$ denotes a basic role which can be either an atomic role or the inverse of an atomic role.

In DL-Lite$_{core}$, an ontology $\mathcal{O} = \langle \mathcal{T}, \mathcal{A} \rangle$ consists of a TBox $\mathcal{T}$ and an ABox $\mathcal{A}$, where $\mathcal{T}$ is a finite set of *concept inclusion assertions* of the form: $B \sqsubseteq C$; and $\mathcal{A}$ is a finite set of *membership assertions* of the form: $A(a)$, $P(a, b)$. DL-Lite$_{FR}$ extends DL-Lite$_{core}$ with inclusion assertions between roles of the form $R \sqsubseteq E$ and functionality on roles (or on their inverses) of the form $(funct R)$ (or $(funct R^-)$). To keep the logic tractable, whenever a role inclusion $R_1 \sqsubseteq R_2$ appears in $\mathcal{T}$, neither $(funct R_2)$ nor $(funct R_2^-)$ can appear in it.

In the following, we call assertions of the form $B_1 \sqsubseteq B_2$ or $R_1 \sqsubseteq R_2$ as positive inclusions (PIs) and $B_1 \sqsubseteq \neg B_2$ or $R_1 \sqsubseteq \neg R_2$ as negative inclusions (NIs).

The semantics of DL-Lite is defined by an interpretation $\mathcal{I} = (\triangle^{\mathcal{I}}, \cdot^{\mathcal{I}})$ which consists of a non-empty domain set $\triangle^{\mathcal{I}}$ and an interpretation function $\cdot^{\mathcal{I}}$, which maps individuals, concepts and roles to elements of the domain, subsets of the domain and binary relations on the domain, respectively. The interpretation function can be extended to arbitrary concept (or role) descriptions and inclusion (or membership) assertions in a standard way[5]. Given an interpretation $\mathcal{I}$ and an assertion $\phi$, $\mathcal{I} \models \phi$ denotes that $\mathcal{I}$ is a *model* of $\phi$. An interpretation is called a *model* of an ontology $\mathcal{O}$, iff it satisfies each assertion in $\mathcal{O}$. An ontology $\mathcal{O}$ logically implies an assertion $\phi$, written $\mathcal{O} \models \phi$, if all models of $\mathcal{O}$ are also models of $\phi$.

Based on the PIs and NIs, we introduce the definitions of $clp(\mathcal{T})$ and $cln(\mathcal{T})$.

**Definition 1.** *Let $\mathcal{T}$ be a DL-Lite$_{FR}$ TBox. We define the PI-closure of $\mathcal{T}$, denoted by $clp(\mathcal{T})$, inductively as follows:*

1. *All positive inclusion assertions (PIs) in $\mathcal{T}$ are also in $clp(\mathcal{T})$.*
2. *If $B_1 \sqsubseteq B_2$ is in $\mathcal{T}$ and $B_2 \sqsubseteq B_3$ is in $clp(\mathcal{T})$, then also $B_1 \sqsubseteq B_3$ is in $clp(\mathcal{T})$.*
3. *If $R_1 \sqsubseteq R_2$ is in $\mathcal{T}$ and $R_2 \sqsubseteq R_3$ is in $clp(\mathcal{T})$, then also $R_1 \sqsubseteq R_3$ is in $clp(\mathcal{T})$.*
4. *If $R_1 \sqsubseteq R_2^-$ is in $\mathcal{T}$ and $R_2 \sqsubseteq R_3^-$ is in $clp(\mathcal{T})$, then also $R_1 \sqsubseteq R_3$ is in $clp(\mathcal{T})$.*
5. *If $R_1 \sqsubseteq R_2$ is in $\mathcal{T}$ and $\exists R_2 \sqsubseteq B$ is in $clp(\mathcal{T})$, then also $\exists R_1 \sqsubseteq B$ is in $clp(\mathcal{T})$.*
6. *If $R_1 \sqsubseteq R_2$ is in $\mathcal{T}$ and $\exists R_2^- \sqsubseteq B$ is in $clp(\mathcal{T})$, then also $\exists R_1^- \sqsubseteq B$ is in $clp(\mathcal{T})$.*

We introduce the definition of NI-closure[5].

**Definition 2.** *Let $\mathcal{T}$ be a DL-Lite$_{FR}$ TBox. We define the NI-closure of $\mathcal{T}$, denoted by $cln(\mathcal{T})$, inductively as follows:*

1. *All negative inclusion assertions (NIs) in $\mathcal{T}$ are also in $cln\mathcal{T}$.*
2. *All functionality assertions in $\mathcal{T}$ are also in $cln\mathcal{T}$.*
3. *If $B_1 \sqsubseteq B_2$ is in $\mathcal{T}$ and $B_2 \sqsubseteq \neg B_3$ or $B_3 \sqsubseteq \neg B_2$ is in $cln(\mathcal{T})$, then also $B_1 \sqsubseteq \neg B_3$ is in $cln(\mathcal{T})$.*
4. *If $R_1 \sqsubseteq R_2$ is in $\mathcal{T}$ and $\exists R_2 \sqsubseteq \neg B$ or $B \sqsubseteq \neg \exists R_2$ is in $cln(\mathcal{T})$, then also $\exists R_1 \sqsubseteq \neg B$ is in $cln(\mathcal{T})$.*
5. *If $R_1 \sqsubseteq R_2$ is in $\mathcal{T}$ and $\exists R_2^- \sqsubseteq \neg B$ or $B \sqsubseteq \neg \exists R_2^-$ is in $cln(\mathcal{T})$, then also $\exists R_1^- \sqsubseteq \neg B$ is in $cln(\mathcal{T})$.*
6. *If $R_1 \sqsubseteq R_2$ is in $\mathcal{T}$ and $R_2 \sqsubseteq \neg R_3$ or $R_3 \sqsubseteq \neg R_2$ is in $cln(\mathcal{T})$, then also $R_1 \sqsubseteq \neg R_3$ is in $cln(\mathcal{T})$.*
7. *If one of the assertions $\exists R \sqsubseteq \neg \exists R$, $\exists R^- \sqsubseteq \neg \exists R^-$ and $R \sqsubseteq \neg R$ is in $cln(\mathcal{T})$, then all three such assertions are in $cln(\mathcal{T})$.*

Finally, let $cl(\mathcal{T}) = clp(\mathcal{T}) \cup cln(\mathcal{T})$, where $cl(\mathcal{T})$ is the closure of $\mathcal{T}$.

We now introduce several important notations closely related to ontology debugging given in [17].

**Definition 3.** *Let $T$ be a DL-Lite$_{FR}$ TBox. A concept $C$ (resp. role $R$) in $T$ is unsatisfiable if and only if for each model $\mathcal{I}$ of $T$, $C^{\mathcal{I}} = \emptyset$ (resp. $R^{\mathcal{I}} = \emptyset$).*

$T$ is incoherent if and only if there is an unsatisfiable concept or role in $\mathcal{O}$.

**Definition 4.** *A TBox $T' \subseteq T$ is a minimal incoherence-preserving sub-TBox (MIPS) of $T$ if and only if $T'$ is incoherent and every sub-TBox $T'' \subsetneq T'$ is coherent.*

*Example 1.* Given a TBox $T$, where $T = \{A \sqsubseteq B, B \sqsubseteq C, B \sqsubseteq D, D \sqsubseteq \neg C, A \sqsubseteq \neg B\}$. According to definition 3, $A$ and $B$ are unsatisfiable concepts. According to definition 4, There are two MIPSs in $T$, one is $\{A \sqsubseteq B, A \sqsubseteq \neg B\}$, the other is $\{B \sqsubseteq C, B \sqsubseteq D, D \sqsubseteq \neg C\}$.

## 2.2   Graph Construction

Let $T$ be a DL-Lite TBox over a signature $\Sigma_T$, containing symbols for atomic elements, i.e., atomic concept and atomic roles. We briefly describe our method to construct a graph from a DL-Lite$_{FR}$ ontology.

Inspired by the work given in [7] and [13], the digraph $\mathcal{G}_T = \langle N, E \rangle$ constructed from TBox $T$ over the signature $\Sigma_T$ is given as follows:

1. for each atomic concept $B$ in $\Sigma_T$, $N$ contains the node $B$.
2. for each atomic role $P$ in $\Sigma_T$, $N$ contains the nodes $P, P^-, \exists P, \exists P^-$.
3. for each concept inclusion $B_1 \sqsubseteq B_2 \in T$, $E$ contains the arc $(B_1, B_2)$.
4. for each concept inclusion $B_1 \sqsubseteq \neg B_2 \in T$, $E$ contains the arc $(B_1, \neg B_2)$ and $N$ contains the node $\neg B_2$.
5. for each role inclusion $R_1 \sqsubseteq R_2 \in T$, $E$ contains the arc $(R_1, R_2)$, arc $(R_1^-, R_2^-)$, arc $(\exists R_1, \exists R_2)$, arc $(\exists R_1^-, \exists R_2^-)$.
6. for each role inclusion $R_1 \sqsubseteq \neg R_2 \in T$, $E$ contains the arc $(R_1, \neg R_2)$, arc $(R_1^-, \neg R_2^-)$, arc $(\exists R_1, \neg \exists R_2)$, arc $(\exists R_1^-, \neg \exists R_2^-)$ and $N$ contains nodes $\neg R_2$, $\neg R_2^-, \neg \exists R_2, \neg \exists R_2^-$.

In items (1)-(6), we construct the graph based on TBox $T$. It has been shown in [13] that the problem of TBox classification in a DL-Lite ontology $\mathcal{O}$ can be done by computing transitive closure of the graph $\mathcal{G}_T$. For simplicity, we call these rules **Construction Rules**, and we use node $C$ (node $R$) to denote the node w.r.t. concept $C$ (role $R$).

In our graph, each node represents a basic concept or a basic role, while each arc represents an inclusion assertion, i.e. the start node of the arc corresponds to the left-hand side of the inclusion assertion and the end node of the arc corresponds to the right-hand side of the inclusion assertion. In order to ensure that the information represented in the TBox is preserved by the graph, we add nodes $R, R^-, \exists R, \exists R^-$ for each role $R$, $arc(R_1, R_2)$, $arc(R_1^-, R_2^-)$, $arc(\exists R_1, \exists R_2)$, $arc(\exists R_1^-, \exists R_2^-)$ for each role inclusion assertion $R_1 \sqsubseteq R_2$.

# 3    A Graph-Based Algorithm

In this section, we first give several theorems which are served as the theoretical basis of our approach, and then we describe our algorithm to computing MIPS.

## 3.1    Theoretical Basis

In order to prove the equivalence of a DL-Lite ontology and a graph w.r.t. reasoning, we give several theorems as follows. At first, we introduce a theorem given in [13].

**Theorem 1.** *Let $T$ be a DL-Lite$_{\mathcal{FR}}$ TBox containing only positive inclusions and let $S_1$ and $S_2$ be two atomic concepts, or two atomic roles. $S_1 \sqsubseteq S_2$ is entailed by $T$ if and only if at least one of the following conditions holds:*

1. *a set $\mathcal{P}$ of positive inclusions exists in $T$, such that $\mathcal{P} \models S_1 \sqsubseteq S_2$*
2. *$T \models S_1 \sqsubseteq \neg S_1$*

We can extend the above theorem to general concepts or roles, meaning that the subsumption relation may be a negative inclusion.

**Theorem 2.** *Let $T$ be a DL-Lite$_{\mathcal{FR}}$ TBox and let $S_1$ and $S_2$ be two atomic concepts, or two atomic roles. $S_1 \sqsubseteq \neg S_2$ is entailed by $T$ if and only if at least one of the following conditions holds:*

1. *a set $\mathcal{P}$ of positive inclusions and a negative inclusion $n \in T$ exist in $T$, such that $\mathcal{P} \cup \{n\} \models S_1 \sqsubseteq \neg S_2$*
2. *$T \models S_1 \sqsubseteq \neg S_1$*

**Proof.** ($\Leftarrow$) This can be easily seen, thus we do not provide a proof.

($\Rightarrow$) Assume $T \models S_1 \sqsubseteq \neg S_2$. Without loss of generality, we suppose that there exists a minimal axiom sequence set $\gamma = \{\phi_1, \phi_2, \ldots, \phi_n\}$ in $TBox$ and $\gamma \models S_1 \sqsubseteq \neg S_2$. Positive inclusion is in the form of $B_1 \sqsubseteq B_2$ or $R_1 \sqsubseteq R_2$, whereas negative inclusion is in the form of $B_1 \sqsubseteq \neg B_2$ or $R_1 \sqsubseteq \neg R_2$ [5], then negative inclusions do not concur in the entailment of a set of positive inclusions. Therefore, there exist at least one negative inclusion in $\gamma$. Let $\phi_i$ be the first negative inclusion that occurs in the $\gamma$ and is of the form $B_1 \sqsubseteq \neg B_2$, then the following cases are considered:

1. $\phi_i$ is the first negative inclusion of $\gamma$, then $\phi_1, \phi_2, \ldots, \phi_{i-1}$ are positive inclusion. if there has no negative inclusion in the $\phi_{i+1}, \phi_{i+2}, \ldots, \phi_n$, axiom set $\gamma$ has a negative inclusion $\phi_i$ and a positive inclusion set $\{\phi_1, \ldots, \phi_{i-1}, \phi_{i+1}, \ldots\}$.
2. Suppose that there have more than one negative inclusion in the $\phi_{i+1}, \phi_{i+2}, \ldots, \phi_n$, we select one negative inclusion and assume that it is of the form $B_1' \sqsubseteq \neg B_2'$. According to the syntax of $DL$-$Lite$[5], the general concept(role) only occurs on the right-hand side of inclusion assertions. Due to the constraint, inference can not be carried out between $\phi_i$ (we assume the form of $\phi_i$ is $B_1 \sqsubseteq \neg B_2$) and $B_1' \sqsubseteq \neg B_2'$, so $\gamma$ is not a minimal set that entails $S_1 \sqsubseteq \neg S_2$. Then we get a contraction.    ❏

Let $\mathcal{T}$ be a DL-Lite$_{\mathcal{FR}}$ TBox and let $\mathcal{G_T} = \langle N, E \rangle$ be the digraph constructed from $\mathcal{T}$ according to the Construction Rules, i.e., each node represents a concept or a role and each edge represents an inclusion assertion. We denote the transitive closure of $\mathcal{G_T}$ by $\mathcal{G_T^*} = \langle N, E^* \rangle$. According to Theorem 1 and Theorem 2, we can see that logic entailment between concepts and roles can be reduced to the graph reachability problem. Consider for example, $\mathcal{T} = \{B_1 \sqsubseteq B_2, B_2 \sqsubseteq B_3\}$ and there are two edges $arc(B_1, B_2)$ and $arc(B_2, B_3)$ in graph $\mathcal{G}$, it is obvious that $B_1 \sqsubseteq B_2 \in cl(\mathcal{T})$ and $arc(B_1, B_3)$ belongs to the transitive closure of $\mathcal{G}$.

In order to prove the equivalence of an ontology and the graph constructed from it w.r.t. classification, we cite a theorem given in [13].

**Theorem 3.** *Let $\mathcal{T}$ be a DL-Lite$_{\mathcal{FR}}$ TBox containing only positive inclusions and let $\mathcal{G_T} = \langle N, E \rangle$ be its digraph representation. Let $S_1, S_2$ be two basic concepts(roles). $S_1 \sqsubseteq S_2 \in cl(\mathcal{T})$ if and only if $arc(S_1, S_2) \in E^*$.*

Then we extend the theorem as follows.

**Theorem 4.** *Let $\mathcal{T}$ be a DL-Lite$_{\mathcal{FR}}$ TBox and let $\mathcal{G_T} = \langle N, E \rangle$ be the digraph constructed from $\mathcal{T}$ according to the Construction Rules. Let $m$ be a basic concept(role) and $n$ be a general concept(role). $m \sqsubseteq n \in cl(\mathcal{T})$ iff $arc(m, n) \in E^*$.*

**Proof.** ($\Leftarrow$) If $arc(m, n) \in E^*$, according to the definition of transitive closure of graph, there exists at least one path that from node m to node n in $G_\mathcal{T}$. Moreover, according to our Graph Construction Rules, for each edge in $G_\mathcal{T}$, it corresponds to at least one inclusion assertion in $\mathcal{T}$ or $cl(\mathcal{T})$ (if concept w.r.t. to node is the form of $\exists R$ ). We assume $\mathcal{S}$ is the set of these assertions and it is obvious that $\mathcal{S} \models m \sqsubseteq n$. Therefore, $m \sqsubseteq n \in cl(\mathcal{T})$.

($\Rightarrow$) If $m \sqsubseteq n \in cl(\mathcal{T})$, we have two cases in the following:

1. If $m \sqsubseteq n \in clp(\mathcal{T})$, we can infer $arc(m, n) \in E^*$ by applying Theorem 3.
2. If $m \sqsubseteq n \in cln(\mathcal{T})$, then $m \sqsubseteq n$ is similar to $B_1 \sqsubseteq \neg B_2$. According to Theorem 2, there exist a set of PIs and a NI $n$ such that $P \cup \{n\} \models B_1 \sqsubseteq \neg B_2$. Without loss of generality, we assume $n = \{B' \sqsubseteq \neg B_2\}$ and $P \models B_1 \sqsubseteq B'$. According to Theorem 1, we can infer that $arc(B_1, B') \in E^*$. Therefore, $arc(B_1, \neg B_2) \in E^*$.

$\square$

For every edge on the graph, it corresponds to an inclusion assertion. Therefore, a path P on the graph will correspond to a set of inclusion assertions that we denote by $\mathcal{S}$. It is obvious that $\mathcal{S} \subseteq cl(\mathcal{T})$. According to Theorem 2 and Theorem 4, there will be a minimal subset $\mathcal{S}' \subseteq \mathcal{T}$ that $\mathcal{S}' \models \mathcal{S}$ and for every $\mathcal{S}'' \subset \mathcal{S}', \mathcal{S}'' \nvDash \mathcal{S}$. We call $\mathcal{S}'$ as the minimal set in $\mathcal{T}$ corresponding to P, denoted with $\mathcal{S}'_{\mathcal{T} \rightarrow P}$.

**Definition 5.** *Let $\mathcal{T}$ be a DL-Lite$_{\mathcal{FR}}$ TBox and let $\mathcal{G_T} = \langle N, E \rangle$ be the digraph constructed from TBox $\mathcal{T}$ according to the Construction Rules. Then for arbitrary node $C \in N$, if there exist two paths that are from node $C$ to node $D$ and from node $C$ to node $\neg D$ respectively in $\mathcal{G_T}$ and there does not exist joint edge in $C \rightarrow D$ and $C \rightarrow \neg D$, we call these two paths as minimal incoherence-preserving path-pair (MIPP).*

We use $mipp(\mathcal{G}_T)$ to denote the set of all MIPP in $\mathcal{G}_T$.

Then we show that the problem to computing MIPS of $T$ can be transformed to finding MIPP in graph $\mathcal{G}_T$. Thus, to compute $mips(T)$, we only need to find $mipp(\mathcal{G}_T)$. The existence of MIPP indicates that the ontology is incoherent.

**Theorem 5.** *Let $T$ be a DL-Lite$_{\mathcal{FR}}$ TBox, $T$ is incoherent iff there exists at least one MIPP in $G_T$.*

**Proof.** ($\Leftarrow$) If there exists a MIPP in $G_T$, according to Definition 5 we assume that there exists two paths, one is $C \rightarrow D$, the other is $C \rightarrow \neg D$. In this case, we can derive that $C \sqsubseteq D$ and $C \sqsubseteq \neg D$ by Theorem 4. Therefore, $T$ contains at least one unsatisfiable concept or role and $T$ is incoherent.

($\Rightarrow$) $T$ is incoherent so that there will be at least one unsatisfiable concept or role in $T$. Without loss of generality, we assume that it is the unsatisfiable concept $C$ and $C \sqsubseteq D \in cl(T)$, $C \sqsubseteq \neg D \in cl(T)$. According to Theorem 4, we can get $arc(C, D) \in G_T$ and $arc(C, \neg D) \in G_T$. It is easy to see that there exist two paths without joint edge in the digraph $G_T$, that is a MIPP by Definition 5. ❑

In the graph transformed from an ontology, MIPP of graph has tightly relationship with MIPS of the ontology.

**Theorem 6.** *Let $T$ be a DL-Lite$_{FR}$ TBox and let $\mathcal{S}'_{T \rightarrow \mathcal{MIPP}}$ be a subset of $T$ that corresponds to a MIPP in $\mathcal{G}_T$. Then $\mathcal{S}'_{T \rightarrow \mathcal{MIPP}}$ is a MIPS in $T$.*

**Proof.** By theorem 5, it is easy to see that $T$ is incoherent. Then we will prove that for any $\mathcal{S}'' \subseteq \mathcal{S}'_{T \rightarrow \mathcal{MIPP}}$, $\mathcal{S}'_{T \rightarrow \mathcal{MIPP}}$ is incoherent but $\mathcal{S}''$ is coherent. Let $\mathcal{S}$ be the set of axioms corresponding to the MIPP. $\mathcal{S}'_{T \rightarrow \mathcal{MIPP}} \models \mathcal{S}$ and $\mathcal{S}'' \not\models \mathcal{S}$ because $\mathcal{S}'_{T \rightarrow \mathcal{MIPP}}$ is the minimal subset of $T$ that $\mathcal{S}'_{T \rightarrow \mathcal{MIPP}} \models \mathcal{S}$. According to our construction rules, there doesn't exists a MIPP in $G_{\mathcal{S}''}$. Then by theorem 4, $\mathcal{S}'_{T \rightarrow \mathcal{MIPP}}$ is incoherent and $\mathcal{S}''$ is coherent. ❑

**Theorem 7.** *Let $T$ be a DL-Lite$_{FR}$ TBox , the set of all $\mathcal{S}'_{T \rightarrow \mathcal{MIPP}}$ in $G_T$ corresponds to the set of all MIPS in $T$.*

**Proof.** For any MIPP in $mipp(\mathcal{G}_T)$, by theorem 6, $\mathcal{S}'_{T \rightarrow \mathcal{MIPP}}$ is a MIPS so that $\mathcal{S}'_{T \rightarrow \mathcal{MIPP}} \in mips(T)$.

For any MIPS $\mathcal{M}$ in $mips(T)$, $\mathcal{M}$ is incoherent and for any $\mathcal{M}' \subset \mathcal{M}$, $\mathcal{M}'$ is coherent. Therefore, there exists a MIPP in $G_{\mathcal{M}}$ and it does not exist in $G_{\mathcal{M}'}$. It is easy to see that $\mathcal{M}$ corresponds to the MIPP in $G_T$ and $\mathcal{M}$ is contained in the set of all $\mathcal{S}'_{T \rightarrow \mathcal{MIPP}}$ corresponding to mipp in $T$.

The case of role R can be proved analogously. ❑

## 3.2    A Graph-Based algorithm

In the following, we describe our graph-based algorithm which will find all MIPP on $\mathcal{G}_{\mathcal{T}}$ and then transform them to all MIPS in $\mathcal{T}$. After that, an example is provided to illustrate the algorithm.

Let $S$ be a concept or a role expression. In steps 3-11 of algorithm 1, we find all descendant nodes of node $S$ and node $\neg S$ on the graph $\mathcal{G}_{\mathcal{T}}$. If there is a same node, which we assume as $S'$, between descendant nodes of node $S$ (Specially, it includes $S$ itself for the case $S \sqsubseteq \neg S$) and those of node $\neg S$, it means that there are two paths $S' \rightarrow S$ and $S' \rightarrow \neg S$ on $\mathcal{G}_{\mathcal{T}}$. While $S' \rightarrow S$ and $S' \rightarrow \neg S$ have no joint edge, then $\{S' \rightarrow S, S' \rightarrow \neg S\}$ is a MIPP. For each edge $arc(B_1, B_2)$ in a MIPP, if $B_1 \sqsubseteq B_2 \in \mathcal{T}$. In steps 12-14 of Algorithm 1, we can get a MIPS that the MIPP corresponding to.

---

**Algorithm 1.**  CompMIPS

---

**Require:** an incoherent TBox $\mathcal{T}$
**Ensure:** set of MIPS in $\mathcal{T}$
 1: $pset \leftarrow \emptyset, mips \leftarrow \emptyset$;
 2: **construct** $\mathcal{G}_{\mathcal{T}} = \langle N, E \rangle$;
 3: **for** each $\neg S \in N$ **do**
 4:     **for** each $n_1 \in$ descendants($\neg S, \mathcal{G}_{\mathcal{T}}$) **do**
 5:         **for** each $n_2 \in$ descendants($S, \mathcal{G}_{\mathcal{T}}$) $\cup \{S\}$ **do**
 6:             **if** $n_1 = n_2$ and not hasJointEdge($path(n_1, S), path(n_2, \neg S)$) **then**
 7:                 $pset \leftarrow pset \cup \{\langle n_1 \rightarrow \neg S, n_2 \rightarrow S \rangle\}$;
 8:             **end if**
 9:         **end for**
10:     **end for**
11: **end for**
12: **for** each $\langle P_1, P_2 \rangle \in pset$ **do**
13:     $mips \leftarrow mips \cup \{\{transToAxioms(P_1, P_2, \mathcal{T})\}\}$;
14: **end for**
    **return** $mips$;

---

Function *hasJointEdge* is applied to check whether there exists a joint edge between two paths and function *transToAxioms* handles the case that $B_1 \sqsubseteq B_2 \notin \mathcal{T}$ but $\mathcal{T} \models B_1 \sqsubseteq B_2$), $B_1 \sqsubseteq B_2$ belongs to a MIPS.

*Example 2.* Given a TBox $\mathcal{T}$, where $\mathcal{T} = \{A \sqsubseteq B, B \sqsubseteq C, C \sqsubseteq D, B \sqsubseteq \neg C, R_1 \sqsubseteq R_2, B \sqsubseteq \exists R_1, \exists R_2 \sqsubseteq \neg D\}$. According to the Construction Rules, we construct a directed graph as shown in Fig.1.

In steps 3-9 of Algorithm 1, we can find two MIPPs in Fig. 1: $\{B \rightarrow \neg C, B \rightarrow C\}$ and $\{B \rightarrow D, B \rightarrow \neg D\}$.

In steps 10-12 of Algorithm 1, function *transToAxioms*($P_1, P_2, \mathcal{T}$) will transform paths $P_1$ and $P_2$ to axioms that these paths correspond to axioms in $\mathcal{T}$. For $\exists R_1 \sqsubseteq \exists R_2 \notin \mathcal{T}$ and $R_1 \sqsubseteq R_2 \models \exists R_1 \sqsubseteq \exists R_2$ , $arc(\exists R_1, \exists R_2)$ corresponds to

---

**Algorithm 2.** TransToAxioms

---

**Require:** $path : P_1, path : P_2, \mathcal{T}$
**Ensure:** a set of axioms
1: $axiomset \leftarrow \emptyset$;
2: **for** $i = 1 : 2$ **do**
3:    **for each** $arc\langle B_1, B_2 \rangle \in P_i$ **do**
4:        **if** $B_1 \sqsubseteq B_2 \in \mathcal{T}$ **then**
5:            $axiomset \leftarrow axiomset \cup \{B_1 \sqsubseteq B_2\}$;
6:        **else if** $B_1 = \exists R_1$ and $B_2 = \exists R_2$ and $R_1 \sqsubseteq R_2 \in \mathcal{T}$ **then**
7:            $axiomset \leftarrow axiomset \cup \{R_1 \sqsubseteq R_2\}$;
8:        **end if**
9:    **end for**
10: **end for**
    **return** $axiomset$;

---

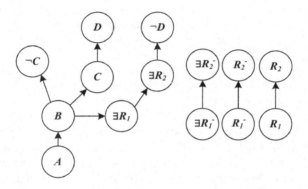

**Fig. 1.** Directed Graph in Example 2

$R_1 \sqsubseteq R_2$ in $\mathcal{T}$. Finally, two MIPSs $\{B \sqsubseteq C, B \sqsubseteq \neg C\}$ and $\{B \sqsubseteq C, C \sqsubseteq D, R_1 \sqsubseteq R_2, B \sqsubseteq \exists R_1, \exists R_2 \sqsubseteq \neg D\}$ will be generated.

**Theorem 8.** *Given an arbitrary incoherent TBox $\mathcal{T}$, the output of Algorithm 1 is the set of all MIPS of $\mathcal{T}$.*

**Proof.** According to Theorem 4, $n \in descendants(S, \mathcal{G}_\mathcal{T})$ denotes $n \sqsubseteq S$. Thus, in the steps 4-6 of algorithm 1, $n_1 = n_2$ means that there exists a path pair from node $n_1 (or \ n_2)$ to disjointness pair$(S, \neg S)$. By definition 4, these two paths consist in a MIPP. By applying algorithm 2, we can transform a MIPP into a MIPS.

For any MIPS in $\mathcal{T}$, there exists a concept(or role) that is subsumed by two disjointness concepts(or roles). Thus, from Theorem 4, we can get two intersectional path derived from two disjointness node in $\mathcal{G}_\mathcal{T}$. Obviously, the intersectional point is the descendant of the two disjointness node. Namely, each MIPS can be calculated by steps 3-11 of algorithm 1. Therefore, the result of algorithm 1 is mips($\mathcal{T}$).

❏

In our algorithm, calculating a MIPP is done by traversing on the directed graph $\mathcal{G}_{\mathcal{T}}$, which can be done in polynomial time with the size of vertexes and arcs. Therefore, a MIPS can be computed in polynomial time with the size of $\mathcal{T}$. However, in the worst case, the computational complexity of algorithm 1 is exponential since there may be exponentially many MIPS for a given TBox [3].

## 4    Experimental Evaluation

Several tableau-based reasoners which can compute MIPS, such as Pellet, FaCT++, etc, have exhibited their excellent performance. In this section, we will compare our graph-based algorithm with these reasoners in computing MIPS. We carry out our experiments by two strategies: one is that we carried experiments on different ontologies to compare efficiency and the other is to compare performance changes on an ontology with scalable number of unsatisfiable concepts or roles.

All experiments have been performed on a *Lenovo* desktop PC with Intel Corei5-2400 3.1 GHz CPU and 4GB of RAM, running Microsoft window 7 operating system, and Java 1.7 with 3GB of heap space.

In our experiments, we stores graph structure of DL-Lite ontology in a Neo4j graph database[16]. Neo4j is an open-source and high-performance graph database supported by Neo Technology[20]. The main ingredients of Neo4j are nodes and relationships. A graph database is used to record data in nodes which have user-defined properties, while nodes are organized by relationships which also have user-defined properties. Users could look up nodes or relationships by index. Cypher is a powerful declarative graph query language[1]. We can leverage its ability to expressive and efficient querying and updating of the graph store. In our implementation, we apply Cypher to find the path between disjointness nodes.

Ontologies used in our experiments have significantly different sizes and structures. In order to fit $DL\text{-}Lite_{FR}$ expressivity, when an expression cannot be expressed by $DL\text{-}Lite_{FR}$, it will be approximated [13]. Since these original ontologies is coherent, we modified them by inserting some "incoherent-generating" axioms randomly, such as negative inclusion. For example, if $\mathcal{T} = \{A \sqsubseteq B, B \sqsubseteq C, B \sqsubseteq D\}$, it is obvious that $\mathcal{T}$ is coherent. To make it incoherent, we may insert $C \sqsubseteq \neg D$ into $\mathcal{T}$. Table 1 lists part of the detailed information about the ontologies used in the experiments.

Table 2 shows the detailed performance comparison on different ontologies. We carried out these experiments on ten ontologies. According to the results, our system outperforms other systems. Especially, for some complex ontologies, such as FMA, our system is one order of magnitude faster than the existing ones. Fig.2 gives another view of the results.

---

[1] http://www.neo4j.org/learn/cypher

**Table 1.** Experimental Ontologies

| Ontology | Axioms | Unsatisfiable Concepts | MIPS |
|---|---|---|---|
| Economy | 803 | 51 | 47 |
| Terrorism | 185 | 14 | 5 |
| Transportation | 1186 | 62 | 36 |
| Aeo | 521 | 49 | 17 |
| CL | 8783 | 59 | 25 |
| DOLCE-LITE | 98 | 33 | 5 |
| Fly-Anatomy | 17735 | 304 | 3 |
| FMA | 160936 | 45 | 12 |
| GO | 43934 | 97 | 18 |
| Plant | 3295 | 45 | 12 |

**Table 2.** Time required to Find All MIPS in milliseconds on different ontologies

| Ontology | Pellet | Hermit | FaCT++ | Jfact | Graph |
|---|---|---|---|---|---|
| Economy | 1286 | 1378 | 619 | 1836 | 601 |
| Terrorism | 388 | 449 | 387 | 455 | 157 |
| Transportation | 1772 | 2735 | 1339 | 4712 | 618 |
| Aeo | 1287 | 1963 | 792 | 1517 | 357 |
| CL | 1465 | 2375 | 1152 | 2269 | 628 |
| DOLCE-Lite | 765 | 1041 | 475 | 813 | 325 |
| Fly-Anatomy | 2019 | 3319 | 1878 | 4901 | 590 |
| FMA | 47262 | 49444 | 45559 | 91776 | 3674 |
| GO | 3765 | 5371 | 3796 | 9315 | 875 |
| Plant | 1543 | 2009 | 935 | 1883 | 513 |

Table 3 gives the result of scalability over adapted Go ontology, with increasing number of unsatisfiable concepts or roles added. We first found all the disjoint concepts in Go ontology, and then increased the number of unsatisfiable concepts by adding common subsumed concept to disjoint concepts randomly (we began with 100 unsatisfiable concepts for clearer comparison). The same strategy is also performed on roles in Go ontology. With the growing number of unsatisfiable concepts and roles, the execution time of systems implementing the state of the art debugging algorithm increases sharply. However, the execution time of our system grows moderately. The results show that our system performs more steadily than other systems in computing MIPS.

# 5   Related Work

The work of ontology debugging has been widely discussed in the literature and many algorithms have been provided for debugging DL-based ontology. These algorithms mainly aim to finding explanations of entailment of incoherence [11]. In general, the approach for debugging ontology can be divided into two categories: glass-box approach and black-box approach.

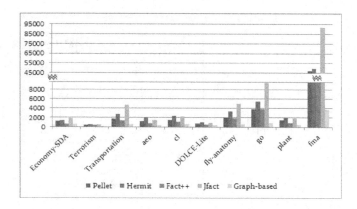

**Fig. 2.** Time required to Find All MIPS

**Table 3.** Time required to Find All MIPS in milliseconds with scalable number of unsatisfiable concepts or roles

| Number | Pellet | Hermit | FaCT++ | Jfact | Graph |
|--------|--------|--------|--------|-------|-------|
| 100 | 4381 | 6090 | 3991 | 8882 | 827 |
| 110 | 4641 | 6280 | 4342 | 10864 | 912 |
| 120 | 4982 | 7155 | 4948 | 12636 | 933 |
| 130 | 6609 | 9771 | 6425 | 17174 | 1016 |
| 140 | 8077 | 10132 | 7305 | 66663 | 1127 |
| 150 | 9449 | 12082 | 8714 | overflow | 1209 |
| 160 | 10210 | 13065 | 9630 | overflow | 1263 |
| 200 | 12889 | 16573 | 12503 | overflow | 1769 |

A glass-box approach is built on an existing tableau-based DL reasoner. The advantage of a glass-box approach is that it can find all MUPS of an incoherent ontology by a single run of a modified reasoner. Most of the glass-box algorithms are obtained as extension of tableau-based algorithms for checking satisfiability of a DL-based ontology. In [17], the authors propose the first tableau-based algorithm for debugging terminologies of an ontology. The algorithm is restricted to *unfoldable ALC TBoxes*, i.e., the left-hand sides of the concept axioms (the defined concepts) are atomic and if the right-hand sides (the definitions) contain no direct or indirect reference to the defined concept. It is realized by attaching label to axioms in order to keep track of which axioms are responsible for assertions generated during the expansion of the completion forests. A general tableau algorithm is proposed in [3], and the concept of "tableau algorithm" is provided in this paper. In the tableau-based algorithm, some blocking techniques cannot be used. In order to resolve this problem and other problems, in [2], an automata-based approach is proposed, which has some excellent theoretical attributes.

A black-box approach uses a DL reasoner to check satisfiability of an ontology. Different from the glass-box approach, a black-box approach does not need to modify the internal reasoner. In the worst case, the black-box approach may call the reasoner an exponential number of times, thus it cannot be used for handling larger ontologies directly. In [12], a bottom-up approach is proposed, which is based on external reasoner, such as RACER and FaCT++. It gets a single MUPS of the unsatisfiable concept first and then it utilizes the Reiter's Hitting Set algorithm to compute all the MUPS. A general DL-based ontology diagnosing approach is given in [6], which is based on Reiter's algorithm and employ a simplified variant of QUICKXPLAIN[10] to generate the HS-tree.

There are several other approaches to debugging ontology. Wang et al. in [19] proposes a heuristic approach, but it cannot get the complete set of MUPS. [21] propose a approach to calculating MUPS. The advantage of this approach is that it calculates the $cln(\mathcal{T})$ off-line by reasoner, and then the rest of process does not depend on reasoner.

Although there have been much work on ontology debugging, most of them are applied to find all the MUPS of an incoherent ontology w.r.t. an unsatisfiable concept. The computation of all the MIPS relies on the computation of all the MUPS of an incoherent ontology w.r.t. all unsatisfiable concept. According to our experimental results, the systems implementing this method are hard to be used to handle large ontologies in DL-Lite. In contrast, our algorithm does not need to compute MUPS and can use optimizations in graph databases. Thus, it is more efficient and scalable than the existing algorithms for computing MIPS.

## 6  Conclusion

In this paper, we have proposed a new approach for debugging incoherent DL-Lite ontologies. It is based on a graph-based algorithm to calculate minimal incoherence preserving subterminology(MIPS). We have implemented the algorithm and conduct experiments over some real ontologies. The experimental results showed that our algorithm outperforms existing algorithms for computing MIPS both in efficiency and stability, especially for ontologies that are very large in size or contain many unsatisfiable concepts.

As a future work, we will develop a semi-automatic ontology repair system in DL-Lite by using the debugging algorithm given in this paper. We also plan to work on ontology merging[15] by adapting the graph-based debugging algorithm.

## References

1. Artale, A., Calvanese, D., Kontchakov, R., Zakharyaschev, M.: The DL-Lite family and relations. Journal of Artificial Intelligence Research 36(1), 1–69 (2009)
2. Baader, F., Peñaloza, R.: Automata-based axiom pinpointing. In: Armando, A., Baumgartner, P., Dowek, G. (eds.) IJCAR 2008. LNCS (LNAI), vol. 5195, pp. 226–241. Springer, Heidelberg (2008)

3. Baader, F., Peñaloza, R., Suntisrivaraporn, B.: Pinpointing in the description logic $\mathcal{EL}$. In: Hertzberg, J., Beetz, M., Englert, R. (eds.) KI 2007. LNCS (LNAI), vol. 4667, pp. 52–67. Springer, Heidelberg (2007)
4. Bienvenu, M., Rosati, R.: Tractable approximations of consistent query answering for robust ontology-based data access. In: Proc. of IJCAI 2013 (2013)
5. Calvanese, D., De Giacomo, G., Lembo, D., Lenzerini, M., Rosati, R.: Tractable reasoning and efficient query answering in description logics: The DL-Lite family. Journal of Automated Reasoning **39**(3), 385–429 (2007)
6. Friedrich, G.E., Shchekotykhin, K.: A general diagnosis method for ontologies. In: Gil, Y., Motta, E., Benjamins, V.R., Musen, M.A. (eds.) ISWC 2005. LNCS, vol. 3729, pp. 232–246. Springer, Heidelberg (2005)
7. Gao, S., Qi, G., Wang, H.: A new operator for ABox revision in DL-Lite. In: Proc. of AAAI 2012, pp. 2423–2324 (2012)
8. Grau, B.C., Horrocks, I., Kazakov, Y., Sattler, U.: Just the right amount: extracting modules from ontologies. In: Proc. of WWW 2007, pp. 717–726 (2007)
9. Ji, Q., Gao, Z., Huang, Z., Zhu, M.: An efficient approach to debugging ontologies based on patterns. In: Pan, J.Z., Chen, H., Kim, H.-G., Li, J., Horrocks, I., Mizoguchi, R., Wu, Z., Wu, Z. (eds.) JIST 2011. LNCS, vol. 7185, pp. 425–433. Springer, Heidelberg (2012)
10. Junker, U.: Quickxplain: preferred explanations and relaxations for over-constrained problems. In: Proc. of AAAI 2004, pp. 167–172 (2004)
11. Kalyanpur, A., Parsia, B., Horridge, M., Sirin, E.: Finding all justifications of OWL DL entailments. In: Aberer, K., Choi, K.-S., Noy, N., Allemang, D., Lee, K.-I., Nixon, L.J.B., Golbeck, J., Mika, P., Maynard, D., Mizoguchi, R., Schreiber, G., Cudré-Mauroux, P. (eds.) ASWC 2007 and ISWC 2007. LNCS, vol. 4825, pp. 267–280. Springer, Heidelberg (2007)
12. Kalyanpur, A., Parsia, B., Sirin, E., Hendler, J.: Debugging unsatisfiable classes in OWL ontologies. Journal of Web Semantics **3**(4), 268–293 (2005)
13. Lembo, D., Santarelli, V., Savo, D.F.: Graph-Based ontology classification in OWL 2 QL. In: Cimiano, P., Corcho, O., Presutti, V., Hollink, L., Rudolph, S. (eds.) ESWC 2013. LNCS, vol. 7882, pp. 320–334. Springer, Heidelberg (2013)
14. Parsia, B., Sirin, E., Kalyanpur, A.: Debugging OWL ontologies. In: Proc. of the WWW 2005, pp. 633–640 (2005)
15. Raunich, S., Rahm, E.: Towards a benchmark for ontology merging. In: Herrero, P., Panetto, H., Meersman, R., Dillon, T. (eds.) OTM-WS 2012. LNCS, vol. 7567, pp. 124–133. Springer, Heidelberg (2012)
16. Robinson, I., Webber, J., Eifrem, E.: Graph Databases, pp. 25–63. O'Reilly Media Inc. (2013)
17. Schlobach, S., Cornet, R.: Non-standard reasoning services for the debugging of description logic terminologies. In: Proc. of IJCAI 2003, pp. 355–362 (2003)
18. Schlobach, S., Huang, Z., Cornet, R., Van Harmelen, F.: Debugging incoherent terminologies. Journal of Automated Reasoning **39**(3), 317–349 (2007)
19. Wang, H., Horridge, M., Rector, A.L., Drummond, N., Seidenberg, J.: Debugging OWL-DL ontologies: a heuristic approach. In: Gil, Y., Motta, E., Benjamins, V.R., Musen, M.A. (eds.) ISWC 2005. LNCS, vol. 3729, pp. 745–757. Springer, Heidelberg (2005)
20. Webber, J.: A programmatic introduction to neo4j. In: Proc. of SPLASH 2012, pp. 217–218 (2012)
21. Zhou, L., Huang, H., Qi, G., Qu, Y., Ji, Q.: An algorithm for calculating minimal unsatisfiability-preserving subsets of ontology in dl-lite. Journal of Computer Research and Development **48**(3), 2334–2342 (2011)

# Reasoning for $\mathcal{ALCQ}$ Extended with a Flexible Meta-Modelling Hierarchy

Regina Motz[2], Edelweis Rohrer[2], and Paula Severi[1](✉)

[1] Department of Computer Science, University of Leicester, Leicester, England
`ps330@leicester.ac.uk`
[2] Instituto de Computación, Facultad de Ingeniería, Universidad de la República,
Montevideo, Uruguay
{`rmotz,erohrer`}`@fing.edu.uy`

**Abstract.** This works is motivated by a real-world case study where it is necessary to integrate and relate existing ontologies through *meta-modelling*. For this, we introduce the Description Logic $\mathcal{ALCQM}$ which is obtained from $\mathcal{ALCQ}$ by adding statements that equate individuals to concepts in a knowledge base. In this new extension, a concept can be an individual of another concept (called *meta-concept*) which itself can be an individual of yet another concept (called *meta meta-concept*) and so on. We define a tableau algorithm for checking consistency of an ontology in $\mathcal{ALCQM}$ and prove its correctness.

**Keywords:** Description logic · Meta-modelling · Meta-concepts · Well founded sets · Consistency · Decidability

## 1 Introduction

Our extension of $\mathcal{ALCQ}$ is motivated by a real-world application on geographic objects that requires to reuse existing ontologies and relate them through meta-modelling [10].

Figure 1 describes a simplified scenario of this application in order to illustrate the meta-modelling relationship. It shows two ontologies separated by a line. The two ontologies conceptualize the same entities at different levels of granularity. In the ontology above the line, rivers and lakes are formalized as individuals while in the one below the line they are concepts. If we want to integrate these ontologies into a single ontology (or into an ontology network) it is necessary to interpret the individual *river* and the concept *River* as the same real object. Similarly for *lake* and *Lake*.

Our solution consists in equating the individual *river* to the concept *River* and the individual *lake* to the concept *Lake*. These equalities are called *meta-modelling axioms* and in this case, we say that the ontologies are related through *meta-modelling*. In Figure 1, meta-modelling axioms are represented by dashed edges. After adding the meta-modelling axioms for rivers and lakes, the concept *HydrographicObject* is now also a *meta-concept* because it is a concept that contains an individual which is also a concept.

© Springer International Publishing Switzerland 2015
T. Supnithi et al.(Eds.): JIST 2014, LNCS 8943, pp. 47–62, 2015.
DOI: 10.1007/978-3-319-15615-6_4

The kind of meta-modelling we consider in this paper can be expressed in OWL Full but it cannot be expressed in OWL DL. The fact that it is expressed in OWL Full is not very useful since the meta-modelling provided by OWL Full is so expressive that leads to undecidability [11].

OWL 2 DL has a very restricted form of meta-modelling called *punning* where the same identifier can be used as an individual and as a concept [7]. These identifiers are treated as different objects by the reasoner and it is not possible to detect certain inconsistencies. We next illustrate two examples where OWL would not detect inconsistencies because the identifiers, though they look syntactically equal, are actually different.

*Example 1.* If we introduce an axiom expressing that *HydrographicObject* is a subclass of *River*, then OWL's reasoner will not detect that the interpretation of *River* is not a well founded set (it is a set that belongs to itself).

*Example 2.* We add two axioms, the first one says that *river* and *lake* as individuals are equal and the second one says that the classes *River* and *Lake* are disjoint. Then OWL's reasoner does not detect that there is a contradiction.

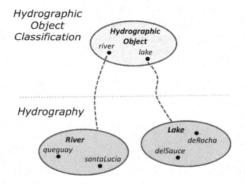

**Fig. 1.** Two ontologies on Hydrography

In this paper, we consider $\mathcal{ALCQ}$ ($\mathcal{ALC}$ with qualified cardinality restrictions) and extend it with *Mboxes*. An Mbox is a set of equalities of the form $a =_m A$ where $a$ is an individual and $A$ is a concept. In our example, we have that $river =_m River$ and these two identifiers are semantically equal, i.e. the interpretations of the individual *river* and the concept *River* are the same. The domain of an interpretation cannot longer consists of only basic objects but it must be any well-founded set. The well-foundness of our model is not ensured by means of fixing layers beforehand as in [8,12] but it is our reasoner which checks for circularities. Our approach allows the user to have any number of levels (or layers) (meta-concepts, meta meta-concepts and so on). The user does not have to write or know the layer of the concept because the reasoner will infer it for him. In this way, axioms can also naturally mix elements of different layers and the user has the flexibility of changing the status of an individual at any point without having to make any substantial change to the ontology.

We define a tableau algorithm for checking consistency of an ontology in $\mathcal{ALCQM}$ by adding new rules and a new condition to the tableau algorithm for $\mathcal{ALCQ}$. The new rules deal with the equalities and inequalities between individuals with meta-modelling which need to be transferred to the level of concepts as equalities and inequalities between the corresponding concepts. The new condition deals with circularities avoiding non well-founded sets. From the practical point of view, extending tableau for $\mathcal{ALCQ}$ has the advantage that one can easily change and reuse the code of existing OWL's reasoners. From the theoretical point of view, we give an elegant proof of correctness by showing an isomorphism between the canonical interpretations of $\mathcal{ALCQ}$ and $\mathcal{ALCQM}$. Instead of re-doing inductive proofs, we "reuse" and invoke the results of correctness of the tableau algorithm for $\mathcal{ALCQ}$ from [1] wherever possible.

*Related Work.* As we mentioned before, OWL 2 DL has a very restricted form of meta-modelling called *punning* [7]. In spite of the fact that the same identifier can be used simultaneously as an individual and as a concept, they are semantically different. In order to use the punning of OWL 2 DL in the example of Figure 1, we could change the name *river* to *River* and *lake* to *Lake*. In spite of the fact that the identifiers look syntactically equal, OWL would not detect certain inconsistencies as the ones illustrated in Examples 1 and 2, and in Example 4 which appears in Section 3. In the first example, OWL won't detect that there is a circularity and in the other examples, OWL won't detect that there is a contradiction. Apart from having the disadvantage of not detecting certain inconsistencies, this approach is not natural for reusing ontologies. For these scenarios, it is more useful to assume the identifiers be syntactically different and allow the user to equate them by using axioms of the form $a =_{\mathsf{m}} A$.

Motik proposes a solution for meta-modelling that is not so expressive as RDF but which is decidable [11]. Since his syntax does not restrict the sets of individuals, concepts and roles to be pairwise disjoint, an identifier can be used as a concept and an individual at the same time. From the point of view of ontology design, we consider more natural to assume that the identifiers for a concept and an individual that conceptualize the same real object (with different granularity) will be syntactically different (because most likely they will live in different ontologies). In [11], Motik also defines two alternative semantics: the context approach and the HiLog approach. The context approach is similar to the so-called punning supported by OWL 2 DL. The HiLog semantics looks more useful than the context semantics since it can detect the inconsistency of Example 2. However, this semantics ignores the issue on well-founded sets. Besides, this semantics does not look either intuitive or direct as ours since it uses some intermediate extra functions to interpret individuals with meta-modelling. The algorithm given in [11, Theorem 2] does not check for circularities (see Example 1) which is one of the main contributions of this paper.

De Giacomo et al. specifies a new formalism, "Higher/Order Description Logics", that allows to treat the same symbol of the signature as an instance, a concept and a role [4]. This approach is similar to punning in the sense that the three new symbols are treated as independent elements.

Pan et al address meta-modelling by defining different "layers" or "strata" within a knowledge base [8,12]. This approach forces the user to explicitly write the information of the layer in the concept. This has several disadvantages: the user should know beforehand in which layer the concept lies and it does not give the flexibility of changing the layer in which it lies. Neither it allows us to mix different layers when building concepts, inclusions or roles, e.g. we cannot express that the intersection of concepts in two different layers is empty or define a role whose domain and range live in different layers.

Glimm et al. codify meta-modelling within OWL DL [5]. This codification consists in adding some extra individuals, axioms and roles to the original ontology in order to represent meta-modelling of concepts. As any codification, this approach has the disadvantage of being involved and difficult to use, since adding new concepts implies adding a lot of extra axioms. This codification is not enough for detecting inconsistencies coming from meta-modelling (see Example 4). The approach in [5] has also other limitations from the point of view of expressibility, e.g. it has only two levels of meta-modelling (concepts and meta-concepts).

*Organization of the Paper.* The remainder of this paper is organized as follows. Section 2 shows a case study and explains the advantages of our approach. Section 3 defines the syntax and semantics of $\mathcal{ALCQM}$. Section 4 proposes an algorithm for checking consistency. Section 5 proves its correctness. Finally, Section 6 sets the future work.

## 2   Case Study on Geography

In this section, we illustrate some important advantages of our approach through the real-world example on geographic objects presented in the introduction.

Figure 2 extends the ontology network given in Figure 1. Ontologies are delimited by light dotted lines. Concepts are denoted by ovals and individuals by small filled circles. Meta-modelling between ontologies is represented by dashed edges. Thinnest arrows denote roles within a single ontology while thickest arrows denote roles from one ontology to another ontology.

Figure 2 has five separate ontologies. The ontology in the uppermost position conceptualizes the politics about geographic objects, defining *GeographicObject* as a meta meta-concept, and *Activity* and *GovernmentOffice* as concepts. The ontology in the left middle describes hydrographic objects through the meta-concept *HydrographicObject* and the one in the right middle describes flora objects through the meta-concept *FloraObject*. The two remaining ontologies conceptualize the concrete natural resources at a lower level of granularity through the concepts *River*, *Lake*, *Wetland* and *NaturalForest*.

Note that horizontal dotted lines in Figure 2 do not represent meta-modelling levels but just ontologies. The ontology "Geographic Object Politics" has the meta meta-concept *GeographicObject*, whose instances are concepts which have also instances being concepts, but we also have the concepts *GovernmentOffice* and *Activity* whose instances conceptualize atomic objects. OWL has only one

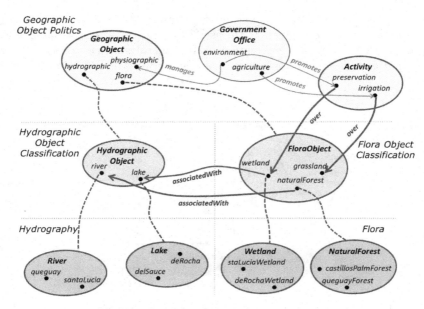

**Fig. 2.** Case Study on Geography

notion of hierarchy which classifies concepts with respect to the inclusion $\sqsubseteq$. Our approach has a new notion of hierarchy, called *meta-modelling hierarchy*, which classifies concepts with respect to the membership relation $\in$. The meta-modelling hierarchy for the concepts of Figure 2 is depicted in Figure 3. The concepts are *GovernmentOffice*, *Activity*, *River*, *Lake*, *Wetland* and *NaturalForest*, the meta-concepts are *HydrographicObject* and *FloraObject*, and the meta meta-concept is *GeographicObject*.

The first advantage of our approach over previous work concerns the reuse of ontologies when the same conceptual object is represented as an individual in one ontology and as a concept in the other. The identifiers for the individual and the concept will be syntactically different because they belong to different ontologies (with different URIs). Then, the ontology engineer can introduce an equation between these two different identifiers. This contrasts with previous approaches where one has to use the same identifier for an object used as a concept and as an individual. In Figure 2, *river* and *River* represent the same real object. In order to detect inconsistency and do the proper inferences, one has to be able to equate them.

The second advantage is about the flexibility of the meta-modelling hierarchy. This hierarchy is easy to change by just adding equations. This is illustrated in the passage from Figure 1 to Figure 2. Figure 1 has a very simple meta-modelling hierarchy where the concepts are *River* and *Lake* and the meta-concept is *HydrographicObject*. The rather more complex meta-modelling hierarchy for the ontology of Figure 2 (see Figure 3) has been obtained by combining the ontologies of Figure 1 with other ontologies and by simply adding some few meta-modelling axioms.

After adding the meta-modelling equations, the change of the meta-modelling hierarchy is *automatic* and *transparent* to the user. Concepts such as *GeographicObject* will automatically pass to be meta meta-concepts and roles such as *associatedWith* will automatically pass to be meta-roles, i.e. roles between meta-concepts.

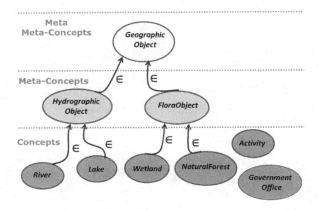

**Fig. 3.** Meta-modelling Hierarchy for the Ontology of Figure 2

The third advantage is that we do not have any restriction on the level of meta-modelling, i.e. we can have concepts, meta-concepts, meta meta-concepts and so on. Figure 1 has only one level of meta-modelling since there are concepts and meta-concepts. In Figure 2, there are two levels of meta-modelling since it has concepts, meta-concepts and meta meta-concepts. If we needed, we could extend it further by adding the equation $santaLucia =_m SantaLucia$ for some concept *SantaLucia* and this will add a new level in the meta-modelling hierarchy: concepts, meta-concepts, meta meta-concepts and meta meta meta-concepts.

Moreover, the user does not have to know the meta-modelling levels, they are transparent for him. Our algorithm detects inconsistencies without burdening the user with syntactic complications such as having to explicitly write the level the concept belongs to.

The fourth advantage is about the possibility of mixing levels of meta-modelling in the definition of concepts and roles. We can build concepts using union or intersection between two concepts of different levels (layers). We can also define roles whose domain and range live in different levels (or layers). For example, in Figure 2, we have: 1) a role *over* whose domain is just a concept while the range is a meta-concept, 2) a role *manages* whose domain is just a concept and whose range is a meta meta-concept. We can also add axioms to express that some of these concepts, though at different levels of meta-modelling, are disjoint, e.g. the intersection of the concept *Activity* and the meta-concept *FloraObject* is empty.

# 3  $\mathcal{ALCQM}$

In this section we introduce the $\mathcal{ALCQM}$ Description Logics (DL), with the aim of expressing meta-modelling in a knowledge base. The syntax of $\mathcal{ALCQM}$ is obtained from the one of $\mathcal{ALCQ}$ by adding new statements that allow us to equate individuals with concepts. The definition of the semantics for $\mathcal{ALCQM}$ is the key to our approach. In order to detect inconsistencies coming from meta-modelling, a proper semantics should give *the same interpretation* to individuals and concepts which have been equated through meta-modelling.

Recall the formal syntax of $\mathcal{ALCQ}$ [2,7]. We assume a finite set of atomic individuals, concepts and roles. If $A$ is an atomic concept and $R$ is a role, the concept expressions $C$, $D$ are constructed using the following grammar:

$C, D ::= A \mid \top \mid \bot \mid \neg C \mid C \sqcap D \mid C \sqcup D \mid \forall R.C \mid \exists R.C \mid\, \geq nR.C \mid\, \leq nR.C$

Recall also that $\mathcal{ALCQ}$-statements are divided in two groups, namely TBox statements and ABox statements, where a TBox contains statements of the form $C \sqsubseteq D$ and an ABox contains statements of the form $C(a)$, $R(a,b)$, $a = b$ or $a \neq b$.

A *meta-modelling axiom* is a new type of statement of the form

$$a =_{\mathsf{m}} A \text{ where } a \text{ is an individual and } A \text{ is an atomic concept.}$$

which we pronounce as *a corresponds to A through meta-modelling.* An *Mbox* is a set $\mathcal{M}$ of meta-modelling axioms. We define $\mathcal{ALCQM}$ by keeping the same syntax for concept expressions as for $\mathcal{ALCQ}$ and extending it only to include MBoxes. An ontology or a knowledge base in $\mathcal{ALCQM}$ is denoted by $\mathcal{O} = (\mathcal{T}, \mathcal{A}, \mathcal{M})$ since it is determined by three sets: a Tbox $\mathcal{T}$, an Abox $\mathcal{A}$ and an Mbox $\mathcal{M}$. The set of all individuals with meta-modelling of an ontology is denoted by $\mathsf{dom}(\mathcal{M})$.

Figure 4 shows the $\mathcal{ALCQM}$-ontologies of Figure 1. In order to check for cycles in the tableau algorithm, it is convenient to have the restriction that $A$ should be a concept name in $a =_{\mathsf{m}} A$. This restriction does not affect us in practice at all. If one would like to have $a =_{\mathsf{m}} C$ for a concept expression $C$, it is enough to introduce a concept name $A$ such that $A \equiv C$ and $a =_{\mathsf{m}} A$.

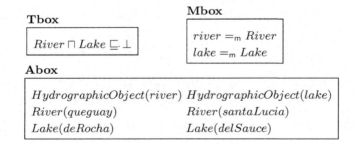

**Fig. 4.** The $\mathcal{ALCQM}$-ontology of Figure 1

**Definition 1** ($S_n$ for $n \in \mathbb{N}$). *Given a non empty set $S_0$ of atomic objects, we define $S_n$ by induction on $\mathbb{N}$ as follows: $S_{n+1} = S_n \cup \mathcal{P}(S_n)$*

The sets $S_n$ are clearly well-founded. Recall from Set Theory that *a relation $R$ is well-founded on a class $X$* if every non-empty subset $Y$ of $X$ has a minimal element. Moreover, *a set $X$ is well-founded* if the set membership relation is well-founded on the the set $X$.

**Definition 2 (Model of an Ontology in $\mathcal{ALCQM}$).** *An interpretation $\mathcal{I}$ is a model of an ontology $\mathcal{O} = (\mathcal{T}, \mathcal{A}, \mathcal{M})$ in $\mathcal{ALCQM}$ (denoted as $\mathcal{I} \models \mathcal{O}$) if the following holds:*

1. *the domain $\Delta$ of the interpretation is a subset of $S_N$ for some $N \in \mathbb{N}$. The smallest $N$ such that $\Delta \subseteq S_N$ is called the* level *of the interpretation $\mathcal{I}$.*
2. *$\mathcal{I}$ is a model of the ontology $(\mathcal{T}, \mathcal{A})$ in $\mathcal{ALCQ}$.*
3. *$\mathcal{I}$ is a model of $\mathcal{M}$, i.e. $\mathcal{I}$ satisfies each statement in $\mathcal{M}$. An interpretation $\mathcal{I}$ satisfies the statement $a =_m A$ if $a^{\mathcal{I}} = A^{\mathcal{I}}$.*

Usually, the domain of an interpretation of an ontology is a set of atomic objects. In the first part of Definition 2 we redefine the domain $\Delta$ of the interpretation, so it does not consists only of atomic objects any longer. The domain $\Delta$ can now contain sets since the set $S_N$ is defined recursively using the power-set operation. A similar notion of interpretation domain is defined in [9, Definition 1] for RDF ontologies.

It is sufficient to require that it is a subset of some $S_N$ so it remains well-founded [1]. Note that $S_0$ does not have to be the same for all models of an ontology. The second part of Definition 2 refers to the $\mathcal{ALCQ}$-ontology without the Mbox axioms. In the third part of the definition, we add another condition that the model must satisfy considering the meta-modelling axioms. This condition restricts the interpretation of an individual that has a corresponding concept through meta-modelling to be equal to the concept interpretation.

*Example 3.* We define a model for the ontology of Figure 4 where

$$S_0 = \{queguay, santaLucia, deRocha, delSauce\}$$

Individuals and concepts equated through meta-modelling are semantically equal:

$$river^{\mathcal{I}} = River^{\mathcal{I}} = \{queguay, santaLucia\}$$
$$lake^{\mathcal{I}} = Lake^{\mathcal{I}} = \{deRocha, delSauce\}$$

**Definition 3 (Consistency of an Ontology in $\mathcal{ALCQM}$).** *We say that an ontology $\mathcal{O} = (\mathcal{T}, \mathcal{A}, \mathcal{M})$ is consistent if there exists a model of $\mathcal{O}$.*

The $\mathcal{ALCQM}$-ontology defined in Figure 4 is consistent.

---

[1] In principle, non well-founded sets are not source of contradictions since we could work on non-well founded Set Theory. The reason why we exclude them is because we think that non well-founded sets do not occur in the applications we are interested in.

*Example 4.* We consider the ontology of Figure 2 and the axioms:

$$River \sqcap Lake \sqsubseteq \bot$$
$$Wetland \equiv NaturalForest$$

and the fact that *associatedWith* is a functional property. Note that we have the following axioms in the Abox:

$$associatedWith(wetland, lake)$$
$$associatedWith(naturalForest, river)$$

As before, the $\mathcal{ALCQ}$-ontology (without the Mbox) is consistent. However, the $\mathcal{ALCQM}$-ontology (with the Mbox) is not consistent.

Example 1 illustrates the use of the first clause of Definition 2. Actually, this example is inconsistent because the first clause of this definition does not hold. Examples 2 and 4 illustrate how the second and third conditions of Definition 2 interact.

**Definition 4 (Logical Consequence from an Ontology in $\mathcal{ALCQM}$).** *We say that $S$ is a* logical consequence *of $\mathcal{O} = (\mathcal{T}, \mathcal{A}, \mathcal{M})$ (denoted as $\mathcal{O} \models S$) if all models of $\mathcal{O}$ are also models of $S$ where $S$ is any of the following ALCQM-statements, i.e. $C \sqsubseteq D$, $C(a)$, $R(a, b)$, $a =_m A$, $a = b$ and $a \neq b$.*

It is possible to infer new knowledge in the ontology with the meta-modelling that is not possible without it as illustrated by Examples 1, 2 and 4.

**Definition 5 (Meta-concept).** *We say that $C$ is a meta concept in $\mathcal{O}$ if there exists an individual $a$ such that $\mathcal{O} \models C(a)$ and $\mathcal{O} \models a =_m A$.*

Then, $C$ is a meta meta-concept if there exists an individual $a$ such that $\mathcal{O} \models C(a)$, $\mathcal{O} \models a =_m A$ and $A$ is a meta-concept. Note that a meta meta-concept is also a meta-concept.

We have some new inference problems:

1. *Meta-modelling.* Find out whether $\mathcal{O} \models a =_m A$ or not.
2. *Meta-concept.* Find out whether $C$ is a meta-concept or not.

Most inference problems in Description Logic can be reduced to satisfiability by applying a standard result in logic which says that a formula $\phi$ is a semantic consequence of a set of formulas $\Gamma$ if and only if $\Gamma \cup \neg\phi$ is not satisfiable. The above two problems can be reduced to satisfiability following this general idea. For the first problem, note that since $a \neq_m A$ is not directly available in the syntax, we have replaced it by $a \neq b$ and $b =_m A$ which is an equivalent statement to the negation of $a =_m A$ and can be expressed in $\mathcal{ALCQM}$.

**Lemma 1.** $\mathcal{O} \models a =_m A$ *if and only if for some new individual $b$, $\mathcal{O} \cup \{a \neq b, b =_m A\}$ is unsatisfiable.*

**Lemma 2.** *$C$ is a meta-concept if and only if for some individual $a$ we have that $\mathcal{O} \cup \{\neg C(a)\}$ is unsatisfiable and for some new individual $b$, $\mathcal{O} \cup \{a \neq b, b =_m A\}$ is unsatisfiable.*

# 4   Checking Consistency of an Ontology in $\mathcal{ALCQM}$

In this section we will define a tableau algorithm for checking consistency of an ontology in $\mathcal{ALCQM}$ by extending the tableau algorithm for $\mathcal{ALCQ}$. From the practical point of view, extending tableau for $\mathcal{ALCQ}$ has the advantage that one can easily change and reuse the code of existing OWL's reasoners.

The tableau algorithm for $\mathcal{ALCQM}$ is defined by adding three expansion rules and a condition to the tableau algorithm for $\mathcal{ALCQ}$. The new expansion rules deal with the equalities and inequalities between individuals with meta-modelling which need to be transferred to the level of concepts as equalities and inequalities between the corresponding concepts. The new condition deals with circularities avoiding sets that belong to themselves and more generally, avoiding non well-founded sets.

**Definition 6 (Cycles).**   *We say that the tableau graph $\mathcal{L}$ has a cycle with respect to $\mathcal{M}$ if there exist a sequence of meta-modelling axioms $A_0 =_m a_0$, $A_1 =_m a_1, \ldots A_n =_m a_n$ all in $\mathcal{M}$ such that*

$$\begin{aligned}
A_1 &\in \mathcal{L}(x_0) & x_0 &\approx a_0 \\
A_2 &\in \mathcal{L}(x_1) & x_1 &\approx a_1 \\
&\ \vdots & &\ \vdots \\
A_n &\in \mathcal{L}(x_{n-1}) & x_{n-1} &\approx a_{n-1} \\
A_0 &\in \mathcal{L}(x_n) & x_n &\approx a_n
\end{aligned}$$

*Example 5.* Suppose we have an ontology $(\mathcal{T}, \mathcal{A}, \mathcal{M})$ with two individuals $a$ and $b$, the individual assignments: $B(a)$ and $A(b)$; and the meta-modelling axioms:

$$a =_m A \qquad b =_m B.$$

The tableau graph $\mathcal{L}(a) = \{B\}$ and $\mathcal{L}(b) = \{A\}$ has a cycle since $A \in \mathcal{L}(b)$ and $B \in \mathcal{L}(a)$.

Initialization for the $\mathcal{ALCQM}$-tableau is nearly the same as for $\mathcal{ALCQ}$. The nodes of the initial tableau graph will be created from individuals that occur in the Abox as well as in the Mbox. After initialization, the tableau algorithm proceeds by non-deterministically applying the **expansion rules** for $\mathcal{ALCQM}$. The expansion rules for $\mathcal{ALCQM}$ are obtained by adding the rules of Figure 5 to the expansion rules for $\mathcal{ALCQ}$.

We explain the intuition behind the new expansion rules. If $a =_m A$ and $b =_m B$ then the individuals $a$ and $b$ represent concepts. Any equality at the level of individuals should be transferred as an equality between concepts and similarly with the difference.

The $\approx$-rule transfers the equality $a \approx b$ to the level of concepts by adding two statements to the Tbox which are equivalent to $A \equiv B$. This rule is necessary to detect the inconsistency of Example 2 where the equality $river = lake$ is transferred as an equality $River \equiv Lake$ between concepts. A particular case of the application of the $\approx$-rule is when $a =_m A$ and $a =_m B$. In this case, the algorithm also adds $A \equiv B$.

$\approx$-rule:    Let $a =_m A$ and $b =_m B$ in $\mathcal{M}$. If $a \approx b$ and $A \sqcup \neg B, B \sqcup \neg A$ does not belong to $\mathcal{T}$ then $\mathcal{T} \leftarrow A \sqcup \neg B, B \sqcup \neg A$.

$\not\approx$-rule:    Let $a =_m A$ and $b =_m B$ in $\mathcal{M}$. If $a \not\approx b$ and there is no $z$ such that $A \sqcap \neg B \sqcup B \sqcap \neg A \in \mathcal{L}(z)$ then create a new node $z$ with $\mathcal{L}(z) = \{A \sqcap \neg B \sqcup B \sqcap \neg A\}$.

**close-rule**: Let $a =_m A$ and $b =_m B$ where $a \approx x$, $b \approx y$, $\mathcal{L}(x)$ and $\mathcal{L}(y)$ are defined. If neither $x \approx y$ nor $x \not\approx y$ are set then equate$(a, b, \mathcal{L})$ or differenciate$(a, b, \mathcal{L})$.

**Fig. 5.** Additional Expansion Rules for $\mathcal{ALCQM}$

The $\not\approx$-rule is similar to the $\approx$-rule. However, in the case that $a \not\approx b$, we cannot add $A \not\equiv B$ because the negation of $\equiv$ is not directly available in the language. So, what we do is to replace it by an equivalent statement, i.e. add an element $z$ that witness this difference.

The rules $\approx$ and $\not\approx$ are not sufficient to detect all inconsistencies. With only these rules, we could not detect the inconsistency of Example 4. The idea is that we also need to transfer the equality $A \equiv B$ between concepts as an equality $a \approx b$ between individuals. However, here we face a delicate problem. It is not enough to transfer the equalities that are in the Tbox. We also need to transfer the semantic consequences, e.g. $\mathcal{O} \models A \equiv B$. Unfortunately, we cannot do $\mathcal{O} \models A \equiv B$. Otherwise we will be captured in a vicious circle [2] since the problem of finding out the semantic consequences is reduced to the one of satisfiability. The solution to this problem is to explicitly try either $a \approx b$ or $a \not\approx b$. This is exactly what the close-rule does. The close-rule adds either $a \approx b$ or $a \not\approx b$ through two new functions *equate* and *differenciate*. It is similar to the choose-rule which adds either $C$ or $\neg C$. This works because we are working in Classical Logic and we have the law of excluded middle. For a model $\mathcal{I}$ of the ontology, we have that either $a^{\mathcal{I}} = b^{\mathcal{I}}$ or $a^{\mathcal{I}} \neq b^{\mathcal{I}}$ (see also Lemma 5). Since the tableau algorithm works with canonical representatives of the $\approx$-equivalence classes, we have to be careful how we equate two individuals or make them different.

Note that the application of the tableau algorithm to an $\mathcal{ALCQM}$ knowledge base $(\mathcal{T}, \mathcal{A}, \mathcal{M})$ changes the Tbox as well as the tableau graph $\mathcal{L}$.

**Definition 7 ($\mathcal{ALCQM}$-Complete).** $(\mathcal{T}, \mathcal{L})$ is $\mathcal{ALCQM}$-complete if none of the expansion rules for $\mathcal{ALCQM}$ is applicable.

The algorithm terminates when we reach some $(\mathcal{T}, \mathcal{L})$ where either $(\mathcal{T}, \mathcal{L})$ is $\mathcal{ALCQM}$-complete, $\mathcal{L}$ has a contradiction or $\mathcal{L}$ has a cycle. The ontology $(\mathcal{T}, \mathcal{A}, \mathcal{M})$ is consistent if there exists some $\mathcal{ALCQM}$-complete $(\mathcal{T}, \mathcal{L})$ such that $\mathcal{L}$ has neither contradictions nor cycles. Otherwise it is inconsistent.

## 5   Correctness of the Tableau Algorithm for $\mathcal{ALCQM}$

In this section we prove termination, soundness and completeness for the tableau algorithm described in the previous section. We give an elegant proof of completeness by showing an isomorphism between the canonical interpretations of $\mathcal{ALCQ}$ and $\mathcal{ALCQM}$.

---

[2] Consistency is the egg and semantic consequence is the chicken.

**Theorem 1 (Termination).** *The tableau algorithm for* $\mathcal{ALCQM}$ *described in the previous section always terminates.*

*Proof.* Suppose the input is an arbitrary ontology $\mathcal{O} = (\mathcal{T}, \mathcal{A}, \mathcal{M})$. We define

$$\text{concepts}(\mathcal{M}) = \bigcup_{a =_m A, b =_m B} \{A \sqcap \neg B \sqcup B \sqcap \neg A, A \sqcup \neg B, B \sqcup \neg A\}$$

Suppose we have an infinite sequence of rule applications:

$$(\mathcal{T}_0, \mathcal{L}_0) \Rightarrow (\mathcal{T}_1, \mathcal{L}_1) \Rightarrow (\mathcal{T}_2, \mathcal{L}_2) \Rightarrow \ldots \tag{1}$$

where $\Rightarrow$ denotes the application of one $\mathcal{ALCQM}$-expansion rule. In the above sequence, the number of applications of the $\approx, \not\approx$ and close-rules is finite as we show below:

1. The $\approx$ and $\not\approx$-rules can be applied only a finite number of times in the above sequence. The $\approx$ and $\not\approx$-rules add concepts to the Tbox and these concepts that can be added all belong to $\text{concepts}(\mathcal{M})$ which is finite. We also have that $\mathcal{T}_i \subseteq \mathcal{T} \cup \text{concepts}(\mathcal{M})$ for all $i$. Besides none of the other rules remove elements from the Tbox.
2. Since the set $\{(a, b) \mid a, b \in \text{dom}(\mathcal{M})\}$ is finite, the close-rule can be applied only a finite number of times. This is because once we set $a \approx b$ or $a \not\approx b$, no rule can "undo" this.

This means that from some $n$ onwards in sequence (1)

$$(\mathcal{T}_n, \mathcal{L}_n) \Rightarrow (\mathcal{T}_{n+1}, \mathcal{L}_{n+1}) \Rightarrow (\mathcal{T}_{n+2}, \mathcal{L}_{n+2}) \Rightarrow \ldots \tag{2}$$

there is no application of the rules $\approx$, $\not\approx$ and close. Moreover, $\mathcal{T}_n = \mathcal{T}_i$ for all $i \geq n$. Now, sequence (2) contains only application of $\mathcal{ALCQ}$-expansion rules. This sequence is finite by [1, Proposition 5.2]. This is a contradiction.

The proof of the following theorem is similar to Soundness for $\mathcal{ALCQM}$ [1]..

**Theorem 2 (Soundness).** *If* $\mathcal{O} = (\mathcal{T}, \mathcal{A}, \mathcal{M})$ *is consistent then the* $\mathcal{ALCQM}$*-tableau graph terminates and yields an* $\mathcal{ALCQM}$*-complete* $(\mathcal{T}_k, \mathcal{L}_k)$ *such that* $\mathcal{L}_k$ *has neither cycles nor contradictions.*

The following definition of canonical interpretation is basically the one in [1, Definition 4.3]. Instead of $<$, we use the idea of descendants.

**Definition 8 ($\mathcal{ALCQ}$-Canonical Interpretation).** *We define the* $\mathcal{ALCQ}$*-canonical interpretation* $\mathcal{I}_c$ *from a tableau graph* $\mathcal{L}$ *as follows.*

$$
\begin{aligned}
\Delta^{\mathcal{I}_c} &= \{x \mid \mathcal{L}(x) \text{ is defined}\} \\
(x)^{\mathcal{I}_c} &= \begin{cases} x & \text{if } x \in \Delta^{\mathcal{I}_c} \\ y & \text{if } x \approx y \text{ and } y \in \Delta^{\mathcal{I}_c} \end{cases} \\
(A)^{\mathcal{I}_c} &= \{x \in \Delta^{\mathcal{I}_c} \mid A \in \mathcal{L}(x)\} \\
(R)^{\mathcal{I}_c} &= \{(x, y) \in \Delta^{\mathcal{I}_c} \times \Delta^{\mathcal{I}_c} \mid R \in \mathcal{L}(x, y), \ x \text{ is not blocked or} \\
& \quad R \in \mathcal{L}(z, y), \text{ where } x \text{ is blocked by } z \text{ and } z \text{ is not blocked }\}
\end{aligned}
$$

Note that the canonical interpretation is not defined on equivalence classes of $\approx$ but by choosing canonical representatives.

**Lemma 3.** *If the tableau algorithm for $\mathcal{ALCQM}$ with input $\mathcal{O} = (\mathcal{T}, \mathcal{A}, \mathcal{M})$ yields an $\mathcal{ALCQM}$-complete $(\mathcal{T}', \mathcal{L})$ such that $\mathcal{L}$ has no contradictions then $\mathcal{I}_c$ is a model of $(\mathcal{T}, \mathcal{A})$.*

*Proof.* We define $\mathrm{rel}(\mathcal{A}_\mathcal{L})$ as follows.

$$\mathrm{rel}(\mathcal{A}_\mathcal{L}) = \{C(x) \mid C \in \mathcal{L}(y), y \approx x\} \cup$$
$$\{R(a, b) \mid R \in \mathcal{L}(x, y), a \approx x, b \approx y, \{a, b\} \subseteq \mathcal{O}\} \cup$$
$$\{x = y \mid x \approx y\} \cup \{x \neq y \mid x \not\approx y\}$$

By [1, Lemma 5.5]), $\mathcal{I}_c$ is a model of $(\mathcal{T}', \mathrm{rel}(\mathcal{A}_\mathcal{L}))$. Since $\mathcal{T} \subseteq \mathcal{T}'$ and $\mathcal{A} \subseteq \mathrm{rel}(\mathcal{A}_\mathcal{L})$, we have that $\mathcal{I}_c$ is a model of $(\mathcal{T}, \mathcal{A})$.

So, how can we now make $\mathcal{I}_c$ into a model of the whole ontology $(\mathcal{T}, \mathcal{A}, \mathcal{M})$? We will transform $\mathcal{I}_c$ into a model of $(\mathcal{T}, \mathcal{A}, \mathcal{M})$ by defining a function set. The following lemma allows us to give a recursive definition of set.

**Lemma 4.** *If the tableau graph $\mathcal{L}$ has no cycles then $(\Delta^{\mathcal{I}_c}, \prec)$ is well-founded where $\prec$ is the relation defined as $y \prec x$ if $y \in (A)^{\mathcal{I}_c}$, $x \approx a$ and $a =_m A \in \mathcal{M}$.*

*Proof.* Suppose $(\Delta^{\mathcal{I}_c}, \prec)$ is not well-founded. Since $\Delta^{\mathcal{I}_c}$ is finite, infinite descendent $\prec$-sequences can only be formed from $\prec$-cycles, i.e. they are of the form

$$y_n \prec y_1 \prec \ldots \prec y_n$$

It is easy to see that this contradicts the fact that $\mathcal{L}$ has no cycles.

**Definition 9 (From Basic Objects to Sets: the function set).** *Let $\mathcal{L}$ a tableau graph without cycles and $\mathcal{I}_c$ be the $\mathcal{ALCQ}$-canonical interpretation from $\mathcal{L}$. For $x \in \Delta^{\mathcal{I}_c}$ we define $\mathrm{set}(x)$ as follows.*

$$\mathrm{set}(x) = \{\mathrm{set}(y) \mid y \in (A)^{\mathcal{I}_c}\} \quad \textit{if } x \approx a \textit{ for some } a =_m A \in \mathcal{M}$$
$$\mathrm{set}(x) = x \quad\quad\quad\quad\quad\quad\quad\quad \textit{otherwise}$$

**Lemma 5.** *Let $\mathcal{L}$ be an $\mathcal{ALCQM}$-complete tableau graph without contradictions. If $a =_m A$ and $a' =_m A'$ then either $a \approx a'$ or $a \not\approx a'$. In the first case, $A^{\mathcal{I}_c} = A'^{\mathcal{I}_c}$ and in the second case, $A^{\mathcal{I}_c} \neq A'^{\mathcal{I}_c}$*

**Lemma 6.** *Let $\mathcal{L}$ be an $\mathcal{ALCQM}$-complete tableau graph that has neither contradictions nor cycles and let $\mathcal{I}_c$ be the canonical interpretation from $\mathcal{L}$. Then, set is an injective function, i.e. $x = x'$ if and only if $\mathrm{set}(x) = \mathrm{set}(x')$.*

*Proof.* We prove first that set is a function. It is enough to consider the case when $x \approx a =_m A$ and $x \approx a' =_m A'$. By Lemma 5, $a \approx a'$ and $(A)^{\mathcal{I}_c} = (A')^{\mathcal{I}_c}$. Hence, $\mathrm{set}(x)$ is uniquely determined.

To prove that set is injective, we do induction on $(\Delta^{\mathcal{I}_c}, \prec)$ which we know that is well-founded by Lemma 4. By Definition of set, we have two cases. The first case is when $\mathsf{set}(x) = x$. We have that $\mathsf{set}(x') = x$ and $x'$ is exactly $x$. This was the base case. In the second case, we have that for $x \approx a$ and $a =_m A$,

$$\mathsf{set}(x) = \{\mathsf{set}(y) \mid y \in (A)^{\mathcal{I}_c}\}$$

Since $\mathsf{set}(x) = \mathsf{set}(x')$, we also have that $x' \approx a'$ and $a' =_m A'$ such that

$$\mathsf{set}(x') = \{\mathsf{set}(y') \mid y' \in (A')^{\mathcal{I}_c}\}$$

Again since $\mathsf{set}(x) = \mathsf{set}(x')$, we have that $\mathsf{set}(y) = \mathsf{set}(y')$. By Induction Hypothesis, $y = y'$ for all $y \in (A)^{\mathcal{I}_c}$. Hence, $(A)^{\mathcal{I}_c} \subseteq (A')^{\mathcal{I}_c}$. Similarly, we get $(A')^{\mathcal{I}_c} \subseteq (A)^{\mathcal{I}_c}$. So, $(A)^{\mathcal{I}_c} = (A')^{\mathcal{I}_c}$. It follows from Lemma 5 that $a \approx a'$. Then, $x = x'$ because the canonical representative of an equivalence class is unique.

We are now ready to define the canonical interpretation for an ontology in $\mathcal{ALCQM}$.

**Definition 10 (Canonical Interpretation for $\mathcal{ALCQM}$).** *Let $\mathcal{L}$ be an $\mathcal{ALCQM}$-complete tableau graph without cycles and without contradictions. We define the canonical interpretation $\mathcal{I}^m$ for $\mathcal{ALCQM}$ as follows:*

$$\begin{aligned}
\Delta^{\mathcal{I}_m} &= \{\mathsf{set}(x) \mid x \in \Delta^{\mathcal{I}_c}\} \\
(a)^{\mathcal{I}_m} &= \mathsf{set}(a) \\
(A)^{\mathcal{I}_m} &= \{\mathsf{set}(x) \mid x \in A^{\mathcal{I}_c}\} \\
(R)^{\mathcal{I}_m} &= \{(\mathsf{set}(x), \mathsf{set}(y)) \mid (x, y) \in (R)^{\mathcal{I}_c}\}
\end{aligned}$$

**Definition 11 (Isomorphism between interpretations of $\mathcal{ALCQ}$).**
*An isomorphism between two interpretations $\mathcal{I}$ and $\mathcal{I}'$ of $\mathcal{ALCQ}$ is a bijective function $f : \Delta \to \Delta'$ such that*

- $f(a^{\mathcal{I}}) = a^{\mathcal{I}'}$
- $x \in A^{\mathcal{I}}$ *if and only if* $f(x) \in A^{\mathcal{I}'}$
- $(x, y) \in R^{\mathcal{I}}$ *if and only if* $(f(x), f(y)) \in R^{\mathcal{I}'}$.

**Lemma 7.** *Let $\mathcal{I}$ and $\mathcal{I}'$ be two isomorphic interpretations of $\mathcal{ALCQ}$. Then, $\mathcal{I}$ is a model of $(\mathcal{T}, \mathcal{A})$ if and only if $\mathcal{I}'$ is a model of $(\mathcal{T}, \mathcal{A})$.*

To prove the previous lemma is enough to show that $x \in C^{\mathcal{I}}$ if and only if $f(x) \in C^{\mathcal{I}'}$ by induction on $C$.

**Theorem 3 (Completeness).** *If $(\mathcal{T}, \mathcal{A}, \mathcal{M})$ is not consistent then the $\mathcal{ALCQM}$-tableau algorithm with input $(\mathcal{T}, \mathcal{A}, \mathcal{M})$ terminates and yields an $\mathcal{ALCQM}$-complete $(\mathcal{T}', \mathcal{L})$ such that $\mathcal{L}$ that has either a contradiction or a cycle.*

*Proof.* By Theorem 1, the $\mathcal{ALCQM}$-tableau algorithm with input $(\mathcal{T}, \mathcal{A}, \mathcal{M})$ terminates. Suppose towards a contradiction that the algorithm yields an $\mathcal{ALCQM}$-complete $(\mathcal{T}', \mathcal{L})$ such that that $\mathcal{L}$ has neither a contradiction nor a cycle. We will prove that $\mathcal{I}_m$ is a model of $(\mathcal{T}, \mathcal{A}, \mathcal{M})$. For this we have to check that $\mathcal{I}_m$ satisfies the three conditions of Definition 2.

1. In order to prove that $\Delta^{\mathcal{I}_m} \subseteq S_N$ for some $S_N$ and $N$, we define $S_0 = \{x \in \Delta^{\mathcal{I}_c} \mid \mathsf{set}(x) = x\}$.
2. We now prove that $\mathcal{I}_m$ is a model of $(\mathcal{T}, \mathcal{A})$. By Lemma 3, the canonical interpretation $\mathcal{I}_c$ is a model of $(\mathcal{T}, \mathcal{A})$. It follows from Lemma 6 that $\mathsf{set} : \Delta^{\mathcal{I}_c} \rightarrow \Delta^{\mathcal{I}_m}$ is a bijective map. It is also easy to show that $\mathcal{I}_c$ and $\mathcal{I}_m$ are isomorphic interpretations in $\mathcal{ALCQ}$. By Lemma 7, $\mathcal{I}_m$ is a model of $(\mathcal{T}, \mathcal{A})$.
3. Finally, we prove that $a^{\mathcal{I}_m} = (A)^{\mathcal{I}_m}$ for all $a =_m A \in \mathcal{M}$. Suppose that $a =_m A \in \mathcal{M}$. Then,

$$
\begin{aligned}
a^{\mathcal{I}_m} &= \mathsf{set}(a) && \text{by Definition 10} \\
&= \{\mathsf{set}(x) \mid x \in (A)^{\mathcal{I}_c}\} && \text{by Definition 9} \\
&= A^{\mathcal{I}_m} && \text{by Definition 10}
\end{aligned}
$$

A direct corollary from the above result is that $\mathcal{ALCQM}$ satisfies the finite model property.

# 6 Conclusions and Future Work

In this paper we present a tableau algorithm for checking consistency of an ontology in $\mathcal{ALCQM}$ and prove its correctness. In order to implement our algorithm, we plan to incorporate optimization techniques such as normalization, absorption or the use of heuristics [2, Chapter 9].

A first step to optimize the algorithm would be to impose the following order on the application of the expansion rules. We apply the rules that create nodes ($\exists$ and $\geq$) only if the other rules are not applicable. We apply the bifurcating rules ($\sqcup$, choose or close-rules) if the remaining rules (all rules except the $\exists$, $\geq$, $\sqcup$, choose and close-rules) are not applicable. One could prove that this strategy is correct similarly to Section 5.

A second step to optimize the algorithm would be to change the $\approx$-rule. Instead of adding $A \sqcup \neg B$ and $\neg A \sqcup B$, we could add $A \equiv B$ and treat this as a trivial case of lazy unfolding.

We would also like to study decidability of consistency for the kind of meta-modelling presented in this paper in more powerful Description Logics than $\mathcal{ALCQM}$.

We believe that consistency in $\mathcal{ALCQM}$ has the same complexity as $\mathcal{ALCQ}$, which is Exp-time complete [13]. We also plan to study worst-case optimal tableau algorithms for $\mathcal{ALCQM}$ [3,6].

**Acknowledgments.** We are grateful to Diana Comesaña for sharing with us the data from the ontology network on geographic objects she is developing in Uruguay [10].

# References

1. Baader, F., Buchheit, M., Hollunder, B.: Cardinality restrictions on concepts. Artif. Intell. **88**(1–2), 195–213 (1996)
2. Baader, F., Calvanese, D., McGuinness, D.L., Nardi, D., Patel-Schneider, P.F. (eds.) The Description Logic Handbook: Theory, Implementation, and Applications. Cambridge University Press (2003)
3. Donini, F.M., Massacci, F.: Exptime tableaux for ALC. Artif. Intell. **124**(1), 87–138 (2000)
4. De Giacomo, G., Lenzerini, M., Rosati, R.: Higher-order description logics for domain metamodeling. In: AAAI (2011)
5. Glimm, B., Rudolph, S., Völker, J.: Integrated metamodeling and diagnosis in OWL 2. In: Patel-Schneider, P.F., Pan, Y., Hitzler, P., Mika, P., Zhang, L., Pan, J.Z., Horrocks, I., Glimm, B. (eds.) ISWC 2010, Part I. LNCS, vol. 6496, pp. 257–272. Springer, Heidelberg (2010)
6. Goré, R., Nguyen, L.A.: Exptime tableaux for ALC using sound global caching. In: Description Logics (2007)
7. Hitzler, P., Krötzsch, M., Rudolph, S.: Foundations of Semantic Web Technologies. Chapman & Hall/CRC (2009)
8. Jekjantuk, N., Gröner, G., Pan, J.Z.: Modelling and reasoning in metamodelling enabled ontologies. Int. J. Software and Informatics **4**(3), 277–290 (2010)
9. Kaushik, S., Farkas, C., Wijesekera, D., Ammann, P.: An algebra for composing ontologies. In: Bennett, B., Fellbaum, C. (eds.) FOIS. Frontiers in Artificial Intelligence and Applications, vol 150, pp. 265–276. IOS Press (2006)
10. Militar, S.G., Catálogo de objetos y símbolos geográficos (2014). http://www.sgm.gub.uy/index.php/component/content/article/49-noticias/novedades/124-catalogoversion1/
11. Motik, B.: On the properties of metamodeling in OWL. In: Gil, Y., Motta, E., Benjamins, V.R., Musen, M.A. (eds.) ISWC 2005. LNCS, vol. 3729, pp. 548–562. Springer, Heidelberg (2005)
12. Pan, J.Z., Horrocks, I., Schreiber, G.: OWL FA: A metamodeling extension of OWL DL. In: OWLED (2005)
13. Tobies, S.: Complexity results and practical algorithms for logics in knowledge representation. PhD thesis, LuFG Theoretical Computer Science, RWTH-Aachen, Germany (2001)

# Employing *DL-Lite$_R$*-Reasoners for Fuzzy Query Answering

Theofilos Mailis$^{(\boxtimes)}$ and Anni-Yasmin Turhan

Chair for Automata Theory, Theoretical Computer Science,
TU Dresden, Dresden, Germany
mailis@tcs.inf.tu-dresden.de

**Abstract.** Fuzzy Description Logics generalize crisp ones by providing membership degree semantics for concepts and roles by fuzzy sets. Recently, answering of conjunctive queries has been investigated and implemented in optimized reasoner systems based on the rewriting approach for crisp DLs. In this paper we investigate how to employ such existing implementations for crisp query answering in *DL-Lite$_R$* over *fuzzy* ontologies. To this end we give an extended rewriting algorithm for the case of fuzzy *DL-Lite$_R$*-ABoxes that employs the one for crisp *DL-Lite$_R$* and investigate the limitations of this approach. We also tested the performance of our proto-type implementation FLITE of this method.

## 1 Introduction

Description Logics (DLs) are a class of knowledge representation languages with well-defined semantics that are widely used to represent the conceptual knowledge of an application domain in a structured and formally well-understood way. Some applications require to describe sets for which there exists no sharp, unambiguous distinction between the members and nonmembers. For example, when classifying numerical sensor values into symbolic classes, a crisp (non-fuzzy), unambiguous distinction between the members and nonmembers is not a natural way of modeling. To represent this kind of information faithfully, fuzzy variants of DLs were introduced. These variants generalize crisp ones by providing membership degree semantics for their concepts and roles by fuzzy sets.

In the last years conjunctive query answering was the main reasoning task investigated for DLs. This reasoning task allows to access data in a flexible way. In order to cope with huge amounts of data, the property of first order (FOL) rewritability of DLs was defined and investigated. This property of a DL allows to implement query answering by a two step procedure: First, the initial query is rewritten such that it captures the information from the TBox. Second, this query is executed over a database capturing the facts from the ABox by means of SQL queries. FOL rewritability is the key feature of the DL-Lite family, which has been proposed and investigated in [2]. It guarantees that query answering can be done efficiently—in the size of the data and in the overall size

---

Partially supported by DFG SFB 912 (HAEC).

T. Supnithi et al.(Eds.): JIST 2014, LNCS 8943, pp. 63–78, 2015.
DOI: 10.1007/978-3-319-15615-6_5

of the corresponding ontology. This is the main reason why $DL\text{-}Lite_R$ is the DL underlying OWL 2 QL, one of the three profiles of OWL 2 language.

So far several fuzzy extensions of DL-Lite have been investigated. In [11,12] the problem of evaluating ranked top-$k$ queries in fuzzy DL-Lite is considered, and a variety of query languages by which a fuzzy DL-Lite knowledge base can be queried is presented in [6]. Though all of these approaches are tractable w.r.t. data complexity, they do not exploit the optimized query rewriting techniques that have been implemented in many systems for the classical case such as QuOnto2 [1,7], Ontop [8], Owlgres [9], and IQAROS [14]. There are also reduction techniques for very expressive DLs such as $\mathcal{SHIQ}$ from fuzzy to crisp [5] for query answering. These techniques are not promising in terms of efficiency, since they don't allow for FOL rewriting-based algorithms that employ relational databases (as the DL-Lite family).

Our approach to answering conjunctive queries over fuzzy $DL\text{-}Lite_R$-ontologies is to use existing optimized crisp DL-Lite reasoners as a black box to obtain an initial rewriting of the conjunctive query. We extend this query by (1) fuzzy atoms and (2) by so-called degree variables that capture the numerical membership degrees, which are used to return the corresponding fuzzy degrees. This straightforward approach allows to employ a standard SQL query engine—as in the crisp case and is thus easy to implement. It gives correct answers for the Gödel family of operators, which is widely used. However, for other families of fuzzy operators, answers concerning the degrees may be incorrect. We give a characterization of such cases and an estimation function for the interval in which the correct degrees lie. We have implemented the query answering engine FLITE based on this approach for fuzzy $DL\text{-}Lite_R$, which uses the Ontop system [8] to obtain the initial crisp rewriting.

The rest of the paper is structured as follows: next, we introduce fuzzy $DL\text{-}Lite_R$. Section 3 presents the algorithms for consistency checking and query answering for fuzzy $DL\text{-}Lite_R$-ontologies. In Section 4 we describe limitations of our approach: we characterize the cases in which incorrect results are obtained and why other fuzzy extensions of DL-Lite-ontologies are problematic. The FLITE system, based on the Ontop framework, is described and evaluated in Section 5. We end with conclusions and future work.

## 2    Preliminaries

We introduce the logic $DL\text{-}Lite_R$, its ontologies and then the fuzzy variant of the latter [6,12]. Starting from a set of concept names $\mathsf{N_C}$ and role names $\mathsf{N_R}$ complex concepts can be build. $DL\text{-}Lite_R$ distinguishes basic concepts represented by $B$, general concepts represented by $C$, basic roles represented by $Q$, and general roles represented by $R$, by the grammar:

$$B \to A \mid \exists Q \qquad C \to \top \mid B \mid \neg B \qquad Q \to P \mid P^- \qquad R \to Q \mid \neg Q$$

where $\top$ is the top concept. A *degree* $d$ is a number from the unit interval $[0, 1]$. The $DL\text{-}Lite_R$-concepts and -roles are used in axioms, which can have the

**Table 1.** Families of fuzzy logic operators

| Family | t-norm $a \otimes b$ | negation $\ominus a$ | implication $\alpha \Rightarrow b$ |
|---|---|---|---|
| Gödel | $\min(a,b)$ | $\begin{cases} 1, & a = 0 \\ 0, & a > 0 \end{cases}$ | $\begin{cases} 1, & a \leqslant b \\ b, & a > b \end{cases}$ |
| Łukasiewicz | $\max(a + b - 1, 0)$ | $1 - a$ | $\min(1 - a + b, 1)$ |
| Product | $a \times b$ | $\begin{cases} 1, & a = 0 \\ 0, & a > 0 \end{cases}$ | $\begin{cases} 1, & a \leqslant b \\ b/a, & a > b \end{cases}$ |

following forms:

$$B \sqsubseteq C \qquad \text{(general concept inclusion axiom)}$$
$$Q \sqsubseteq R \qquad \text{(role inclusion axiom)}$$
$$\text{funct}(Q) \qquad \text{(functionality axiom)}$$

A TBox $\mathcal{T}$ is a finite set of axioms. The set $\mathsf{N_I}$ is the set of individual names. Let $a, b \in \mathsf{N_I}$ and $d$ be a degree, then a *fuzzy assertion* is a statement of the form:

$$\langle B(a), d \rangle \qquad \text{(fuzzy concept assertion)}$$
$$\langle P(a, b), d \rangle \qquad \text{(fuzzy role assertion)}$$

An ABox $\mathcal{A}$ is a finite set of fuzzy assertions. A fuzzy DL-Lite ontology $\mathcal{O} = (\mathcal{T}, \mathcal{A})$ consists of a TBox $\mathcal{T}$ and an ABox $\mathcal{A}$. The crisp $DL\text{-}Lite_R$-ontologies (ABoxes) are a special case, where only degrees $d = 1$ are admitted.

The semantics of fuzzy $DL\text{-}Lite_R$ are provided via the different families of fuzzy logic operators depicted in Table 1 and interpretations. An *interpretation* for fuzzy $DL\text{-}Lite_R$ is a pair $\mathcal{I} = (\Delta^{\mathcal{I}}, \cdot^{\mathcal{I}})$ where $\Delta^{\mathcal{I}}$ is the interpretation domain and $\cdot^{\mathcal{I}}$ is an interpretation function mapping every individual $a$ onto an element $a^{\mathcal{I}} \in \Delta^{\mathcal{I}}$, every concept name $A$ onto a *concept membership function* $A^{\mathcal{I}} : \Delta^{\mathcal{I}} \to [0,1]$, every atomic role $P$ onto a *role membership function* $P^{\mathcal{I}} : \Delta^{\mathcal{I}} \times \Delta^{\mathcal{I}} \to [0,1]$.

Let $\delta, \delta'$ denote elements of $\Delta^{\mathcal{I}}$ and $\ominus$ denote fuzzy negation (Table 1), then the semantics of concepts and roles are inductively defined as follows:

$$(\exists Q)^{\mathcal{I}}(\delta) = \sup_{\delta' \in \Delta^{\mathcal{I}}} Q^{\mathcal{I}}(\delta, \delta') \qquad (\neg B)^{\mathcal{I}}(\delta) = \ominus B^{\mathcal{I}}(\delta) \qquad \top^{\mathcal{I}}(\delta) = 1$$
$$P^{-\mathcal{I}}(\delta, \delta') = P^{\mathcal{I}}(\delta', \delta) \qquad (\neg Q)^{\mathcal{I}}(\delta, \delta') = \ominus Q^{\mathcal{I}}(\delta, \delta')$$

We say an interpretation $\mathcal{I}$ *satisfies* a

- concept inclusion axiom $B \sqsubseteq C$ iff $B^{\mathcal{I}}(\delta) \leqslant C^{\mathcal{I}}(\delta)$ for every $\delta \in \Delta^{\mathcal{I}}$,
- role inclusion axiom $Q \sqsubseteq R$ iff $Q^{\mathcal{I}}(\delta, \delta') \leqslant R^{\mathcal{I}}(\delta, \delta')$ for every $\delta, \delta' \in \Delta^{\mathcal{I}}$,
- functionality axiom $\text{func}(Q)$ iff for every $\delta \in \Delta^{\mathcal{I}}$ there is a unique $\delta' \in \Delta^{\mathcal{I}}$ such that $Q^{\mathcal{I}}(\delta, \delta') > 0$.

We say that an interpretation $\mathcal{I}$ is a *model of a TBox* $\mathcal{T}$, i.e. $\mathcal{I} \models \mathcal{T}$, iff it satisfies all axioms in $\mathcal{T}$. $\mathcal{I}$ satisfies a fuzzy concept assertion $\langle B(a), d \rangle$ iff $B^{\mathcal{I}}(a^{\mathcal{I}}) \geqslant d$, and a fuzzy role assertion $\langle P(a,b), d \rangle$ iff $P^{\mathcal{I}}(a^{\mathcal{I}}, b^{\mathcal{I}}) \geqslant d$. $\mathcal{I}$ is a *model of an ABox* $\mathcal{A}$, i.e. $\mathcal{I} \models \mathcal{A}$, iff it satisfies all assertions in $\mathcal{A}$. Finally an interpretation $\mathcal{I}$ is a model of an ontology $\mathcal{O} = (\mathcal{T}, \mathcal{A})$ iff it is a model of $\mathcal{A}$ and $\mathcal{T}$.

Based on the formal semantics several reasoning problems can be defined for DLs. A *DL-Lite$_R$*-concept or TBox is *satisfiable* iff it has a model. Likewise, a *DL-Lite$_R$*-ontology is *consistent* iff it has a model, otherwise it is *inconsistent*. Given a TBox $\mathcal{T}$ and two concepts $C$ and $D$, $C$ is *subsumed by* $D$ w.r.t. $\mathcal{T}$ (denoted $C \sqsubseteq_{\mathcal{T}} D$), iff for all models $\mathcal{I}$ of $\mathcal{T}$ $C^{\mathcal{I}}(\delta) \leq D^{\mathcal{I}}(\delta)$ holds. The reasoning problem we want to address in this paper is answering of (unions of) conjunctive queries, which allows retrieval of tuples of individuals from the ontology by the use of variables. Let $N_V$ be a set of variable names and let $t_1, t_2 \in N_I \cup N_V$ be terms (either individuals or variable names). An *atom* is an expression of the form: $C(t_1)$ (*concept atom*) or $P(t_1, t_2)$ (*role atom*). Let $\mathbf{x}$ and $\mathbf{y}$ be vectors over $N_V$, then $\phi(\mathbf{x}, \mathbf{y})$ is a conjunction of atoms of the form $A(t_1)$ and $P(t_1, t_2)$. A *conjunctive query* (CQ) $q(\mathbf{x})$ over an ontology $\mathcal{O}$ is a first-order formula $\exists \mathbf{y}.\phi(\mathbf{x}, \mathbf{y})$, where $\mathbf{x}$ are the *answer variables*, $\mathbf{y}$ are *existentially quantified variables* and the concepts and roles in $\phi(\mathbf{x}, \mathbf{y})$ appear in $\mathcal{O}$. Observe, that the atoms in a CQ do not contain degrees. A *union of conjunctive queries* (UCQ) is a finite set of conjunctive queries that have the same number of answer variables.

Given a CQ $q(\mathbf{x}) = \exists \mathbf{y}.\phi(\mathbf{x}, \mathbf{y})$, an interpretation $\mathcal{I}$, a vector of individuals $\boldsymbol{\alpha}$ with the same arity as $\mathbf{x}$, we define the mapping $\pi$ that maps: i) each individual $a$ to $a^{\mathcal{I}}$, ii) each variable in $\mathbf{x}$ to a corresponding element of $\boldsymbol{\alpha}^{\mathcal{I}}$, and iii) each variable in $\mathbf{y}$ to a corresponding element $\delta \in \Delta^{\mathcal{I}}$. Suppose that for an interpretation $\mathcal{I}$, $\Pi$ is the *set of mappings* that comply to these three conditions. Computing the $t$-norm $\otimes$ of all atoms: $A^{\mathcal{I}}(\pi(t_1))$ and $P^{\mathcal{I}}(\pi(t_1), \pi(t_2))$ yields the degree of $\phi^{\mathcal{I}}(\boldsymbol{\alpha}^{\mathcal{I}}, \pi(\mathbf{y}))$. A tuple of individuals $\boldsymbol{\alpha}$ is a *certain answer* to $q(\mathbf{x})$, over $\mathcal{O}$, with a degree greater or equal than $d$ (denoted $\mathcal{O} \models q(\boldsymbol{\alpha}) \geqslant d$), if for every model $\mathcal{I}$ of $\mathcal{O}$:

$$q^{\mathcal{I}}(\boldsymbol{\alpha}^{\mathcal{I}}) = \sup_{\pi \in \Pi} \{\phi^{\mathcal{I}}(\boldsymbol{\alpha}, \pi(\mathbf{y}))\} \geqslant d.$$

We denote the set of certain answers along with degrees, to a query $q(\mathbf{x})$ w.r.t. an ontology $\mathcal{O}$ with $ans(q(\mathbf{x}), \mathcal{O})$:

$$ans(q(\mathbf{x}), \mathcal{O}) = \{(\boldsymbol{\alpha}, d) \mid \mathcal{O} \models q(\boldsymbol{\alpha}) \geqslant d$$
$$\text{and there exists no } d' > d \text{ such that } \mathcal{O} \models q(\boldsymbol{\alpha}) \geqslant d'\}.$$

A special case of CQs and UCQs are those with an empty vector $\mathbf{x}$ of answer variables. These queries return only a degree of satisfaction and are called *degree queries*. An ontology *entails* a degree query $q$ to a degree of $d$, if $\mathcal{O} \models q() \geqslant d$ and $\mathcal{O} \not\models q() \geqslant d'$ for every $d' > d$. In the crisp case, these queries are *Boolean queries* and return true of false. A crisp ontology entails a Boolean query $q$, if $\mathcal{O} \models q()$.

*Example 1.* To illustrate the expressiveness of fuzzy *DL-Lite$_R$*, we give an example from the operating systems domain focusing on information about servers.

The first two concept inclusions in the TBox $\mathcal{T}_{ex}$ state that each server has a part that is a CPU. The functional restriction states that no CPU can belong to more than one server. The ABox $\mathcal{A}_{ex}$ provides information about the connections between servers and CPUs and each CPU's degree of overutilization.

$$\mathcal{T}_{ex} := \{\text{Server} \sqsubseteq \exists\text{hasCPU},\ \exists\text{hasCPU}^- \sqsubseteq \text{CPU},\ \text{func}(\text{hasCPU}^-)\}$$
$$\mathcal{A}_{ex} := \{\langle\text{Server}(\text{server}_1), 1\rangle,\ \langle\text{hasCPU}(\text{server}_1, \text{cpu}_1), 1\rangle,$$
$$\langle\text{OverUtilized}(\text{cpu}_1), 0.6\rangle,\ \langle\text{hasCPU}(\text{server}_1, \text{cpu}_2), 1\rangle,$$
$$\langle\text{OverUtilized}(\text{cpu}_2), 0.8\rangle \qquad\qquad\qquad \}$$

Based on the ontology $\mathcal{O}_{ex} = (\mathcal{T}_{ex}, \mathcal{A}_{ex})$ we can formulate the following queries:

$$q_1(x) = \text{CPU}(x) \tag{1}$$
$$q_2(x, y) = \text{hasCPU}(x, y) \wedge \text{OverUtilized}(y) \tag{2}$$
$$q_3(x) = \exists y\ \text{hasCPU}(x, y) \wedge \text{OverUtilized}(y) \tag{3}$$

The query $q_1$ asks for all the CPUs of our system. The query $q_2$ asks for pairs of Servers and CPUs with an overutilized CPU. The query $q_3$ asks for Servers for which there exists an overutilized CPU. If conjunction and negation are interpreted based on the Gödel family of operators, the certain answers for each of the queries w.r.t. $\mathcal{O}_{ex}$ are:

$$ans(q_1(x), \mathcal{O}_{ex}) = \{(\text{cpu}_1, 1),\ (\text{cpu}_2, 1)\}$$
$$ans(q_2(x, y), \mathcal{O}_{ex}) = \{(\text{server}_1, \text{cpu}_1, 0.6),\ (\text{server}_2, \text{cpu}_2, 0.8)\}$$
$$ans(q_3(x), \mathcal{O}_{ex}) = \{(\text{server}_1, 0.8)\}.$$

## 3   Fuzzy Reasoning by Extending Crisp Rewritings

Let $q(\mathbf{x})$ be the conjunctive query that the user has formulated over the vocabulary of the $DL\text{-}Lite_R$ ontology $\mathcal{O} = (\mathcal{T}, \mathcal{A})$. The main idea underlying the classic $DL\text{-}Lite_R$ reasoning algorithms is to rewrite the query with the information from the TBox and then apply the resulting UCQ to the ABox $\mathcal{A}$ alone. The reasoning algorithm rewrites $q(\mathbf{x})$ by the use of $\mathcal{T}$ into a UCQ $q_{\mathcal{T}}(\mathbf{x})$, called the *rewriting* of $q$ w.r.t. $\mathcal{T}$. For $DL\text{-}Lite_R$-ontologies it is well-known that $\mathcal{O} \models q(\boldsymbol{\alpha})$ iff $\mathcal{A} \models q_{\mathcal{T}}(\boldsymbol{\alpha})$ for any ABox $\mathcal{A}$ and any tuple of individuals in $\mathcal{A}$ holds [2,4]. The PerfectRef$(q, \mathcal{T})$ algorithm, described in [4], computes the rewriting, i.e., the corresponding UCQ.

In order to perform consistency checking for a given $DL\text{-}Lite_R$-ontology $\mathcal{O} = (\mathcal{T}, \mathcal{A})$ the system rewrites the information from $\mathcal{T}$ into a Boolean UCQ $q_{\mathcal{T}}^{\text{unsat}}()$ that contains only existentially quantified variables by the Consistent $(\mathcal{O})$ algorithm, described in [4]. It holds that: an ontology $\mathcal{O} = (\mathcal{T}, \mathcal{A})$ is inconsistent iff $\mathcal{A} \models q_{\mathcal{T}}^{\text{unsat}}()$.

For fuzzy DLs we adopt the same approach for reasoning. The main difference is that the degrees of ABox assertions must also be taken into account here. The extensive investigation on the crisp algorithms for $DL\text{-}Lite_R$ [3,11] and the readily available optimized reasoner systems motivate our investigation on how to employ the classic $DL\text{-}Lite_R$ algorithm as a black box procedure to perform reasoning for the fuzzy case as well. The main idea is to apply the $DL\text{-}Lite_R$ rewriting algorithm on the crisp part of the ontology, i.e., by considering assertions as crisp and treating the degrees in a separate form of atoms in a second rewriting step. We apply this idea for satisfiability checking and query answering, extending the classical $\text{CONSISTENT}(\mathcal{O})$ and $\text{PERFECTREF}(q, \mathcal{T})$ algorithms to the fuzzy setting.

Before presenting the algorithm, we need introduce some additional notation to accommodate the degrees. For each concept name $A$ we introduce the binary predicate $A_f$ and for each role name $P$ we introduce the ternary predicate $P_f$. Intuitively, a fuzzy assertion of the form $A(a) \geqslant d$ (or $P(a, b) \geqslant d$) can be represented by a predicate assertion of the form $A_f(a, d)$ (or $P_f(a, b, d)$), where $d \in [0, 1]$. The $A_f$, $P_f$ predicates can be stored as tables in a database. Similarly to the relational database tables $\text{tab}_A$, $\text{tab}_r$ of arity 2 and 3 respectively, presented in [12].

Now, to have the fuzzy connectors implemented by the SQL engine correctly, degree variables and degree predicates are needed, which represent the fuzzy operators in the resulting query. These degree variables and predicates are used in the rewritings and enrich the query format used by our algorithms internally. Let $\mathsf{N}_{\mathsf{Vd}}$ be a set of *degree variables*. Such degree variables $x_d$, $y_d \in \mathsf{N}_{\mathsf{Vd}}$ can only be mapped to a value in $[0, 1]$. By using degree variables in conjunctive queries, we obtain again crisp UCQs with the fuzzy part represented by an additional answer variable $x_d$.

In order to represent fuzzy conjunction and negation by the $t$-norm and negation operator described in Table 1, we consider the *degree predicates* $\Phi_>, \Phi_\ominus, \Phi_\otimes$ such that for every $\alpha, \beta, \beta_1, \ldots, \beta_n \in \mathsf{N}_{\mathsf{Vd}}$:

$$\Phi_>(\alpha, \beta) = \{(\alpha, \beta) \mid \alpha > \beta\} \tag{4}$$

$$\Phi_\ominus(\alpha, \beta) = \{(\alpha, \beta) \mid \alpha = \ominus\beta\} \tag{5}$$

$$\Phi_\otimes(\alpha, \beta_1, \beta_2) = \{(\alpha, \beta_1, \beta_2) \mid \alpha = \beta_1 \otimes \beta_2\} \tag{6}$$

$$\Phi_\otimes(\alpha, \beta_1, \ldots, \beta_n) = \{(\alpha, \beta_1, \ldots, \beta_n) \mid \alpha = \beta_1 \otimes \ldots \otimes \beta_n\} \tag{7}$$

We call an expression formed over a degree predicate and a tuple of degree variables a *degree atom*. The degree predicates can be materialized in a query language such as SQL or SPARQL by standard mathematical functions and comparison operators. Depending on the family of operators used for fuzzy $DL\text{-}Lite_R$, the degree predicates $\Phi_\ominus$ and $\Phi_\otimes$ are instantiated according to Table 1.

In the remainder of the paper we use $_f$ to distinguish between the fuzzy and the crisp version of the algorithms and the parameters. For example, the $\text{CONSISTENT}$ algorithm used for classic $DL\text{-}Lite_R$ is extended to the fuzzy case in the $\text{CONSISTENT}_f$ algorithm, similarly we use the predicates $A$ and $A_f$.

## 3.1    The Consistent$_f$ Algorithm

The CONSISTENT$_f$ method depicted in Algorithm 1 first computes the query $q_T^{\text{unsat}}()$ used for consistency checking in the crisp case. A second rewriting step by REWRITEWITHDEGREES introduces CQs with degree variables and atoms to the query $q_{Tf}^{\text{unsat}}()$ to take into account the degrees from the ABox. The idea is that each CQ in $q_T^{\text{unsat}}()$ corresponds to a different type of inconsistency that may appear in our ontology: line 5 of Function REWRITEWITHDEGREES ensures that no functional restriction is violated, line 7 that no inverse functional restriction is violated, line 9 that no subsumption of the form $A \sqsubseteq_T \neg A'$ is violated, line 11 that no subsumption of the form $A \sqsubseteq_T \neg \exists P$ is violated and so on. Since these are all forms of clashes that can occurr in *DL-Lite$_R$*, the crisp CONSISTENT algorithm produces the UCQ $q_T^{\text{unsat}}()$, which covers all possible cases. The correctness of the method can be shown based on the semantics of fuzzy and crisp *DL-Lite$_R$*.

*Example 2.* According to the TBox $\mathcal{T} = \{\text{OverUtilized} \sqsubseteq \neg\text{UnderUtilized}\}$ a CPU cannot be in both states of utilization in the crisp case. Therefore, if the conjunctive query $q_{\mathcal{T}_{ex}}^{\text{unsat}}() = \exists x.\text{OverUtilized}(x) \wedge \text{UnderUtilized}(x)$ is entailed, our ontology is inconsistent. However, for the fuzzy case, the degree of OverUtilization should also be taken into account. The query $q_{\mathcal{T}_{ex}}^{\text{unsat}}()$ is rewritten to:

$$q_{\mathcal{T}_{ex}f}^{\text{unsat}}() = \exists x, y_{d_1}, y_{d_2}, y_{d_3}.\text{OverUtilized}_f(x, y_{d_1}) \wedge \text{UnderUtilized}_f(x, y_{d_2}) \wedge$$
$$\Phi_>(y_{d_1}, y_{d_3}) \wedge \Phi_\ominus(y_{d_3}, y_{d_2}).$$

This query asks, if there exists a CPU such that its degree of over-utilization is greater than the negation of its degree of under-utilization. In such a case an entailment $\mathcal{O} \models q_{\mathcal{T}_{ex}f}^{\text{unsat}}()$ would only be given, if $\mathcal{O}$ is inconsistent.

## 3.2    The PerfectRef$_f$ Algorithm for Answering Conjunctive Queries

Suppose, the conjunctive query $q(\mathbf{x}) = \exists \mathbf{y}.\phi(\mathbf{x}, \mathbf{y})$, where $\phi(\mathbf{x}, \mathbf{y})$ is a conjunction of concept and role atoms containing variables from $\mathbf{x}, \mathbf{y}$, is to be answered. Based on the crisp *DL-Lite$_R$* PERFECTREF algorithm, the CQ $q(\mathbf{x})$ is rewritten to the $q_T(\mathbf{x})$. This UCQ $q_T(\mathbf{x})$ contains atoms of the form $A(t_1)$ and $P(t_1, t_2)$, where $t_1, t_2$ are variables in $\mathbf{x}, \mathbf{y}$ or individuals from $\mathcal{O}$. For each CQ $q'(\mathbf{x})$ in the UCQ $q_T(\mathbf{x})$, each atom $A(t_1)$, $P(t_1, t_2)$ is replaced by $A_f(t_1, y_{d'})$, $P_f(t_1, t_2, y_{d'})$ respectively, where $y_{d'}$ is a new degree variable. Likewise, the $t$-norms of all the degree variables $y_{d'}$ appearing in $A_f(t_1, y_{d'})$ and $P_f(t_1, t_2, y_{d'})$ are added in the extended rewriting in form of degree predicates $\Phi_\otimes$. The actual computation of the degree values takes place, when the query is evaluated over the ABox. This idea is made precise in Algorithm 2. This algorithm returns a UCQ that, if answered w.r.t. the ABox $\mathcal{A}$, results in tuples of individuals, along with the degree by which they satisfy the query. If the same tuple of individuals is returned as an answer, but with a different degree, then only the answer with the highest degree is kept.

**Algorithm 1.** The CONSISTENT$_f$ algorithm

---

1: **function** CONSISTENT$_f(\mathcal{O})$
    ▷ $\mathcal{O}$ is a fuzzy $DL\text{-}Lite_{R_\mathcal{A}}$ ontology $\mathcal{O} = (\mathcal{T}, \mathcal{A})$.
2:     $q_\mathcal{T}^{\text{unsat}}() := $ CONSISTENT(remove-degrees($\mathcal{O}$))
    ▷ The query $q_\mathcal{T}^{\text{unsat}}()$ is obtained from the crisp CONSISTENT algorithm.
3:     **if** $ans($REWRITEWITHDEGREES$(q_\mathcal{T}^{\text{unsat}}()), \mathcal{A}) = \emptyset$ **then**
4:         **return** true
5:     **else**
6:         **return** false
7:     **end if**
8: **end function**

1: **function** REWRITEWITHDEGREES$(q_\mathcal{T}^{\text{unsat}}())$
2:     $q_{\mathcal{T}f}^{\text{unsat}}() := \emptyset$
    ▷ $q_{\mathcal{T}f}^{\text{unsat}}()$ is an initially empty crisp UCQ.
3:     **for all** CQs $q$ in $q_\mathcal{T}^{\text{unsat}}()$ **do**
4:         **if** $q$ has the form $\exists x, y_1, y_2.P(x,y_1) \wedge P(x,y_2) \wedge y_1 \neq y_2$ **then**
5:             $q_f := \exists x, y_1, y_2, y_{d_1}, y_{d_2}.P_f(x,y_1,y_{d_1}) \wedge P_f(x,y_2,y_{d_2}) \wedge y_1 \neq y_2 \wedge$
                 $\Phi_>(y_{d_1}, 0) \wedge \Phi_>(y_{d_2}, 0)$
    ▷ $q_f$ the extension of $q$ for querying fuzzy ABoxes.
6:         **else if** $q$ has the form $\exists x_1, x_2, y.P(x_1,y) \wedge P(x_2,y) \wedge x_1 \neq x_2$ **then**
7:             $q_f := \exists x_1, x_2, y, y_{d_1}, y_{d_2}.P_f(x_1,y,y_{d_1}) \wedge P_f(x_2,y,y_{d_2}) \wedge x_1 \neq x_2 \wedge$
                 $\Phi_>(y_{d_1}, 0) \wedge \Phi_>(y_{d_2}, 0)$
8:         **else if** $q$ has the form $\exists x.A(x) \wedge A'(x)$ **then**
9:             $q_f := \exists x, y_{d_1}, y_{d_2}, y_{d_3}.A_f(x,y_{d_1}) \wedge A'_f(x,y_{d_2}) \wedge \Phi_>(y_{d_1}, y_{d_3}) \wedge$
                 $\Phi_\ominus(y_{d_3}, y_{d_2})$
10:        **else if** $q$ has the form $\exists x, y.A(x) \wedge P(x,y)$ **then**
11:            $q_f := \exists x, y, y_{d_1}, y_{d_2}, y_{d_3}.A_f(x,y_{d_1}) \wedge P_f(x,y,y_{d_2}) \wedge$
                 $\Phi_>(y_{d_1}, y_{d_3}) \wedge \Phi_\ominus(y_{d_3}, y_{d_2})$
12:        **else if** $q$ has the form $\exists x, y.A(x) \wedge P(y,x)$ **then**
13:            $q_f := \exists x, y, y_{d_1}, y_{d_2}, y_{d_3}.A_f(x,y_{d_1}) \wedge P_f(y,x,y_{d_2}) \wedge$
                 $\Phi_>(y_{d_1}, y_{d_3}) \wedge \Phi_\ominus(y_{d_3}, y_{d_2})$
14:        **else if** $q$ has the form $\exists x, y_1, y_2.P(x,y_1) \wedge P'(x,y_2)$ **then**
15:            $q_f := \exists x, y_1, y_2, y_{d_1}, y_{d_2}, y_{d_3}.P_f(x,y_1,y_{d_1}) \wedge P'_f(x,y_2,y_{d_2}) \wedge$
                 $\Phi_>(y_{d_1}, y_{d_3}) \wedge \Phi_\ominus(y_{d_3}, y_{d_2})$
16:        **else if** $q$ has the form $\exists x, y_1, y_2.P(x,y_1) \wedge P'(y_2,x)$ **then**
17:            $q_f := \exists x, y_1, y_2, y_{d_1}, y_{d_2}, y_{d_3}.P_f(x,y_1,y_{d_1}) \wedge P'_f(y_2,x,y_{d_2}) \wedge$
                 $\Phi_>(y_{d_1}, y_{d_3}) \wedge \Phi_\ominus(y_{d_3}, y_{d_2})$
18:        **else if** $q$ has the form $\exists x, y_1, y_2.P(y_1,x) \wedge P'(y_2,x)$ **then**
19:            $q_f := \exists x, y_1, y_2, y_{d_1}, y_{d_2}, y_{d_3}.P_f(y_1,x,y_{d_1}) \wedge P'_f(y_2,x,y_{d_2}) \wedge$
                 $\Phi_>(y_{d_1}, y_{d_3}) \wedge \Phi_\ominus(y_{d_3}, y_{d_2})$
20:        **else if** $q$ has the form $\exists x, y.P(x,y) \wedge P'(x,y)$ **then**
21:            $q_f := \exists x, y, y_{d_1}, y_{d_2}, y_{d_3}.P_f(x,y,y_{d_1}) \wedge P'_f(x,y,y_{d_2}) \wedge$
                 $\Phi_>(y_{d_1}, y_{d_3}) \wedge \Phi_\ominus(y_{d_3}, y_{d_2})$
22:        **else if** $q$ has the form $\exists x, y.P(x,y) \wedge P'(y,x)$ **then**
23:            $q_f := \exists x, y, y_{d_1}, y_{d_2}, y_{d_3}.P_f(x,y,y_{d_1}) \wedge P'_f(y,x,y_{d_2}) \wedge$
                 $\Phi_>(y_{d_1}, y_{d_3}) \wedge \Phi_\ominus(y_{d_3}, y_{d_2})$
24:        **end if**
25:        $q_f^{\text{unsat}} := q_f^{\text{unsat}} \cup \{q_f\}$
26:    **end for**
27: **return** $q_f^{\text{unsat}}$
28: **end function**

---

**Algorithm 2.** The PERFECTREF$_f$ algorithm

---

1: **function** PERFECTREF$_f(q(\mathbf{x}), \mathcal{T})$
2:     $q_{\mathcal{T}}(\mathbf{x}) := \text{PERFECTREF}(q(\mathbf{x}), \mathcal{T})$
3:     $q_{\mathcal{T}}^f(\mathbf{x}) := \emptyset$
4:     **for all** CQs $q'(\mathbf{x}) = \exists \mathbf{y}.\phi(\mathbf{x}, \mathbf{y})$ in $q_{\mathcal{T}}(\mathbf{x})$ **do**
5:         $\mathbf{y_d} := ()$
         ▷ $\mathbf{y_d}$ is a vector that keeps the existentially quantified degree variables.
6:         $\phi_f(\mathbf{x}, \mathbf{y}) := \emptyset$
         ▷ $\phi_f(\mathbf{x}, \mathbf{y})$ is a conjunction of atoms corresponding to the fuzzy version of $\phi(\mathbf{x}, \mathbf{y})$.
7:         **for all** $A(t)$ in $q'(\mathbf{x})$ **do**
8:             Add the degree variable $y_{d'}$ to the vector $\mathbf{y_d}$
         ▷ $y_{d'}$ is a fresh degree variable name.
9:             $\phi_f(\mathbf{x}, \mathbf{y}) := \phi_f(\mathbf{x}, \mathbf{y}) \land A_f(t, y_{d'})$
10:         **end for**
11:         **for all** $P(t_1, t_2)$ in $q'(\mathbf{x})$ **do**
12:             Add the degree variable $y_{d'}$ to the vector $\mathbf{y_d}$
13:             $\phi_f(\mathbf{x}, \mathbf{y}) := \phi_f(\mathbf{x}, \mathbf{y}) \land P_f(t_1, t_2, y_{d'})$
14:         **end for**
15:         $q_f'(\mathbf{x}, x_d) := \exists \mathbf{y}, \mathbf{y_d}.\phi_f(\mathbf{x}, \mathbf{y}) \land \Phi_{\otimes}(x_d, \mathbf{y_d})$
16:         $q_{\mathcal{T}}^f(\mathbf{x}, x_d) := q_{\mathcal{T}}^f(\mathbf{x}, x_d) \cup \{q_f'(\mathbf{x}, x_d)\}$
17:     **end for**
18:     **return** $q_{\mathcal{T}}^f(\mathbf{x})$
19: **end function**

---

*Example 3.* Based on $\mathcal{O}_{ex} = (\mathcal{T}_{ex}, \mathcal{A}_{ex})$ from Example 1 we illustrate the application of PERFECTREF$_f$ algorithm to the queries $q_1, q_2, q_3$ from Example 1. Initially, $q_1, q_2, q_3$ are rewritten, by the crisp PERFECTREF algorithm to the following UCQs:

$$q_{1\mathcal{T}_{ex}}(x) = \{\text{CPU}(x), \quad \exists y.\text{hasCPU}(y, x)\}$$
$$q_{2\mathcal{T}_{ex}}(x, y) = \{\text{hasCPU}(x, y) \land \text{OverUtilized}(y)\}$$
$$q_{3\mathcal{T}_{ex}}(x) = \{\exists y.\text{hasCPU}(x, y) \land \text{OverUtilized}(y)\}$$

In the next step, the PERFECTREF$_f$ algorithm extends the queries with degree variables and atoms, so that the corresponding degrees can be returned:

$$q_{1\mathcal{T}_{ex}}^f(x, x_d) = \{\text{CPU}(x, x_d), \quad \exists y.\text{hasCPU}(y, x, x_d)\}$$
$$q_{2\mathcal{T}_{ex}}^f(x, y, x_d) = \{\text{hasCPU}(x, y, y_{d_1}) \land \text{OverUtilized}(y, y_{d_2}) \land \Phi_{\otimes}(x_d, y_{d_1}, y_{d_2})\}$$
$$q_{3\mathcal{T}_{ex}}^f(x, x_d) = \{\exists y.\text{hasCPU}(x, y, y_{d_1}) \land \text{OverUtilized}(y, y_{d_2}) \land \Phi_{\otimes}(x_d, y_{d_1}, y_{d_2})\}$$

For the ABox $\mathcal{A}_{ex}$ we get the following set of answers to each of the queries:

$$ans(q_{1\mathcal{T}_{ex}}^f(x, x_d), \mathcal{A}_{ex}) = \{(cpu_1, 1), (cpu_2, 1)\}$$
$$ans(q_{2\mathcal{T}_{ex}}^f(x, y, x_d), \mathcal{A}_{ex}) = \{(server_1, cpu_1, 0.6), (server_1, cpu_2, 0.8)\}$$
$$ans(q_{3\mathcal{T}_{ex}}^f(x, x_d), \mathcal{A}_{ex}) = \{(server_1, 0.6), (server_1, 0.8)\}$$

Finally, for each answer to a query, only the one with the highest degree is kept per (tuple of) individual(s):

$$ans(q1^f_{T_{ex}}(x, x_d), \mathcal{A}_{ex}) = \{(cpu_1, 1), (cpu_2, 1)\}$$
$$ans(q2^f_{T_{ex}}(x, x_d), \mathcal{A}_{ex}) = \{(server_1, cpu_1, 0.6), (server_1, cpu_2, 0.8)\}$$
$$ans(q3^f_{T_{ex}}(x, x_d), \mathcal{A}_{ex}) = \{(server_1, 0.8)\}$$

Unfortunately, this practical approach does not always yield correct results. The simplifications made during the rewriting step by the crisp algorithms PERFECTREF and CONSISTENT are correct for the crisp, but not for the fuzzy case. Specifically, a conjunctive query that contains the atom $A(x)$ repeatedly is simplified in the crisp case to a conjunctive query containing the same atom only once—an obvious optimization for the crisp case. However, in the fuzzy case, such simplification causes our algorithm to become unsound, since for every $A^\mathcal{I}(o) \in (0, 1)$ it applies that $A^\mathcal{I}(o) > A^\mathcal{I}(o) \otimes A^\mathcal{I}(o)$ for the Łukasiewicz and product families of operators. Similarly, each time two atoms are unified during the rewriting, one contribution degree is lost. These effects are better illustrated by the following example.

*Example 4.* Suppose that $\otimes$ is the product ($\times$) $t$-norm and our ontology has the following TBox and ABox:

$$\mathcal{T} = \{A_1 \sqsubseteq A_2, \ A_3 \sqsubseteq A_4\} \qquad \mathcal{A} = \{A_1(a) \geqslant 0.8, \ A_3(a) \geqslant 0.9\}.$$

Then the conjunctive query $q(x) = A_1(x) \wedge A_2(x) \wedge A_3(x) \wedge A_4(x)$ has $a$ as an answer with degree $\geq 0.5184$, since $A_1^\mathcal{I}(a^\mathcal{I}) \times A_1^\mathcal{I}(a^\mathcal{I}) \times A_3^\mathcal{I}(a^\mathcal{I}) \times A_3^\mathcal{I}(a^\mathcal{I}) = 0.5184$. Now, the crisp algorithm returns the following UCQ as rewriting:

$$q_{\mathcal{T}}(x) = \{A_1(x) \wedge A_2(x) \wedge A_3(x) \wedge A_4(x), \ A_1(x) \wedge A_3(x) \wedge A_4(x),$$
$$A_1(x) \wedge A_2(x) \wedge A_3(x), \ A_1(x) \wedge A_3(x)\}$$

For the crisp case there is no difference between the answers to the conjunctive queries $A_1(x) \wedge A_3(x)$ or $A_1(x) \wedge A_1(x) \wedge A_3(x) \wedge A_3(x)$. If we apply our rewriting technique for fuzzy queries to the last query, we get a fuzzy conjunctive query of the form:

$$q_{\mathcal{T}}^f(x, x_d) = \exists y_{d_{A_1}}, y_{d_{A_3}}.A_{1f}(x, y_{d_{A_1}}) \wedge A_{3f}(x, y_{d_{A_3}}) \wedge \Phi_\times(x_d, y_{d_{A_1}}, y_{d_{A_3}}) \quad (8)$$

and the answer for the variables $x$ and $x_d$ is $(a, 0.72)$, i.e., $a$ is an answer with a degree $\geq 0.72$ instead of 0.5184 which is the correct degree.

To conclude, our pragmatic approach for query answering over a fuzzy ontology, that uses the rewritings obtained during crisp query answering, yields sound results fuzzy semantics with idempotent operators such as the Gödel family of operators. For other families of operators, that are not idempotent, the algorithm need not be sound in the sense that the degree of a result returned may be greater than the actual degree.

# 4  Limitations of the Approach

## 4.1  Identifying and Assessing Unsound Results for Non-idempotent Fuzzy DLs

Since our approach for conjunctive query answering is sound for the Gödel family of operators, a natural question is when a case that might yield an unsound result is encountered. To this end we present a straightforward idea for identifing unsound results for the degrees and to give a narrowed down interval for the missed degrees. Recall that the *DL-Lite$_R$*-CQs have concept or role atoms, whereas the UCQs returned from our algorithms have degree atoms in addition. Let $|q(\mathbf{x})|_{\mathsf{CR}}$ denote the number of concept and role atoms of a CQ (degree atoms are not taken into account), and let $q_T^f(\mathbf{x})$ be the UCQ that the algorithm PERFECTREF$_f$ returns. A property of the crisp *DL-Lite$_R$* algorithm is that

$$|q(\mathbf{x})|_{\mathsf{CR}} \geqslant |q_f'(\mathbf{x}, x_d)|_{\mathsf{CR}} \quad \text{for every } q_f'(\mathbf{x}, x_d) \in q_T^f(\mathbf{x}).$$

This property allows to infer: if $|q(\mathbf{x})|_{\mathsf{CR}} = |q_f'(\mathbf{x}, x_d)|_{\mathsf{CR}}$ for every CQ $q_f'(\mathbf{x}, x_d)$ in $q_T^f(\mathbf{x})$, then no atom simplification has been applied and thus our algorithm gave a correct result.

In case $|q(\mathbf{x})|_{\mathsf{CR}} \geqslant |q_f'(\mathbf{x}, x_d)|_{\mathsf{CR}}$ for some $q_f'(\mathbf{x}, x_d) \in q_T^f(\mathbf{x})$, a pessimistic estimation for the not correctly calculated degrees can be computed in the following way. Suppose that $|q(\mathbf{x})|_{\mathsf{CR}} = n$, while $|q_f'(\mathbf{x}, x_d)|_{\mathsf{CR}} = m$ with $n > m$. Based on the PERFECTREF algorithm, each concept and role atom in $q(\mathbf{x})$ can be mapped to some corresponding 'fuzzy' atom in $q_f'(\mathbf{x}, x_d)$. Since $n > m$, there is at least one atom in $q_f'(\mathbf{x}, x_d)$ to which several atoms in $q(\mathbf{x})$ map to. Thus a simplification has taken place and the degree variables of some of the atoms in $q_f'(\mathbf{x}, x_d)$ are not calculated correctly. In fact, exactly $n - m$ occurrences of degree variables are ignored. Since $|q_f'(\mathbf{x}, x_d)|_{\mathsf{CR}} = m$, the query $q_f'(\mathbf{x}, x_d)$ contains the predicate $\Phi_\otimes(x_d, y_{d1}, \ldots, y_{dm})$, where $x_d, y_{d1}, \ldots, y_{dm} \in \mathsf{N_{Vd}}$. Each such simplification step causes a predicate atom $A_f(t_i, y_{di})$ or $P_f(t_i, t_i', y_{di})$ with $1 \leq i \leq n$ occurring in $q(\mathbf{x})$ being omitted when computing the membership degree for the conjunction in $q_f'(\mathbf{x}, x_d)$ by evaluating $\Phi_\otimes(x_d, y_{d1}, \ldots, y_{dm})$. Since it is unknown which of the predicate atoms and consequently which degree variable is missing, we consider the most pessimistic case, i.e., that the variable taking the lowest degree in each answer has not been calculated. This minimum value is represented in the variable $y_\lambda : \Phi_{\min}(y_\lambda, y_{d1}, \ldots, y_{dm})$ (the predicate $\Phi_{\min}$ corresponds to the predicate $\Phi_\otimes$ in Equation 7 where $\otimes$ is replaced by the min $t$-norm). The membership degree for the pessimistic case can be calculated by changing the line 15 of Algorithm 2, so that the value of the degree variable $x_d$ in the query is calculated by:

$$\Phi_\otimes(x_d, \mathbf{y_d}, \underbrace{y_\lambda, \ldots, y_\lambda}_{n - m \text{ times}}) \wedge \Phi_{\min}(y_\lambda, y_{d1}, \ldots, y_{dm}).$$

The difference of the value returned by the algorithm and the value from the pessimistic estimation, give an estimate how close the returned answer is to the correct answer.

*Example 5.* Extending Example 4, the query to acquire a pessimistic degree estimation is:

$$q_{\mathcal{T}}^f(x, x_d) = \exists y_{d_A}, y_{d_B}, y_\lambda . A_f(x, y_{d_A}) \wedge B_f(x, y_{d_B}) \wedge$$
$$\wedge\, \Phi_\times(x_d, y_{d_A}, y_{d_B}, y_\lambda, y_\lambda) \wedge \Phi_{\min}(y_\lambda, y_{d_A}, y_{d_B}).$$

In the single pessimistic answer returned, $y_\lambda$ takes the value of 0.8 and the estimation is that $a$ is an answer to the query with a degree $\geq 0.4608$. This estimation is very close to the correct one, i.e., $a$ is the answer to the query with a degree $\geq 0.5184$. Now, with the pessimistic answer $(a, 0.4608)$ and the unsound answer $(a, 0.72)$, we know that the correct degree is between the two values.

## 4.2   Extended Use of Fuzzy Information

Our pragmatic approach can only handle fuzzy information in ABox assertions. Sometimes it can be useful to have also concept inclusion axioms with degrees or to extend conjunctive queries by fuzzy information.

**Fuzzy *DL-Lite$_R$* with Degrees in Concept Inclusions:** So far we have only considered concept inclusions of the form $B \sqsubseteq C$ in the extended rewriting approach. To extend our approach to the general case of fuzzy concept inclusions, i.e., $\langle B \sqsubseteq C, d \rangle$, is not straightforward. Such concept inclusions are satisfied by an interpretation $\mathcal{I}$ iff for every $\delta \in \Delta^{\mathcal{I}}$ and the implication operator from Table 1:

$$(B^{\mathcal{I}}(\delta) \Rightarrow C^{\mathcal{I}}(\delta)) \geqslant d.$$

We present here the intuition what the obstacles are. Suppose that our algorithm contains the concept inclusion $\langle B \sqsubseteq C, d \rangle$ and the corresponding CQ contains only the atom $C(x)$. During the rewriting, the replacement of $C_f(x, y_C)$ by $B_f(x, y_B)$ takes place and the degree $d$ should also to be calculated, i.e., the CQ returned after the replacement should be $\exists y_B . B_f(x, y_B) \wedge \Phi_\otimes(x_d, y_B, d)$, where $d$ is a degree and not a degree variable. Unfortunately, this cannot be done by the crisp rewriting algorithm since it does not keep track of the degrees in fuzzy concept inclusions.

One could introduce a new set of concept names corresponding to the $\alpha$-cuts of each concept, similar to the reduction technique presented in [10]. Here, the concept $B_{\geqslant 0.3}$ represents the set of elements that belong to the concept $B$ with a degree greater or equal than 0.3 and the concept inclusion $\langle B \sqsubseteq C, d \rangle$ can be replaced by the set of concept inclusions: $\langle B_{\geqslant d \otimes d'} \sqsubseteq C_{\geqslant d'}, 1 \rangle$ for each degree $d'$ in $\mathcal{T}$. Then in the final query each concept atom $B_{f \geqslant d}(x, y_{d_B})$ is replaced by $B_f(x, y_{d_B})$ and the degree $d$ is simply used in the predicate atom $\Phi_\otimes(\dots)$.

This procedure would remedy the above problem, but it would not yield optimized queries for the following reasons:

– Simplifications, optimizations and variable unifications are not performed since the crisp DL-Lite algorithm lacks the information that $B_{\geqslant 0.3}$ and $B_{\geqslant 0.4}$ are different $\alpha$-cuts of the same concept.

– If there are $n$ nested replacements in the rewriting, then the algorithm would need to compute all possible products of $n$ factors for the Łukasiewicz and product families of operators.

Therefore this method needs to be further investigated regarding its applicability and effectiveness.

**Fuzzy $DL\text{-}Lite_R$ with Generalized Query Component:** A generalized form of fuzzy CQs are those queries in which a score of a query is computed via a monotone scoring function. Such kind of queries have already been investigated in [6,12] and the question is whether our black box approach can be applied to answer them as well. Extending Example 4, we can express via a scoring function that the parameter $A_3$ is more important than $A_4$ which in turn is more important than $A_1$ and $A_2$:

$$q(x) = 0.2 \cdot A_1(x) + 0.1 \cdot A_2(x) \wedge 0.4 \cdot A_3(x) + 0.3 \cdot A_4(x). \qquad (9)$$

Again, due to the simplifications taking place in the crisp rewriting step, some of the atoms may be merged and therefore after this step the initial weight corresponding to the merged atoms are unknown. For equation 9, the crisp PERFECTREF algorithm returns an UCQ containing, among others, the CQ $A_1(x) \wedge A_3(x)$. For this CQ, one cannot guess correctly how to assign the weights $0.2, 0.1, 0.4, 0.3$ to the two remaining atoms.

**Fuzzy $DL\text{-}Lite_R$ Threshold Queries:** Another interesting form of queries w.r.t. to a fuzzy ontology, are threshold queries. These queries ask for all individuals that satisfy each atom with at least a certain degree. Threshold conjunctive queries may take the following form:

$$q(x) = \mathsf{Server}(x) \geqslant 1 \wedge \mathsf{hasPart}(x,y) \geqslant 1 \wedge \mathsf{CPU}(y) \geqslant 1 \wedge \mathsf{Overutilized}(y) \geqslant 0.4$$

Again, due to the simplifications taking place, threshold queries cannot be handled directly by employing the crisp rewritings first.

## 5  Practical Implementation and Performance Test

### 5.1  The FLite Reasoner

We have developed a reasoner for conjunctive query answering with respect to a TBox $\mathcal{T}$ and a fuzzy ABox $\mathcal{A}$ for $DL\text{-}Lite_R$. FLITE (Fuzzy $DL\text{-}Lite_R$ query engine) implements the query answering algorithm presented in Section 3 and it builds on the rewriting algorithms for crisp $DL\text{-}Lite_R$ implemented in the Ontop framework [8] developed at the Free University of Bozen Bolzano.

Figure 1 illustrates the whole query answering pipeline and the components involved. The initial input is a conjunctive query $q(\mathbf{x})$ represented in the form of a SPARQL query. The Ontop framework requires that the ABox $\mathcal{A}$ is stored in a relational database. A mapping

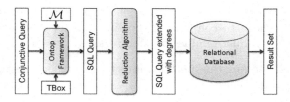

**Fig. 1.** FLITE implementation

$\mathcal{M}$, in the form of multiple SQL queries, translates the Tables of the relational database to ABox assertions. By combining the mapping $\mathcal{M}$ with the TBox assertions, the Ontop framework rewrites the initial query to a UCQ $q_{\mathcal{T}}(\mathbf{x})$, in the form of an SQL query. The rewritten query is post-processed by FLITE, as described in Section 3, resulting in the UCQ $q_{\mathcal{T}}^{f}(\mathbf{x}, x_d)$ that additionally asks for the associated degree of each answer by means of degree variables. The final SQL query is then evaluated over the relational database returning the corresponding result set with degrees.

*Example 6.* Let's consider again the three conjunctive queries and the ontology $\mathcal{O}_{ex}$ from Example 1 and assume that ABox $\mathcal{A}_{ex}$ is stored in a relational database. The mapping $\mathcal{M}$ is used to map the set of answers to: i) the query `select Server_id from Servers` to instances of the concept *Server*, ii) the query `select Server_id,CPU_id from CPUs` to instances of the role *hasCPU*, iii) the query `select CPU_id,Degree from Overutilized` to instances of the concept OverUtilized along with their corresponding degree. In this example, only the concept OverUtilized is fuzzy. It is represented by rows in the Table Overutilized stating the *CPU* and its degree of over-utilization. For an entry $(cpu_1, 0.6)$ in Table Overutilized we have that OverUtilized$(cpu_1) \geqslant 0.6$, all the other concepts are crisp and therefore have a degree of 1.0. Next the Ontop framework transforms the CQ in equation 2 to the following SQL query (in black), which is augmented by our extended rewriting algorithm (in gray).

```
SELECT    QVIEW1.Server_id AS x, QVIEW1.CPU_id AS y,
QVIEW2.Degree AS d
FROM   CPUs QVIEW1,Overutilized QVIEW2
WHERE    QVIEW1.Server_id IS NOT NULL AND QVIEW1.
    CPU_id IS NOT NULL AND (QVIEW1.CPU_id = QVIEW2.
    CPU_id)
```

## 5.2   An Initial Performance Evaluation

We have evaluated the performance of FLITE on an ontology. The current version of the HAEC fuzzy $DL\text{-}Lite_R$ ontology contains 311 TBox axioms, 178 concepts, 39 roles, together with 15 conjunctive queries. We performed our evaluation for a complicated query containing 13 concept and role atoms. Out of these 13 atoms, 9 were about fuzzy concepts, thus the extended SQL contained 9 additional degree variables. Out of the 10 relational database tables used to store the fuzzy ABox information 4 contained fuzzy information. Thus, about

40% of the ABox assertions were fuzzy. We evaluated the performance of our approach by comparing FLITE to the standard Ontop framework for the classic *DL-Lite$_R$* language by simply ignoring the degrees in concept assertions.[1]

The evaluation of the system was performed on a MacBook Pro laptop with *2.6 GHz Intel Core i7 Processor, 8 GB 1600 MHz DDR3 Memory*, running a *PostgreSQL 9.3.4 on x86_64-apple-darwin* database. Figure 2 depicts the comparison between Ontop and its extension FLITE in terms of running time. The graph shows the performance of the two query engines w.r.t. the number of assertions in the ABox. As we can see the over-

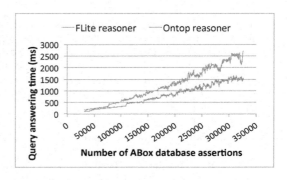

**Fig. 2.** Running times: Ontop - FLITE

head of adding degrees and answering queries containing degrees can be handled well by our algorithm and the database. In fact, FLITE answered the queries having to examine up to $326,340$ ABox assertions within only $1,519$ *ms* for the crisp and within $2,717$ *ms* for the fuzzy case.

# 6   Conclusions

We presented a pragmatic approach for answering conjunctive queries over ontologies with fuzzy ABoxes. Our approach uses rewritings obtained by the algorithm for answering crisp queries. Although described here for *DL-Lite$_R$*, our approach can be extended to other DLs that enjoy FOL rewritability. Our algorithm is sound for those t-norms that have idempotent operators, such as the Gödel t-norm. This does not need be for other t-norms. We devised a method by which unsound answers can be identified and the correct degrees estimated. We implemented our approach in the FLITE system and evaluated it against the Ontop framework. Our initial experiments suggest that the overhead for handling fuzzy information does not crucially affect the overall performance.

Our extended rewriting approach cannot be extended in straight-forward way to other interesting forms of queries such as threshold queries. To answer these kind of queries one would have to implement an algorithm from scratch [6,13] or extend the source code of an existing rewriting implementation. A thorough investigation of this subject remains future work.

---

[1] A comparison of the performance of FLITEwith SoftFacts [13] –an ontology mediated database system based on the DLR-Lite language–would have been more appropriate, but the system could not be set up.

# References

1. Acciarri, A., Calvanese, D., De Giacomo, G., Lembo, D., Lenzerini, M., Palmieri, M., Rosati, R.: Quonto: querying ontologies. In: AAAI, pp. 1670–1671 (2005)
2. Artale, A., Calvanese, D., Kontchakov, R., Zakharyaschev, M.: The DL-lite family and relations. Journal of Artificial Intelligence Research **36**(1), 1–69 (2009)
3. Calvanese, D., De Giacomo, G., Lembo, D., Lenzerini, M., Poggi, A., Rodriguez-Muro, M., Rosati, R.: Ontologies and databases: the *DL-Lite* approach. In: Tessaris, S., Franconi, E., Eiter, T., Gutierrez, C., Handschuh, S., Rousset, M.-C., Schmidt, R.A. (eds.) Reasoning Web. LNCS, vol. 5689, pp. 255–356. Springer, Heidelberg (2009)
4. Calvanese, D., De Giacomo, G., Lembo, D., Lenzerini, M., Rosati, R.: Tractable reasoning and efficient query answering in description logics: The DL-lite family. Journal of Automated Reasoning **39** (2007)
5. Mailis, T., Peñaloza, R., Turhan, A.-Y.: Conjunctive query answering in finitely-valued fuzzy description logics. In: Kontchakov, R., Mugnier, M.-L. (eds.) RR 2014. LNCS, vol. 8741, pp. 124–139. Springer, Heidelberg (2014)
6. Pan, J.Z., Stamou, G.B., Stoilos, G., Thomas, E.: Expressive querying over fuzzy DL-Lite ontologies. In: Description Logics (2007)
7. Poggi, A., Rodriguez, M., Ruzzi, M.: Ontology-based database access with DIG-Mastro and the OBDA plugin for protégé. In: Proc. of OWLED (2008)
8. Rodríguez-Muro, M., Kontchakov, R., Zakharyaschev, M.: Ontology-based data access: *Ontop* of databases. In: Alani, H., Kagal, L., Fokoue, A., Groth, P., Biemann, C., Parreira, J.X., Aroyo, L., Noy, N., Welty, C., Janowicz, K. (eds.) ISWC 2013, Part I. LNCS, vol. 8218, pp. 558–573. Springer, Heidelberg (2013)
9. Stocker, M., Smith, M.: Owlgres: a scalable OWL reasoner. In: OWLED, vol. 432 (2008)
10. Straccia, U.: Transforming fuzzy description logics into classical description logics. In: Alferes, J.J., Leite, J. (eds.) JELIA 2004. LNCS (LNAI), vol. 3229, pp. 385–399. Springer, Heidelberg (2004)
11. Straccia, U.: Answering vague queries in fuzzy DL-Lite. In: Proceedings of the 11th International Conference on Information Processing and Management of Uncertainty in Knowledge-Based Systems, (IPMU 2006), pp. 2238–2245 (2006)
12. Straccia, U.: Towards top-k query answering in description logics: the case of DL-Lite. In: Fisher, M., van der Hoek, W., Konev, B., Lisitsa, A. (eds.) JELIA 2006. LNCS (LNAI), vol. 4160, pp. 439–451. Springer, Heidelberg (2006)
13. Straccia, U.: Softfacts: a top-k retrieval engine for ontology mediated access to relational databases. In: 2010 IEEE International Conference on Systems Man and Cybernetics (SMC), pp. 4115–4122. IEEE (2010)
14. Venetis, T., Stoilos, G., Stamou, G.: Query extensions and incremental query rewriting for OWL 2 QL ontologies. Journal on Data Semantics pp. 1–23 (2014)

# Ontology Based Inferences Engine for Veterinary Diagnosis

H. Andres Melgar S.[1,2]([⊠]), Diego Salas Guillén[1], and Jacklin Gonzales Maceda[1]

[1] Grupo de Reconocimiento de Patrones e Inteligencia Artificial Aplicada, Pontificia Universidad Católica del Perú, Lima, Peru
amelgar@pucp.edu.pe
http://inform.pucp.edu.pe/~grpiaa/
[2] Sección de Ingeniería Informática, Departamento de Ingeniería, Pontificia Universidad Católica del Perú, Lima, Peru

**Abstract.** Motivated on knowledge representation and veterinary domain this project aims at using semantic technologies to develop a tool which supports veterinary diagnosis. For this purpose an ontology based inference engine was developed following the diagnosis task definition provided by CommonKADS methodology. OWL was the language used for representing the ontologies, they were built using Protégé and processed using the Jena API. The inference engine was tested with two different ontologies. This shows the versatility of the developed tool that can easily be used to diagnose different types of diseases. This is an example of the application of CommonKADS diagnosis template using ontologies. We are currently working to make diagnoses in other domains of knowledge.

**Keywords:** Ontology · Inferences · Veterinary diagnosis · Common-KADS

## 1 Introduction

Pets give us an undervalued benefit. Keeping our pets healthy contributes both to our physical and mental health [1]. Veterinary medicine applies scientific and technical knowledge to improve animal health and diseases prevention. Computer science can be used as support of some veterinary activities like education [2], epidemiology control [3] and information retrieval [4].

This research aims to applying Knowledge Engineering (KE) methods and Semantic Web (SW) technology for the development of a prototype of diagnosis system that supports veterinary diagnosis. The diagnosis is performed using ontologies and inference methods. In this kind of systems the knowledge typically is stored in a knowledge database that could be constructed using different representation techniques like a raw database, simple rules or ontologies. We used ontologies as knowledge database.

The inference methods was developed using CommonKADS task taxonomy [5]. CommonKADS is a methodology for KE which offers templates and

© Springer International Publishing Switzerland 2015
T. Supnithi et al.(Eds.): JIST 2014, LNCS 8943, pp. 79–86, 2015.
DOI: 10.1007/978-3-319-15615-6_6

definitions of knowledge intensive task, including diagnosis[6]. According CommonKADS, the knowledge has a stable internal structure that allows us to analyze and distinguish specific types of knowledge. In this project we developed a diagnostic engine in an independent way of the knowledge database. We used two different ontologies to test the diagnosis engine. The first one contains dog skin diseases and the second gastrointestinal diseases. The diagnosis was limited, as suggested by CommonKADS methodology, to discarding and reinforcing different possibilities by causal relation between system features and system observable [6].

This paper is structured in six sections. At first is presented the introduction and work motivation. Secondly, it is presented a brief summary of the CommonKADS methodology. The diagnosis tool prototype implementation is detailed on section three. The results are presented on section four and in the section five the results and future works are discussed. Finally, in the section six, the conclusion is presented.

## 2   CommonKADS for Diagnosis

CommonKADS presents a comprehensive methodology that gives a basic but thorough insight into the related disciplines of KE and Knowledge Management, all the way to knowledge-intensive systems design and implementation, in an integrated fashion [6]. Unlike other KE approaches, CommonKADS provides a clear link to modern object-oriented development and uses notations compatible with UML [7]. CommonKADS product has been developed over some 15 years [8–10] and is now being used in wide variety of application projects and is the in-house standard of a growing number of companies [11,12].

It has some task templates that provide a common type of a reusable combination of model elements [5]. A task template supplies the knowledge engineer with inferences and tasks that are typical for solving a problem of particular type. Diagnosis task is a subtype of analytic tasks. It is concerned with finding a malfunction that causes deviant system behavior. A diagnosis task is a task that tackles a diagnostic problem. Although in theory, "problem" and "task" are distinct entities, in practice these have the same meaning [6]. In recent years, CommonKADS have been used in the academic community to develop systems supporting the diagnosis [13,14].

## 3   A Tool for Veterinary Diagnosis Based on Ontology Inferences

The prototype was built using: i) the OWL language[1], ii) the ontology modeling tool Protégé[2] and iii) the JENA API[3].

---

[1] http://www.w3.org/2001/sw/wiki/Main_Page
[2] http://protege.stanford.edu/
[3] https://jena.apache.org/

To perform the diagnostic process CommonKADS methodology was applied [6]. CommonKADS allows building Knowledge-Based Systems on large scale, structured, in a controllable and repeatable way. Is characterized by the separation of flow-of-control design from selection of representations and techniques [15] distinguishing among the problem-solving agent, the knowledge of the world about which the agent is doing problem solving, and knowledge about how to solve problems [5].

CommonKADS methodology defines a template for the diagnosis task which was adapted to work with ontologies [16]. According to CommonKADS the diagnosis task consists in finding a fault that causes the malfunctioning of a system [6]. The template for the diagnosis task suggests defining a system behavior model but states it is possible to reduce the task to a classification problem replacing this model by a direct association within symptoms and faults. The algorithm 1 presents the diagnosis task proposed by CommonKADS template.

**Input**: complaint: Finding that initiates the diagnostic process
**Output**:
fault: the faults that could have caused the complaint;
evidence: the evidence gathered during diagnostic
**Data**:
INFERENCES: cover, select, specify, verify;
TRANSFER-FUNCTION: obtain;
`/* differential: active candidate solutions;`
`hypothesis: candidate solution;`
`result: boolean indicating result of the test;`
`expected-finding: data one would normally expected to find;`
`actual-finding: the data actually observed in practice;`            `*/`
CONTROL-STRUCTURE:
**while** *NEW-SOLUTION cover(complaint → hypothesis)* **do**
 | differential ← hypothesis ADD differential
**end**
**repeat**
  |  select(differential → hypothesis);
  |  specify(hypothesis → observable);
  |  obtain(observable → finding);
  |  evidence ← finding ADD evidence;
  |  **for** *hypothesis IN differential* **do**
  |   |  verify(hipothesis + evidence → result);
  |   |  **if** *result==false* **then**
  |   |   | differential ← differential SUBTRACT hypothesis;
  |   |  **end**
  |  **end**
**until** *SIZE differential ≤ 1 ∧ 'No more observable left'*;
faults ← differential;

**Algorithm 1.** CommonKADS diagnosis task specification. Adapted from [6]

The task starts with a complaint, an observable which causes malfunctioning. For the problem at hand the complaint is considered a symptom. The output is the faults found and the evidence that supports the results. These are the diseases or disease found and the evidence used to discard other options. The control structure is the method used to obtain the outputs. CommonKADS proposes a causal covering method which will go cover the rules defined on the ontology.

The intermediate roles defined are: differential, hypothesis, result, expected-finding, actual finding. The differential is a set of hypotheses that are candidate to become a fault. A hypothesis is a possible fault that must be tested against evidence. A result is the outcome of a test to discard or reaffirm a hypothesis. A finding is a piece of new information that could be expected or actual, depending if it was internally inferred or external. In the control structure rightwards arrows represent a result obtained and leftward arrows represent assignations.

The following inference functions are defined: cover, select, specify and verify. There is also a transference function, obtain, which is used to get external information about an observable. Picture 1 shows these functions interaction to obtain the results. The `cover` inference takes a complaint and builds the differential with all the possible hypothesis associated to that complaint. In this case, all the diseases associated to a symptom. The `select` inference takes an hypothesis from the differential based on some criteria related to the knowledge representation. Diseases with unique or rare symptoms have higher priority in order to reduce the differential. The `specify` inference takes the selected hypothesis and returns a related observable for testing. The transference function, `obtain`, gets external information about the specified observable. The `verify` inference takes the new finding obtained and validates the differential against the evidence discarding the hypothesis that do not match the evidence. This process follows until the differential has only the desired number of hypothesis, usually one, or when there is no more evidence. In this last case the resulting differential is returned.

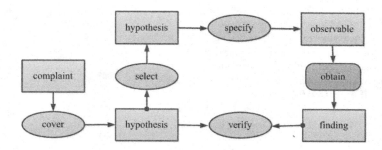

**Fig. 1.** CommonKADS inference method structure. Adapted from [6]

# 4   Results

The tool, built following CommonKADS diagnosis template, allows to diagnose a disease based on a complaint and external information about the presence of symptoms. The rules used for this purpose are encapsulated in an ontology which the inference engine queries searching for symptoms to easily discard diseases. Here is an example of the functioning tool.

The process starts by identifying a representative symptom and providing the number of desired disease suggestions. Once validated, a representative symptom is selected and its presence is verified through external input from the user. If the symptom is present, the related diseases are discarded, else, the evidence of the related diseases is strengthened. In both cases, feedback messages are shown to the user. This process repeats until there is no more evidence or the desired number of results is reached. This allows the user to interactively discard or validate present symptoms related to the current hypothesis considered on the process of diagnosis. After each validation, the user can check which related diseases were affected by the new evidence.

We will show the flow for a test case corresponding to the gastrointestinal diseases ontology. We chose a common initial complaint to show the process of discarding.

Given the disease `Pancreatitis` and its medical description: *"Acute pancreatitis is characterized by the abrupt onset of vomiting and severe pain in the abdomen. The dog may have a tucked-up belly and assume a prayer position. Abdominal pain is caused by the release of digestive enzymes into the pancreas and surrounding tissue. Diarrhea, dehydration, weakness, and shock may ensue"*. With the initial complaint `Vomiting`, the following results were obtained:

**Table 1.** Pancreatitis disease diagnosis

| Question | Answer |
| --- | --- |
| Does the patient present the symptom Frequent Heartburn? | No |
| Does the patient present the symptom Palpable Abdominal Mass? | No |
| Does the patient present the symptom Weakness? | Yes |
| Does the patient present the symptom Polyphagia? | No |
| Does the patient present the symptom Tucked Up Belly? | Yes |
| Does the patient present the symptom Peritonitis? | No |

The system reports that the disease obtained was `pancreatitis`.

This illustrates the process of discarding and reinforcing of each symptom based on the relations established on the ontology. All mapped diseases were tested successfully according to the specifications for each of them. Almost two thirds of the cases, took very few questions, between one or two.

The system was validated using the current gastrointestinal diseases ontology against the criteria of a veterinarian. After choosing two diseases and answering the questions based on his experience, three answers were obtained, the two diseases he choose and a third one which he recognized as valid according to his symptom selection.

The feedback obtained pointed to develop a more flexible interface for interaction with the knowledge. The system guiding the diagnosis is too strict in terms of combining symptoms. It is recommended to let the veterinarian choose a set of symptoms a try multiple combinations so the systems shows possible diseases and symptoms suggestions. The overall opinion of the veterinarian was optimistic given the lack of automatization of peruvian veterinary clinics processes and the possibility of developing a support tool which could broaden the spectrum of possibilities in diagnosis. There were also indications on usability, the veterinarian must interact with the dog during diagnosis so there should be a more simple interaction layer, typing the symptoms can be impractical.

The system was tested against 30 historical cases and it successfully diagnosed 70% based on the information obtained by a veterinarian for each case. The next step is to improve the ontology with the help of a veterinarian, adapt it to present a more flexible interface and resume the testing.

## 5    Discussion and Future Works

CommonKADS methodology and the disease template lead the inference tool development, it offered a structure for the knowledge database and the inference engine. CommonKADS has templates for various knowledge intensive tasks. These tasks are classification, assessment, monitoring, synthesis, configuration design, assignment, planning and scheduling. Any of them would make an interesting research topic to work with.

There are various methods to represent knowledge, ontologies were chosen given their potential reutilization in various projects, saving time and effort modeling knowledge. It would be interesting to test the same method on different domains, this is as simple as expressing the domain knowledge in terms of the ontology created. Not only different disease types can be explored, there are other options like mechanic tools malfunctioning or networks troubleshoot.

If the structure of the ontology is modified, the ontology access methods should reflect the changes. The relations and elements of the ontology are loaded independently so there is no problem with the ontology loader. The inference engine flow depends on the information received by the ontology access methods, it should be adapted to any changes made to the ontology. It is important to maintain the causal relation between a system state and a system feature as this is the basic structure suggested by CommonKADS diagnosis template.

The two formats evaluated for the construction of the knowledge database were OBO and OWL. We choose OWL because it had better documentation and it is a web standard according to W3C. There were two candidate tools to work with on the ontology processing, Jena and BaseVIsor. Jena was chosen because it offered and interesting set of classes to work with ontologies but BaseVIsor seems like a fair alternative too.

The designed tool lacks the potential to be launched as a veterinary tool, one of the future works is to improve the tool and the knowledge database. The first step is to reduce response time when working with huge ontologies. The ontology built proved to be manageable but there are modifications pending to reduce processing time when the knowledge encapsulated is vast.

The inference functions can also be optimized by adding probabilistic values to the relations stated on the ontology. There is a need for veterinary experts to get involved in the development of a complete domain ontology. This could be done incrementally, step by step for different types of disease. Jena also modifications to the ontology so an interface could be built to make this job easier to the experts.

The scope of the project is the development of a complete software for veterinary diagnosis support. Including an ontology editor and dynamic sets of rules, allowing the veterinary to edit the ontology or replace it on the parameters input window. Another interesting option is an online query tool. jOWL is a jQuery plugin which supports ontologies and it offers similar features as Jena.

# 6 Conclusion

The motivation for this project was a personal interest for knowledge representation, simulation of rational thinking and veterinary medicine. The main goal was proving that ontologies make a good alternative for representing knowledge necessary for diagnosis. The tools and the process to develop an ontology based inference engine have been proposed and tested. The prototype build opens a new path into developing a complete tool which could be useful not only for clinical diagnosis but also for educational purposes.

An application of the diagnosis template proposed by CommonKADS has been structured and a functional application is being developed. The prototype proves it is possible to built a complete tool with improved inference functions. It is interesting that, despite not considering ontologies during the development of CommonKADS methodology, the knowledge model proposed was easily adapted to work with ontologies.

Other tools used on the process were the Jena API which allowed loading and querying the ontology, accessing to classes, individuals and properties. Protégé which was used to build and edit the ontology.

The main benefit of using ontologies to represent knowledge is that they make the knowledge structure reusable for various domains. We run the diagnosis process using two different ontologies and obtained positive results. Actually, any diagnosis problem similar to differential diagnosis in medicine could be encapsulated and processed by the same engine.

The tool has some drawbacks with respect to data size and knowledge complexity, solving this problems is the main goal right now. It is a good start point in the development of a complete functional tool for diagnosis based on ontologies and CommonKADS diagnosis task template.

# References

1. Stanley, I.H., Conwell, Y., Bowen, C., Van Orden, K.A.: Pet ownership may attenuate loneliness among older adult primary care patients who live alone. Aging & Mental Health **18**, 394–399 (2014). PMID: 24047314
2. Iotti, B., Valazza, A.: A reliable, low-cost picture archiving and communications system for small and medium veterinary practices built using open-source technology. Journal of Digital Imaging, 1–8 (2014)
3. Volkova, S., Hsu, W.: Computational knowledge and information management in veterinary epidemiology. In: 2010 IEEE International Conference on Intelligence and Security Informatics (ISI), pp. 120–125 (2010)
4. Tangtulyangkul, P., Hocking, T., Fung, C.C.: Intelligent information mining from veterinary clinical records and open source repository. In: IEEE Region 10 Conference TENCON 2009, pp. 1–6 (2009)
5. Chandrasekaran, B., Josephson, J.R., Benjamins, R.: Ontology of tasks and methods. In: AAAI Spring Symposium (1997)
6. Schreiber, G.: Knowledge engineering and Management: the CommonKADS methodology. MIT Press (2000)
7. Surakratanasakul, B., Hamamoto, K.: CommonKADS's knowledge model using UML architectural view and extension mechanism, pp. 59–63 (2011). Cited by (since 1996) 1
8. Schreiber, A., Terpstra, P.: Sisyphus-VT: A CommonKADS solution. International Journal of Human-Computer Studies **44**, 373–402 (1996)
9. Van de Velde, W.: Knowledge and software engineering. International Journal of Engineering Intelligent Systems for Electrical Engineering and Communications **3**, 3–8 (1995). Cited by (since 1996) 0
10. Schreiber, G., Wielinga, B., de Hoog, R., Akkermans, H., Van de Velde, W.: CommonKADS: a comprehensive methodology for KBS development. IEEE Expert **9**, 28–37 (1994)
11. Tsai, Y.H., Ko, C.H., Lin, K.C.: Using CommonKADS method to build prototype system in medical insurance fraud detection. Journal of Networks **9**, 1798–1802 (2014)
12. Sriwichai, P., Meksamoot, K., Chakpitak, N., Dahal, K., Jengjalean, A.: The effectiveness of knowledge management system in research mentoring using knowledge engineering. International Education Studies **7**, 25 (2014)
13. Yang, Y.: Design and implementation of knowledge system of wireless communication equipment fault diagnosis. Advanced Materials Research **945**, 1108–1111 (2014)
14. Lopes, L.F., Lopes, M.C., Fialho, F.A.P., Gonçalves, A.L.: Knowledge system for acupuncture diagnosis: a modeling using CommonKADS. Gestão & Produção **18**, 351–366 (2011)
15. Kingston, J.K.C.: Designing knowledge based systems: the CommonKADS design model. Knowledge-Based Systems **11**, 311–319 (1998)
16. Schreiber, G., Crubézy, M., Musen, M.A.: A case study in using Protégé-2000 as a tool for CommonKADS. In: Dieng, R., Corby, O. (eds.) EKAW 2000. LNCS (LNAI), vol. 1937, pp. 33–48. Springer, Heidelberg (2000)

# An Information Literacy Ontology
# and Its Use for Guidance Plan Design –
# An Example on Problem Solving

Kouji Kozaki[1(✉)], Hiroko Kanoh[2], Takaaki Hishida[3], and Motohiro Hasegawa[4]

[1] The Institute of Scientific and Industrial Research, Osaka University,
8-1 Mihogaoka, Osaka, Ibaraki 567-0047, Japan
kozaki@ei.sanken.osaka-u.ac.jp
[2] Institute of Arts and Sciences, Yamagata University,
1-4-12 Koshirakawamachi, Yamagata-shi, Yamagata 990-8560, Japan
kanoh@pbd.kj.yamagata-u.ac.jp
[3] Department of Information Science, Aichi Institute of Technology,
1247, Yakusa-cho, Toyota, Aichi 470-0392, Japan
hishida@aitech.ac.jp
[4] Department of Global and Media Studies, Kinjogakuin University,
2-1723, Omori, Moriyama-ku, Nagoya, Aichi 463-8521, Japan
ghase@kinjo-u.ac.jp

**Abstract.** Recently, it is very important to educate about information literacy since information techniques are rapidly developed. However, common view and definition on information literacy are not established enough. Therefore, it is required to systematize concepts related to information literacy. This article discusses an experimental development of Information Literacy Ontology and its use for guidance plan design.

**Keywords:** Ontology · Information literacy · Education · Guidance plan

## 1 Introduction

Recently, a variety of information technology devices such as tablets and smart phones in addition to PCs and mobile phones are widespread use. According to such developments of information technologies, it becomes more important to educate about information literacy to deal with a vast of information appropriately. For example, Kanoh discusses information literacy which should be studied by young people so called net-generations [1].

However, necessary skills and considerable problems for appropriate managements of information are very complicated because of rapid developments of information technologies and devices. It seems to cause a situation that there is not enough common understanding and definition of information literacy. It is an important first step for considering education of information literacy to overcome this situation by systematizing concepts related to information literacy and clarifying relationships among them.

© Springer International Publishing Switzerland 2015
T. Supnithi et al. (Eds.): JIST 2014, LNCS 8943, pp. 87–93, 2015.
DOI: 10.1007/978-3-319-15615-6_7

The purpose of our study is to develop an information literacy ontology and classifying goals for education of information literacy based on it. In this paper, we discusses the information literacy ontology we built and how it can be used for design guidance plan.

In section 2, we discuss requirements for an information literacy ontology, then we outlines an ontology on problem solving as an example of our information literacy ontology. Section 4 shows how the ontology is used for guidance design and we conclude this paper with some future works in the section 5.

## 2     Requirements

We define information literacy as abilities to appropriately read, understand and analyze phenomenon, make a judgment, represent and communicate it by acquiring a way of looking and thinking based on information science [1]. We aim to design goal setting and education curriculum for 12 years at elementary, junior high and high schools.

In the current Japanese education curriculum, only high schools teach information technology as an independent subject while some related topics are taught as parts of other subjects. So, we started to extract important terms form textbooks of information technology for high school and classified them in order to develop an information literacy ontology. Then, we added terms in the education guideline by the Ministry of Education, Culture, Sports, Science and Technology, Japan.

After that, we reorganized the classification based on the developmental discussion method proposed Norman R. F. Maier [2]. We conducted a discussion by five researchers for six hours. Through some coordination, we introduced 7 top-level categories *information and communications technology*, i*nformation system*, *problem solving*, information analysis, history of information technology, and operation of IT devices as the result.

## 3     An Information Literacy Ontology

Considering the above requirements, we developed an information literacy ontology using Hozo[1] [3]. It consists of 1,117 concepts and will be available at the URL https://informationliteracy2.wordpress.com/ in Hozo and OWL formats.   In this paper, we outline ontologies on problem solving and information system as examples [4].

Problem solving discusses methods that detect a problem and find solution for it through collection and analysis of various information. It is one of important topic must be taught in education of information technology. Fig. 1[2] shows a part of the ontology.

*Content*, *Representation* and *Representation Form* (see. Fig. 1) are concepts which represent information related to problem solving based on the ontology of representation in a top-level ontology YAMATO [4].   As shown in Fig. 1, Basic concepts such as   *Data*, *Problem*, concepts which represent content (e.g. *Processing procedure*) are

---

[1] http://www.hozo.jp

[2] The whole otology are available at https://informationliteracy2.wordpress.com/

defined. In addition to them, it defines concepts which represents them in a form such as *Model*, *Data structure*, and *Program* and Representation Forms which are used to represent them. These are main concepts which are referred not only in the ontology of problem solving but also in other ontologies.

**Fig. 1.** A part of the ontology on problem solving (Content, Representation, etc.)

On the other hands, Fig. 2 shows definitions of *Acts* which related to problem solving. A problem solving is usually conducted as a compound process which includes sub-processes such as *Discovery* of problem, *Analysis* of the problem, *Discovery* of solution for it, *Optimization* of the solution, *Feedback*, and *Evaluation*. *Partial acts* represent that a problem solving is achieved by these acts. Each *Act* is defined by its *Target object*, *Actor*, *Method*, *Use software,* and etc. Though most acts appeared in the education of

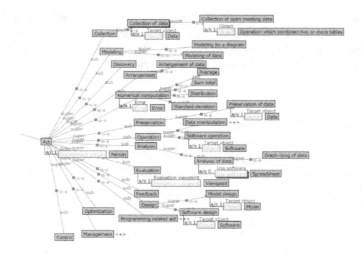

**Fig. 2.** A part of the ontology on problem solving (Act)

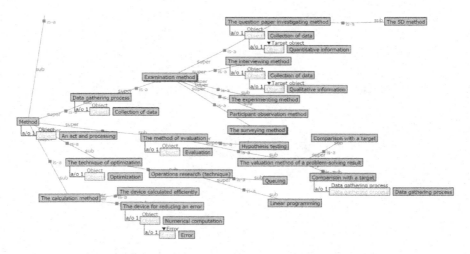

**Fig. 3.** A part of the ontology on problem solving (Method)

information technology are targeted at data such like *Collection of data* and *Analysis of data*, we introduced more general act such as *Collection* and *Analysis* in order to represent abstract concepts which are common to other domains.

Methods for problem solving are defined in another *is-a* hierarchy. For example, *Examination method* for data gathering includes sub-concepts such as *The question paper investigating method*, *The interviewing method*, *The experimenting method*, *Participant observation method*, *The surveying method* (see. Fig. 3) .

# 4     Design Example of Guidance Plan Based on the Ontology

In this section, we discusses how teachers design guidance plan for their class based on the information literacy ontology. The ontology shows concepts which students should study in their class. Therefore, teachers can refer concepts in the ontology when they design their guidance plans.

Table. 1 shows a design example of a guidance plan based on the ontology on problem solving discussed in the previous section. Its topic is to learn a difference between the question paper investigating method and the interviewing method. We suppose a 45 minutes class for around 10 years old in an elementary school. The goal of study is to learn features of each methods and to explain which methods are suitable for collecting which kinds of data.

**Table 1.** An example of guidance plan to learn a diffrence between the question paper investigating method and the interviewing method

| Flow | Study Topic |
|---|---|
| Introduction 10 min. | The teacher introduces two kinds of **Examination methods**; **the question paper investigating method** and **the interviewing method** as examples to examine familiar topics as favorite foods and how to spend at a holiday by illustrate with the following methods; A) Collecting the answers by show of hands for presented choices. B ) Collecting the answers from conversations with students. |
| Deployment [Group Discussion] 20 min. | The teacher assigns supposed themes such as "decide where we should go for school trip" and "decide casts for a drama performed in a school festival" to discuss in each group. Then, each group discusses which ways are good for reasonable decisions. In the discussion processes, the teacher give an advice to consider whether they should take the above method A) or b), and what questions are effective for the purpose. |
| Summary 10 min. | Each group presents the result of their discussions and summarizes which method (the above A) or B)) is suitable for which themes on a blackboard. |
| 5 min. | The teacher summarizes the class through the results written on the blackboard. Then, each student write the summary on his/her notebook. |

In the beginning of the class, the teacher introduce the question paper investigating method and the interviewing method through familiar examples. Then, the teacher gives supposed themes to each students group for discussing about which methods are good and what questions are effective to collect good data. After group discussions, students explain the results of their discussions.

In the Table. 1, concepts defined in the ontology are denoted by underline. It shows that not so much concepts are appeared in guidance plan. This is because the role of the information literacy ontology is to provide main concepts which students should study. That is, strategies for education is out of scope of the ontology while it are discussed by some researches in e-learning domains [6]. In other words, it is important for the information literacy ontology to cover enough wide range of concepts. This is why we introduces 7 categories as discussed in the section 2.

## 5    Concluding Remarks and Future Works

In this paper, we shows an information literacy ontology to classify goal of education and design guidance plans through an example on problem solving filed.  Many concepts are related to problem solving according to kinds of a target problem and its features. In particular, there are much more concepts related to acts and their methods for problem solving than the current ontology we built. Therefore, we suppose that we have to extend our ontology when we consider education plans for 12 years.

The first version of ontologies on other top-level categories are also available. We already have tried to design some sample guidance plans for other categories. Through these experience, we suppose these ontologies are a good first step for systematizing concepts which teachers should consider to teach information literacy.

As the next step, we plan to develop a support system to design goal settings and guidance plans for information literacy based on the information literacy ontology. Because the ontology provides necessary concepts for the education, the users can easily understand which topics should be covered by their plans. That is, the information literacy ontology can be used as a check list to compare learning goals each plans cover. We suppose to develop the system as a web based system using linked data techniques since designed guidance plans should be shared among teachers across schools so that they improve quality of education. After that, we plan to evaluate our ontology and developed system through user tests.

Another important future work is to conduct international comparison of education for information literacy. We suppose to compare education guidelines provided by governments in each country based on the information literacy ontology.

**Acknowledgement.** This work was supported by JSPS KAKENHI Grant Number 25282031.

# References

1. Kanoh, H.: Power of analyzing information:What is the information literacy which net-generations need to study most? In: IADIS International Conference e-Learning 2012 (2012)
2. Maier, N.R.F., Hoffman, L.R.: Using trained "developmental'' discussion leaders to improve further the quality of group decisions. Journal of Applied Psychology **44**(4), 247–251 (1960)
3. Mizoguchi, R., Sunagawa, E., Kozaki, K., Kitamura, Y.: A Model of Roles within an Ontology Development Tool: Hozo. J. of Applied Ontology **2**(2), 159–179 (2007)
4. Mizoguchi, R.: YAMATO: Yet Another More Advanced Top-level Ontology.
   http://www.ei.sanken.osaka-u.ac.jp/hozo/onto_library/upperOnto.htm
5. Kozaki, K., Kanoh, H., Hishida, T., Hasegawa, M.: An ontology of information literacy on "problem solving" and "information system". In: Proc. of the 38th Annual Meeting of Japan Society for Science Education (2014)
6. Hayashi, Y., Bourdeau, J., Mizoguchi, R.: Strategy-centered Modeling for Better Understanding of Learning/Instructional Theories. International Journal of Knowledge and Web Intelligence **1**(3/4), 187–208 (2010)

# Linked Data

# A Roadmap for Navigating the Life Sciences Linked Open Data Cloud

Ali Hasnain[1]([✉]), Syeda Sana e Zainab[1], Maulik R. Kamdar[1],
Qaiser Mehmood[1], Claude N. Warren Jr.[2], Qurratal Ain Fatimah[1,2,3],
Helena F. Deus[3], Muntazir Mehdi[1], and Stefan Decker[1]

[1] Insight Center for Data Analytics, National University of Ireland, Galway, Ireland
{ali.hasnain,syeda.sanaezainab,maulik.kamdar,qaiser.mehmood,
muntazir.mehdi,stefan.decker}@insight-centre.org
[2] Xenei.com, Denver, CO, USA
claude@xenei.com
[3] Foundation Medicine Inc., Cambridge, MA, USA
hdeus@foundationmedicine.com

**Abstract.** Multiple datasets that add high value to biomedical research have been exposed on the web as a part of the Life Sciences Linked Open Data (LSLOD) Cloud. The ability to easily navigate through these datasets is crucial for personalized medicine and the improvement of drug discovery process. However, navigating these multiple datasets is not trivial as most of these are only available as isolated SPARQL endpoints with very little vocabulary reuse. The content that is indexed through these endpoints is scarce, making the indexed dataset opaque for users. In this paper, we propose an approach for the creation of an active Linked Life Sciences Data Roadmap, a set of configurable rules which can be used to discover links (roads) between biological entities (cities) in the LSLOD cloud. We have catalogued and linked concepts and properties from 137 public SPARQL endpoints. Our Roadmap is primarily used to dynamically assemble queries retrieving data from multiple SPARQL endpoints simultaneously. We also demonstrate its use in conjunction with other tools for selective SPARQL querying, semantic annotation of experimental datasets and the visualization of the LSLOD cloud. We have evaluated the performance of our approach in terms of the time taken and entity capture. Our approach, if generalized to encompass other domains, can be used for road-mapping the entire LOD cloud.

**Keywords:** Linked Data (LD) · SPARQL · Life Sciences (LS) · Semantic web · Query federation

## 1 Introduction

A considerable portion of the Linked Open Data cloud is comprised of datasets from Life Sciences Linked Open Data (LSLOD). The significant contributors includes the Bio2RDF project[1], Linked Life Data[2], Neurocommons[3], Health care

---

[1] http://bio2rdf.org/ (l.a.: 2014-03-31 )
[2] http://linkedlifedata.com/ (l.a.: 2014-07-16 )
[3] http://neurocommons.org/page/Main_Page (l.a.: 2014-07-16 )

© Springer International Publishing Switzerland 2015
T. Supnithi et al.(Eds.): JIST 2014, LNCS 8943, pp. 97–112, 2015.
DOI: 10.1007/978-3-319-15615-6_8

and Life Sciences knowledge base[4] (HCLS Kb) and the W3C HCLSIG Linking
Open Drug Data (LODD) effort[5]. The deluge of biomedical data in the last
few years, partially caused by the advent of high-throughput gene sequencing
technologies, has been a primary motivation for these efforts. There had been
a critical requirement for a single interface, either programmatic or otherwise,
to access the Life Sciences (LS) data. Although publishing datasets as RDF
is a necessary step towards unified querying of biological datasets, it is not
sufficient to retrieve meaningful information due to data being heterogeneously
available at different endpoints [2,14]. Despite the popularity and availability
of bio-ontologies through ontology registry services[6], it is still very common
for semantic web experts to publish LS datasets without reusing vocabularies
and terminologies. The popularity of bio-ontologies has also led to the use of
overlapping standards and terminologies, which in turn has led to a low adoption
of standards [14]. For example, the exact term "Drug" is matched in 38 bioportal
ontologies[7] - it is not clear which of these should be chosen for publishing LD.

In the LS domain, LD is extremely heterogeneous and dynamic [8,16]; also
there is a recurrent need for *ad hoc* integration of novel experimental datasets
due to the speed at which technologies for data capturing in this domain are
evolving. As such, integrative solutions increasingly rely on federation of queries
[5–7]. With the standardization of SPARQL 1.1, it is now possible to assemble
federated queries using the "SERVICE" keyword, already supported by multiple
tool-sets (SWobjects, Fuseki and dotNetRDF). To assemble queries encompass-
ing multiple graphs distributed over different places, it is necessary that all
datasets should be query-able using the same global schema [17]. This can be
achieved either by ensuring that the multiple datasets make use of the same
vocabularies and ontologies, an approach previously described as *"a priori integration"* or conversely, using *"a posteriori integration"*, which makes use of mapping rules that change the topology of remote graphs to match the global schema
[6] and the methodology to facilitate the latter approach is the focus of this
paper. Moreover for LD to become a core technology in the LS domain, three
issues need to be addressed: *i)* dynamically discover datasets containing data
on biological entities (e.g. Proteins, Genes), *ii)* retrieve information about the
same entities from multiple sources using different schemas, and *iii)* identify, for
a given query, the highest quality data.

To address the aforementioned challenges, we introduce the notion of an
active Roadmap for LS data – a representation of entities as *"cities"* and the links
as the *"roads"* connecting these *"cities"*. Such a Roadmap would not only help
understand which data exists in each LS SPARQL endpoint, but more importantly enable assembly of multiple source-specific federated SPARQL queries. In
other words, the ability to assemble a SPARQL query that goes from **A** (e.g.
a neuroreceptor) to **B** (e.g. a drug that targets that neuroreceptor), requires a

---

[4] http://www.w3.org/TR/hcls-kb/ (l.a.: 2014-07-16 )

[5] http://www.w3.org/wiki/HCLSIG/LODD (l.a.: 2014-07-16 )

[6] http://bioportal.bioontology.org/ (l.a.: 2014-07-12)

[7] http://bioportal.bioontology.org/search?query=Drug (l.a.: 2014-07-12)

Roadmap that clarifies all the possible *"roads"* or *"links"* between those two entities. Our initial exploratory analysis of several LS endpoints revealed that they not only use different URIs but also different labels for similar concepts (e.g. Molecule vs Compound). Our methodology for developing the active Roadmap consisted of two steps: *i)* catalogue development, in which metadata is collected and analyzed, and *ii)* links creation and Roadmap development, which ensures that concepts and properties are properly mapped to a set of Query Elements (*Qe*) [19]. Hasnain et. al [9] described the Link Creation mechanism, linking approaches as well as the linking statistics and in this paper we focus primarily on the Cataloguing mechanism that facilitates linking as a second step. We assumed in this work that federated queries are assembled within a context – as such, our Roadmap relies on identifying the global schema onto which entities should be mapped. This entails the initial identification of a set of $Qe^8$, in the context of cancer chemoprevention[19], identified by the domain experts participating in the EU GRANATUM project[9].

The rest of this paper is organized as follows: In Section 2, we discuss the related research carried out towards integrating heterogeneous LD. In Section 3, we introduce the catalogue and link generation methodologies to build Roadmap. In Section 4, we showcase four applications of the generated Roadmap - notably a domain-specific federated query engine which reasons over the Roadmap to query the LSLOD. We evaluate the performance of Cataloguing Mechanism in terms of time taken and entity capture in Section 5.

## 2    Related Work

Approaches to facilitate the *"A posteriori integration"* is currently an area of active research. One approach is through the use of available schema: semantic information systems have used ontologies to represent domain-specific knowledge and enable users to select ontology terms in query assembly [13]. BLOOMS, for example, is a system for finding schema-level links between LOD datasets using the concept of ontology alignment [11], but it relies mainly on Wikipedia. Ontology alignment typically relies on starting with a single ontology, which is not available for most SPARQL endpoints in the LOD cloud and therefore could not be applied in our case. Furthermore, ontology alignment does not make use of domain rules (e.g. if two sequences are the same, they map to the same gene) nor the use of URI pattern matching for alignment – these issues had already been discussed by Hasnain et. al [9]. Other approaches such as the VoID [1] and the SILK Framework [18] enable the identification of rules for link creation, but require extensive knowledge of the data prior to links creation. Query federation approaches have developed some techniques to meet the requirements of efficient query computation in the distributed environment. FedX [15], a project which extends the Sesame Framework [3] with a federation layer, enables efficient query processing on distributed LOD sources by relying on the assembly of a catalogue

---

[8] http://srvgal78.deri.ie/RoadMapEvaluation/#Query_Elements(l.a.: 2014-07-19)

[9] http://www.granatum.org(l.a.: 2014-07-05)

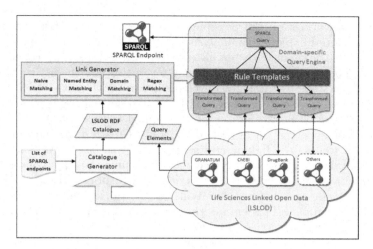

**Fig. 1.** Roadmap and Query Engine Architecture

of SPARQL endpoints but does not use domain rules for links creation. Our approach for link creation towards Roadmap development is a combination of the several linking approaches as already explained by Hasnain et. al [9]: *i)* similarly to ontology alignment, we make use of label matching to discover concepts in LOD that should be mapped to a set of *Qe, ii)* we create *"bags of words"* for discovery of schema-level links similar to the approach taken by BLOOMS, and *iii)* as in SILK, we create domain rules that enable the discovery of links.

## 3   Methodology

We developed an active Roadmap for navigating the LSLOD cloud. Our methodology consists of two stages namely catalogue generation and link generation. Data was retrieved from 137 public SPARQL endpoints[10] and organized in an RDF document - the LSLOD Catalogue. The list of SPARQL endpoints was captured from publicly available Bio2RDF datasets and by searching for datasets in CKAN[11] tagged *"life science"* or *"healthcare"*.

### 3.1   Methodology for Catalogue Development

Connecting the different concepts i.e. making links between them, is an ultimate goal of this research to facilitate the navigation across the LSLOD Cloud. As an example from a drug discovery scenario, it is often necessary to find the links between *"cancer chemopreventive agent"* and *"publication"*. For enabling this, a preliminary analysis of multiple SPARQL Endpoints containing data from both the Life Sciences and the Health care domains was undertaken. A semi-automated method was devised to retrieve all classes (concepts) and associated properties (attributes) available through any particular endpoint by probing

---

[10] http://goo.gl/ZLbLzq
[11] http://wiki.ckan.org/Main_Page (l.a.: 2014-05-05)

data instances. The workflow definition for probing instances through endpoint analysis using the W3C SPARQL algebra notation[12] is as follows:

1. For every SPARQL endpoint $S_i$, find the distinct Classes $C(S_i)$ :

$$C(S_i) = Distinct\ (Project\ (?class\ (toList\ (BGP\ (triple\ [\ ]\ a\ ?class\ ))))) \quad (1)$$

2. Collect the Instances for each Class $C_j(S_i)$ :

$$I_i : C_j(S_i) = Slice\ (Project\ (?I\ (toList\ (BGP\ (triple\ ?a\ a\ < C_j(S_i) > )))),\ rand())\quad (2)$$

3. Retrieve the Predicate/Objects pairs for each $Ii : C_j(S_i)$:

$$I_i(P, O) = Distinct\ (Project\ (?p,\ ?o\ (toList\ (BGP\ (triple\ < I_i : C_j(S_i) > ?p\ ?o\ ))))\quad (3)$$

4. Assign Class $C_j(S_i)$ as domain of the Property $P_k$ :

$$Domain(P_k) = C_j(S_i) \quad (4)$$

5. Retrieve Object type $(O_T)$ and assign as a range of the Property $P_k$ :

$$Range(P_k) = O_T; O_T = \begin{cases} rdf : Literal & \text{if } (O_k\ is\ String) \\ dc : Image & \text{if } (O_k\ is\ Image) \\ dc : InteractiveResource & \text{if } (O_k\ is\ URL) \\ Project\ (?R\ (toList\ (BGP & \\ (triple\ < O_k >\ rdf : type\ ?R))) & \text{if } (O_k\ is\ IRI) \end{cases}$$

$$(5)$$

It is worth noting that step 2 is heuristic – performing step 3 on a list of random instances is only necessary to avoid query timeout as the alternative (`triple [ ] ?p ?o`) would generally retrieve too many results. Step 5 effectively creates links between two entities ($C_j(S_i)$ and the Object Type $O_T$), but only when the object of the triples ($O_k$) retrieved in step 3 are URIs. We found that the content-type of properties can take any of the following formats:

1. Literal (i.e non-URI values e.g: *"Calcium Binds Troponin-C"*)
2. Non-Literal; these can further be divided into one of following types:
   (a) URL (e.g.: <http://www.ncbi.nlm.nih.gov/pubmed/1002129>) which is not equivalent to a URI because is cannot retrieve structured data.
   (b) Images (e.g: <http://www.genome.jp/Fig/drug/D00001.gif>)
   (c) URI (e.g.: <http://bio2rdf.org/pubchem:569483>); the most common types of URI formats that we have discovered were:
      i. Bio2RDF URIs (e.g.: <http://bio2rdf.org/gi:23753>)
      ii. DBpedia URIs (e.g.: <http://dbpedia.org/resource/Ontotext>)
      iii. Frei e Universität Berlin URIs e.g.:
         <http://www4.wiwiss.fu-berlin.de/drugbank/resource/drugs/DB00339>
      iv. Other URIs (e.g.: <http://purl.org/ontology/bibo/Journal>)

RDFS, Dublin Core[13] and VoID[14] vocabularies were used for representing the data in the LSLOD catalogue. A slice of the catalogue is presented as follows[15]:

---

[12] http://www.hpl.hp.com/techreports/2005/HPL-2005-170.pdf (l.a.: 2014-07-30)
[13] http://dublincore.org/documents/dcmi-terms/ (l.a.: 2014-07-12)
[14] http://vocab.deri.ie/void (l.a.: 2014-07-12)
[15] In this paper, we omit URI prefixes for brevity. All prefixes can be looked up at http://prefix.cc/ (l.a.: 2014-07-31 )

**Listing 1.** An Extract from the LSLOD Catalogue for KEGG dataset

```
<http://kegg.bio2rdf.org/sparql> a void:Dataset ;
void:class <http://bio2rdf.org/ns/kegg#Enzyme> ;
void:sparqlEndpoint <http://kegg.bio2rdf.org/sparql>,
<http://s4.semanticscience.org:12014/sparql> .
<http://bio2rdf.org/ns/kegg#Enzyme> rdfs:label "Enzyme";
void:exampleResource <http://bio2rdf.org/ec:3.2.1.161>.
<http://bio2rdf.org/ns/kegg#xSubstrate> a rdf:Property;
rdfs:label "#xSubstrate" ;
voidext:domain <http://bio2rdf.org/ns/kegg#Enzyme> ;
voidext:range   <http://bio2rdf.org/kegg_resource:Compound>.
<http://bio2rdf.org/kegg_resource:Compound>
void:exampleResource <http://bio2rdf.org/cpd:C00001>;
void:uriRegexPattern "^http://bio2rdf\\.org/cpd:.*" ;
voidext:sourceIdentifier "cpd" .
```

The RDF above is an illustrative example of a portion of the catalogue generated for the KEGG SPARQL endpoint[16]. VoID is used for describing the dataset and for linking it with the catalogue entries: the void#Dataset being described in this catalogue entry is "KEGG" SPARQL endpoint. In cases where SPARQL endpoints were available through mirrors (e.g. most Bio2RDF endpoints are available through Carleton Mirror URLs) or mentioned using alternative URLs (e.g. http://s4.semanticscience.org:12014/sparql), these references were also added as a second value for the void#sparqlEndpoint property. Listing 1 also includes one identified Class (http://bio2rdf.org/ns/kegg#Enzyme),and one property using that class as a domain (http://bio2rdf.org/ns/kegg#xSubstrate). Classes are linked to datasets using the void#classproperty; the labels were collected usually from parsing the last portion of the URI and probed instances were also recorded (http://bio2rdf.org/ec:3.2.1.161)as values for void#example Resource. Properties (http://bio2rdf.org/ns/kegg#xSubstrate) collected by our algorithm (steps 3,4,5) were classified as rdfs:property.

When the object of a predicate/object pair is of type URI, (e.g. as for KEGG shown in Listing 1) the algorithm attempts to perform link traversal in order to determine its object type ($O_T$). In most cases, however, the URI was not dereferenceable and in such cases an alternative method relies on querying the SPARQL endpoint for the specific "type" of the instance. In example above, dereferencing the object URI <http://bio2rdf.org/cpd:C00001> resulted in class <http://bio2rdf.org/kegg_resource:Compound>. We call this as a "range class" used as the range of the property <http://bio2rdf.org/ns/kegg#xSubstrate>. Actual object URI <http://bio2rdf.org/cpd:C00001> is classified as void#exampleResource of <http://bio2rdf.org/kegg_resource:Compound>and the URI regular expression pattern is recorded under void#uriRegexPattern. We found that, in many cases, the \sourceIdentifier" or the identifier that appears before the ":" symbol in case of many URIs could be used for discovering the appropriate type for the non-dereferenceable URI when none was provided.

---

[16] http://kegg.bio2rdf.org/sparql (l.a.: 2014-02-01)

Although this is not a standardised method, we found it to be useful in mapping classes nonetheless. For non-dereferenceable URIs with no actual class as $O_T$ (termed Orphan URIs), a new $O_T$ is created using UUID, which is classified as voidext#OrphanClass (Listing 2).

**Listing 2.** Orphan Classes captured in Catalogue

```
<http://bio2rdf.org/ns/kegg#xSubstrate> a rdf:Property;
voidext:domain <http://bio2rdf.org/ns/kegg#Reaction> ;
voidext:range roadmap:CLASS2c2ab5b75a454f678a9056dfc1d1214.
roadmap:CLASS2c2ab5b75a454f678a9056dfc1d1214
a voidext:OrphanClass;
void:exampleResource <http://bio2rdf.org/cpd:c00890> ;
void:uriRegexPattern "http://bio2rdf\\.org/cpd:*" ;
voidext:sourceIdentifier "cpd" .
```

### 3.2  Methodology for Link Generation

During this phase subClassOf and subPropertyOf links were created amongst different concepts and properties to facilitate *"a posteriori integration"*. The creation of links between identified entities (both chemical and biological) is not only useful for entity identification, but also for discovery of new associations such as protein/drug, drug/drug or protein/protein interactions that may not be obvious by analyzing datasets individually. A link is a property of an entity that takes URI as a value. The following RDF statement is both a property of the chemo-prevention agent *acetophenone* as well as a link between that chemoprevention agent and a publication:`<pubmed_id:18991637>dc:hasPart"acetophenone"`.
Leveraging the class descriptions and its properties in the LSLOD catalogue, links were created (discussed previously in [9]) using several approaches: *i)* Naïve Matching/ Syntactic Matching/ Label Matching, *ii)* Named Entity Matching, *iii)* Domain dependent/ unique identifier Matching, and *iv)* Regex Matching.

**Regular Expression Matching.** Regular expression matching can be considered as a special case of "Naïve Matching". The regular expressions for all those URIs that may or may not be dereferenced (Orphan URIs) were captured during catalogue generation phase. By looking to the similar regular expressions it can be concluded that two distinct URIs belong to the same class. Considering the same regular expressions of instances of Orphan and non-Orphan URIs, an Orphan URI can safely be linked with a non-Orphan URIs.

## 4  Roadmap Applications

The Roadmap is exposed as a SPARQL endpoint[17] and relevant information is also documented[18]. As of $31^{st}$ May 2014, the Roadmap consists of 263731 triples representing 1861 distinct classes, 3299 distinct properties and 13027 distinct Orphan Classes catalogued from 137 public SPARQL endpoints.

---

[17] http://srvgal78.deri.ie:8006/graph/Roadmap (l.a.: 2014-07-31)
[18] Roadmap Homepage: https://code.google.com/p/life-science-roadmap/

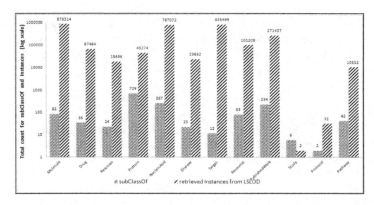

**Fig. 2.** Number of retrieved Instances and Subclasses linked to any $Qe$

## 4.1 Domain-Specific Query Engine

Fundamentally, the Domain-specific Query Engine (DSQE) is a SPARQL query engine that transforms the expressions from one vocabulary into those represented using vocabularies of known SPARQL endpoints and combines those expressions into a single query using SPARQL "SERVICE" calls. The engine executes the resulting statement and returns the results to the user (Fig. 1). DSQE is implemented on top of the Apache Jena query engine and extends it by intercepting and rewriting the SPARQL algebra[19]. Most of the algebra is unmodified and we concentrate on the base graph patterns (BGP) which are effectively the triples found in the original SPARQL query.

DSQE comprises of two major components: the SPARQL Algebra rewriter and the Roadmap. The algebra rewriter examines each segment of the BGP triples and attempts to expand the terms based on the vocabulary mapping into terms of the endpoint graphs and stores the result for each. Hence, it effectively builds the BGPs for all known graphs in parallel. In the final stage, the rewriter wraps the BGPs with SERVICE calls, using the Roadmap to determine the "relevent" endpoint, and unions them together along with the original BGP. The result is passed back into the standard Jena query process and the execution continues normally. Each endpoint is therefore accessed via a SERVICE call and the relevant underlying data is incorporated in the result. Early implementations [6] have shown such algebraic rewrite to be functional and efficient. Our enhancement was two fold - $i)$ Based on the presence of OWL2 `hasKey` properties, we added the ability to identify similar subjects with different URIs, $ii)$ we added the ability to rewrite generic queries for different endpoints.

**Identifying Subjects with Different URIs.** For a given query:
SELECT ?p ?o WHERE { ?x a <T>; ?p ?o } and two endpoints E1 and E2 with synonymous topics T1 and T2 for T and K1 and K2 synonyms for K. We want to expand the query as shown in Listing 3. To ensure that there exists a triple that matches `?alias <K> ?key` in the local graph, we use a temporary graph that

---

[19] http://www.w3.org/TR/sparql11-query/#sparqlAlgebra (l.a.: 2014-07-30)

only exists for the duration of the query and is merged with the normal local graph during execution. Hence, when DSQE executes the query we retrieve the necessary information from the remote SPARQL endpoints to merge the results together even if they use different values for the Subject.

**Listing 3.** SPARQL Construct

```
SELECT ?p ?o WHERE { { SERVICE <E1> {
    SELECT (?Ap as ?p) (?Ao as ?o) (?Akey as ?key)
    WHERE { ?Ax a <T1> ; <K1> ?Akey ; ?Ap ?Ao  } }
  FILTER ( insertKeyFilter( K, ?key ))   }
UNION { SERVICE <E2> {
    SELECT (?Bp as ?p) (?Bo as ?o) (?Bkey as ?key)
    WHERE { ?Bx a <T2> ; <K2> ?Bkey ; ?Bp ?Bo }   }
    FILTER ( inserKeyFilter( K, ?key ))
} ?alias <K> ?key ?p ?o  }
```

An instance[20] of the DSQE is deployed in the context of cancer chemoprevention drug discovery [10]. To facilitate the user to build federated SPARQL queries for execution against the LSLOD, the DSQE provides a 'Standard' and a 'Topic-based' query builder. The user can select a topic of interest (e.g. Molecule) and a list of associated $Qe$ are automatically generated. The user can also select relevant filters (numerical and textual) through the 'Topic-based' builder. Our implementation exposes a REST API and provides a visual query system, named ReVeaLD [12], for intuitive query formulation and domain-specific uses [10]. The catalogued subclasses of few $Qe$ and as a result the total number of distinct instances retrieved per $Qe$ while querying using DSQE is shown in Fig. 2.

### 4.2 Roadmap for Drug Discovery

Mining LD for drug discovery can become possible once domain experts are able to discover and integrate the relevant data necessary to formulate their hypothesis [10]. For domain users, it is not always obvious where the data is stored or what are the appropriate terminologies used to retrieve/publish it. Although multiple SPARQL endpoints contain molecular information, not all of them use the same terminology, as we have shown in our LSLOD catalogue. For example, one of the most intensive uses of LD is to find links which translate between Disease → Drugs → Protein targets → Pathways. Such data is not available at a single source and new datasets relevant for such query needs to be added *ad hoc*. Our Roadmap provides a possible solution to this dilemma by enabling the dynamic discovery of the SPARQL endpoints that should be queried and the links between these concepts (Listing 4).

**Listing 4.** Roadmap Links between concepts relevant for drug discovery

```
?disease  a  :Disease ; :treatedWith ?Drug .
?Drug     a  :Drug ; :interactsWith ?target .
?target   a  :Target ; :involvedIn ?pathway .
?pathway  a  :Pathway .
```

---

[20] http://srvgal78.deri.ie:8007/graph/Granatum

Consider $Qe$ to be any of the possible query elements and $XQe$ to be an instance of that query element, the following can be used by a SPARQL 1.1 engine to list all possible $XQe$ regardless of where they are stored:

$$C_i \sqsubseteq Qe \;\&\&\; XC_i : C_i \Rightarrow XQe, XC_i : Qe \tag{6}$$

**Listing 5.** SPARQL Construct to list all possible instances

```
CONSTRUCT { ?x a [QeRequested] } WHERE {
?SparqlEndpoint void:class [QeRequested]} UNION
{?QeMatched rdfs:subClassOf [QeRequested].}
SERVICE ?SparqlEndpoint {?x a ?QeMatched. } }
```

In Listing 5, [QeRequested] can be Disease, Drug, Protein or Pathway. Similarly, to discover which properties link two $Qe$ together, the Roadmap can be used to create new graphs that map elements to those available in the chosen set of query elements. For any "linked" $Qe$, there is a set of incoming links (properties that use $Qe$ as its rdfs:domain) and a set of outgoing links (properties that use $Qe$ as its rdfs:range). We denote the collection of incoming and outgoing links from a particular concept as $C_i(IC_i, OC_i)$. When concepts from available graphs are mapped to $Qe$, it becomes possible to create new incoming ($I_i$) and outgoing ($O_i$) links connected to $Qe$. This is illustrated by the following principle:

$$C_i(IC_i, \; OC_i) \;\&\&\; C_i \sqsubseteq Qe \Rightarrow Qe(\{IC_i, IQe\}, \; \{OC_i, IQe\}) \tag{7}$$

We exemplify the above principle by the following SPARQL CONSTRUCT:

**Listing 6.** SPARQL Construct to create new incoming and outgoing links

```
CONSTRUCT {?DomainQe ?PropertyDomainRange ?RangeQe } WHERE {
?PropertyDomainRange rdfs:domain [DomainQe]  .
?PropertyDomainRange rdfs:range [RangeQe]
{ FILTER (?DomainQe == [DomainQe]) } UNION
{ ?DomainQe rdfs:subclassOf [DomainQe] }
{ FILTER (?RangeQe == [RangeQe]) } UNION
{ ?RangeQe rdfs:subclassOf [RangeQe] }}
```

## 4.3 SPARQL Endpoint Selection Based on Availability

While probing instances through SPARQL endpoint analysis, there was clear evidence that a particular class or a property may be present at multiple SPARQL endpoints. We also noticed the problems of service disruption and perennial unavailability of some of the endpoints throughout. Our Roadmap has opened an avenue for easy, continuous and uninterrupted accessibility of data from several SPARQL endpoints, in cases where the same data may be available from multiple underlying data sources. Mondeca Labs provides a service[21] which monitors the availability status of SPARQL endpoints [4]. The status of any endpoint can be - *i)* Operating normally, *ii)* Available but problems within last 24 hours, *iii)* Service disruption, or *iv)* Still alive? Exploiting the knowledge available in

---

[21] http://labs.mondeca.com/sparqlEndpointsStatus.html (l.a.: 2014-07-30)

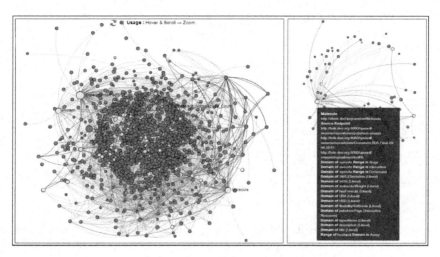

**Fig. 3.** Visualizing the mapped LSLOD Cloud

our Roadmap in conjunction with this service, we can direct our query to only that endpoint which is currently available, quick to respond and has similar data. In Listing 7, we present a sample SPARQL query determining the availability status of SPARQL endpoints providing data on **Protein**-similar concepts in the LSLOD.

**Listing 7.** Availability of SPARQL Endpoints for Protein-similar concepts

```
SELECT * WHERE { SERVICE
<http://hcls.deri.org:8080/openrdf-sesame/repositories/
roadmap>
{{ ?sparqlEndpoint void:class ?proteinClass .
FILTER (?proteinClass = granatum:Protein)}
UNION { ?sparqlEndpoint void:class ?proteinClass .
    ?proteinClass rdfs:subClassOf granatum:Protein . }}
SERVICE <http://labs.mondeca.com/endpoint/ends> {
        ?dataset void:sparqlEndpoint ?sparqlEndpoint .
        ?dataset ends:status ?status .
        ?status dcterms:date ?statusDate .
        ?status ends:statusIsAvailable ?isAvailable . }}
ORDER BY DESC(?statusDate)
```

### 4.4   Visualization of the Mapped LSLOD Cloud

A Visualization interface[22] is also developed to enable the domain users for intu-itively navigating the LSLOD Cloud using the generated Roadmap (Fig. 3). The linked concepts and literals are displayed as nodes arranged in a force-directed concept map representation, as previously introduced in [12]. The $Qe$ used for linking are displayed as *Light Brown*-colored nodes, whereas catalogued concepts

---

[22] http://srvgal78.deri.ie/roadmapViz/ (l.a.: 2014-07-31)

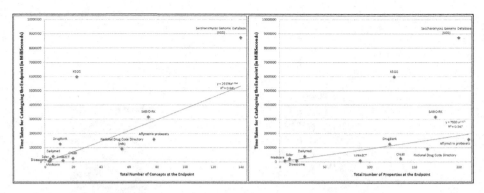

**Fig. 4.** Time taken to catalogue 12 SPARQL endpoints

(*Blue*-colored nodes) are linked to these $Qe$ using the `voidext:subClassOf`. Properties with the content-types `rdf:Literal` and `dc:InteractiveResource` are shown as *Red* and *Green*-colored nodes respectively, whereas URI-type properties are represented as *Black*-colored edges between the domain and the range concepts. The final result is a densely-clustered network graph representing the mapped LSLOD. Hovering over any particular node reduces this graph to display only the first-level associations. An information box is also displayed (Fig. 3), which provides additional details to the user regarding the source SPARQL endpoint from which the concept was catalogued, its super classes and the list of properties for which the selected concept may act as a domain/range, along with the name of the associated node or the $O_T$ (Literal, Interactive Resource).

### 4.5   Google Refine Tool

Drug compounds and molecular data are available in machine-readable formats from many isolated SPARQL endpoints. In order to harvest the benefits of their availability, an extension[23] to the popular Google Refine[24] tool was devised for providing researchers with an intuitive, integrative interface where they can use the Roadmap to semantically annotate their experimental data. Researchers load a list of molecules into the application and link them to URIs. The extension is able to make use of the URIs in conjunction with the Roadmap to collect all possible literal properties that are associated with the "Molecule" instances and help users enhance their datasets with extra molecular properties.

## 5   Evaluation

We have previously evaluated the performance of our Link Generation methodology by comparing it against the popular linking approaches [9].

### 5.1   Experimental Setup

We evaluated the performance of our catalogue generation methodology (shown in Section 3.1). The aim of this evaluation was to determine if we could catalogue

---

[23] http://goo.gl/S809N2 (l.a.: 2014-07-31)
[24] http://refine.deri.ie (l.a.: 2014-07-31)

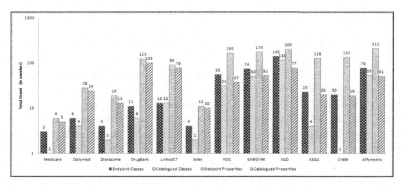

**Fig. 5.** Comparative plot of available versus catalogued Query Elements

any SPARQL endpoint and link it to the Roadmap in a desirable amount of time with a satisfactory percentage capture (catalogued elements/total elements). We proceeded by recording the times taken to probe instances through endpoint analysis of 12 different endpoints whose underlying data sources were considered relevant for drug discovery - Medicare, Dailymed, Diseasome, DrugBank, LinkedCT, Sider, National Drug Code Directory (NDC), SABIO-RK, Saccharomyces Genome Database (SGD), KEGG, ChEBI and Affymetrix probesets. The cataloguing experiments were carried out on a standard machine with 1.60Ghz processor, 8GB RAM using a 10Mbps internet connection. We recorded the total available concepts and properties at each SPARQL endpoint as well as those actually catalogued in our Roadmap (Fig. 5). Total number of triples exposed at each of these SPARQL endpoints and the total time taken for cataloguing was also recorded. We attempted to select those SPARQL endpoints which have a better latency for this evaluation, as the availability and the uptime of the SPARQL endpoint is an important factor for cataloguing. Best fit regression models were then calculated.

### 5.2  Evaluation Results

As shown in Fig. 4, our methodology took less than 1000000 milliseconds (<16 minutes) to catalogue seven of the SPARQL endpoints, and a gradual rise with the increase in the number of available concepts and properties. We obtained two power regression models ($T = 29206 * C_n^{1.113}$ and $T = 7930 * P_n^{1.027}$) to help extrapolate time taken to catalogue any SPARQL endpoint with a fixed set of available concepts ($C_n$) and properties ($P_n$), with $R^2$ values of 0.641 and 0.547 respectively. Using these models and knowing the total number of available concepts/properties, a developer could determine the approximate time (ms) as a vector combination. A comparative plot of the total number of concepts and properties available at the endpoints against those catalogued by our methodology was also prepared. We obtained a >50% concept capture for 9 endpoints, and a >50% property capture for 6 endpoints. KEGG and SGD endpoints had taken an abnormally large amount of time for cataloguing than the trendline. The reason for this may include endpoint timeouts or network delays.

# 6 Discussion

There is great potential in using semantic web and LD technologies for drug discovery. However, in most cases, it is not possible to predict *a priori* where the relevant data is available and its representation. An additional consequence of the deluge of data in life sciences in the past decade is that datasets are too large (in the order of terabytes) to be made available through a single instance of a triple store, and many of the source providers are now enabling native SPARQL endpoints. Sometimes only a fraction of the dataset is necessary to answer a particular question – the decision on "which" fraction is relevant will vary on the context of the query. In this paper we describe the concept and methodology for devising an active Linked Life Sciences Data Roadmap. Our methodology relies on systematically issuing queries on various life sciences SPARQL endpoints and collecting its results in an approach that would otherwise have to be encoded manually by domain experts or those interested in making use of the web of data for answering meaningful biological questions. In an effort to explore how often concepts are reused, we define a methodology that maps concepts to a set of $Qe$, which were defined by domain experts as relevant in a context of drug discovery.

## 6.1 Catalogue Development and Links Creation

The number of classes per endpoint varied from a single class to a few thousands. Our initial exploration of the LSLOD revealed that only 15% of classes are reused. However, this was not the case for properties, of which 48.5% are reused. Most of the properties found were domain independent (e.g. `type`, `xref`, `sameAs`, `comment`, `seeAlso`); however, these are not relevant for the Roadmap as they cannot increase the richness of information content. These properties can be more easily resolved through the concepts that are used as domain/range. In class matching, most orphan classes, which were created in cases where object URI were non-dereferenceable, could be mapped to $Qe$ through matching URI regular expression patterns and source identifiers. From these results, we found that maintaining consistency of the catalogue even when the URIs were non-dereferenceable is critical as the merging of data with other sources enable the classification of the instances and identification of "roads" between different SPARQL endpoints. We faced multiple challenges during catalogue development which can hinder the applicability of our approach:

- Some endpoints return timeout errors when a simple query (`SELECT DISTINCT ?Concept WHERE {[ ] a ?Concept}`) is issued.
- Some endpoints have high downtime and cannot be generally relied.
- Many endpoints provide non-deferenceable URI and some derefenceable URI do not provide a "type" for the instance.

Nevertheless, we still found the Roadmap approach highly applicable for solving complex biological problems in drug discovery [10]. Although a very low percentage of linking becomes possible through naïve matching or manual/domain matching, the quality of links created are highly trusted [9]. For Orphan classes 34.9% of classes were linked by matching the URI regex patterns. It is also worth noticing that 23% of identified classes, and 56.2% of the properties remained unlinked, either because they are out of scope or cannot match any $Qe$. This means that the quality as well as the quantity of links created is highly dependent on the set of $Qe$ used: if the $Qe$ `gr:Gene` is available, `kegg:Gene` would be mapped to it as opposed to the current case where it is mapped to `gr:nucleicAcid`. In such cases, additions to the $Qe$ could be suggested. It is worth noting that these $Qe$ were created to fit the drug discovery scenario, and can be replaced in alternative contexts e.g *"Protein-Protein Interaction"*. Changing the $Qe$ will result in different Roadmaps. The aim of the LSLOD Roadmap is to enable *"a posteriori integration"* e.g. adding the relation `R1:{kegg:Drug voidext:subClassOf gr:Drug}` ensures that DSQE would infer all instances of `kegg:Drug` as instances of `gr:Drug` but not vice versa.

### 6.2   Semantic Inconsistencies and Future Work

In some cases, we found that catalogued classes would be better described as properties e.g. term `Symbol` is used both as a property and as a class. Since RDFS semantics does not allow a Class to be made a subclass of an entity of type "Property", these could not match using our methods. Moreover some URIs are used both as a class and as a property e.g. `http://bio2rdf.org/ncbi_resource:gene`. This can cause an inconsistency since a reasoning engine would either accept a URI as a property or a class. The algorithm used to create catalogue covers only classes containing instances and therefore our catalogue may not be capturing uninstantiated classes in any endpoints. Since our Roadmap was aimed only at linking classes for which roads can be discovered, we considered uninstantiated classes to be out of scope of our Roadmap. Class and property labels, used for discovering links, were generally obtained from the URI itself; however, there may be cases where labels for those entities may have been made available as part of ontologies or through external SPARQL endpoints. Those labels were not investigated in the Roadmap presented – however, Bioportal has exposed their annotated ontologies as a SPARQL endpoint and therefore in future we expect to make use of the labels provided by the original creators of the data.

## 7   Conclusion

Our preliminary analysis of existing SPARQL endpoint reveals that most Life Sciences and bio-related data cannot be easily mapped together. In fact, in the majority of cases there is very little ontology and URI reuse. Furthermore, many datasets include orphan URI - instances that have no "type"; and multiple URIs that cannot be dereferenced. Our Roadmap is a step towards cataloguing and linking the LSLOD through several different techniques. We evaluated the proposed Roadmap in terms of cataloguing time and entity capture and also showcased a few applications - namely query federation, selective SPARQL querying, drug discovery, semantic annotation and LSLOD visualization.

**Acknowledgements.** This research has been supported in part by Science Foundation Ireland under Grant Number SFI/12/RC/2289 and SFI/08/CE/I1380 (Lion 2). The authors would like to acknowledge Gofran Shukair for developing Google Refine Tool.

# References

1. Alexander, K., Hausenblas, M.: Describing linked datasets-on the design and usage of void, the'vocabulary of interlinked datasets. In: Linked Data on the Web Workshop (LDOW 09), in conjunction with WWW09. Citeseer (2009)
2. Bechhofer, S., Buchan, I., De Roure, D., Missier, P., et al.: Why linked data is not enough for scientists. Future Generation Computer Systems **29**(2), 599–611 (2013)
3. Broekstra, J., Kampman, A., van Harmelen, F.: Sesame: a generic architecture for storing and querying RDF and RDF schema. In: Horrocks, I., Hendler, J. (eds.) ISWC 2002. LNCS, vol. 2342, pp. 54–68. Springer, Heidelberg (2002)
4. Buil-Aranda, C., Hogan, A., Umbrich, J., Vandenbussche, P.-Y.: SPARQL web-querying infrastructure: ready for action? In: Alani, H., et al. (eds.) ISWC 2013, Part II. LNCS, vol. 8219, pp. 277–293. Springer, Heidelberg (2013)
5. Cheung, K.H., Frost, H.R., Marshall, M.S., et al.: A journey to semantic web query federation in the life sciences. BMC bioinformatics **10**(Suppl 10), S10 (2009)
6. Deus, H.F., Prud'hommeaux, E., Miller, M., Zhao, J., Malone, J., Adamusiak, T., et al.: Translating standards into practice-one semantic web API for gene expression. Journal of biomedical informatics **45**(4), 782–794 (2012)
7. Deus, H.F., Zhao, J., Sahoo, S., Samwald, M.: Provenance of microarray experiments for a better understanding of experiment results (2010)
8. Goble, C., Stevens, R., Hull, D., et al.: Data curation+ process curation= data integration+ science. Briefings in bioinformatics **9**(6), 506–517 (2008)
9. Hasnain, A., Fox, R., Decker, S., Deus, H.F.: Cataloguing and linking life sciences LOD cloud. In: 1st International Workshop on Ontology Engineering in a Data-driven World collocated with EKAW12 (2012)
10. Hasnain, A., et al.: Linked biomedical dataspace: lessons learned integrating data for drug discovery. In: Mika, P., et al. (eds.) ISWC 2014, Part I. LNCS, vol. 8796, pp. 114–130. Springer, Heidelberg (2014)
11. Jain, P., Hitzler, P., Sheth, A.P., Verma, K., Yeh, P.Z.: Ontology alignment for linked open data. In: Patel-Schneider, P.F., Pan, Y., Hitzler, P., Mika, P., Zhang, L., Pan, J.Z., Horrocks, I., Glimm, B. (eds.) ISWC 2010, Part I. LNCS, vol. 6496, pp. 402–417. Springer, Heidelberg (2010)
12. Kamdar, M.R., Zeginis, D., Hasnain, A., Decker, S., Deus, H.F.: ReVeaLD: A user-driven domain-specific interactive search platform for biomedical research. Journal of Biomedical Informatics **47**, 112–130 (2014)
13. Petrovic, M., Burcea, I., Jacobsen, H.A.: S-ToPSS: semantic toronto publish/subscribe system. In: Proceedings of the 29th international conference on Very large data bases, vol. 29, pp. 1101–1104. VLDB Endowment (2003)
14. Quackenbush, J.: Standardizing the standards. Molecular systems biology **2**(1) (2006)
15. Schwarte, A., Haase, P., Hose, K., Schenkel, R., Schmidt, M.: FedX: a federation layer for distributed query processing on linked open data. In: Antoniou, G., Grobelnik, M., Simperl, E., Parsia, B., Plexousakis, D., De Leenheer, P., Pan, J. (eds.) ESWC 2011, Part II. LNCS, vol. 6644, pp. 481–486. Springer, Heidelberg (2011)
16. Stein, L.D.: Integrating biological databases. Nature Reviews Genetics **4**(5), 337–345 (2003)
17. Studer, R., Grimm, S., Abecker, A.: Semantic web services: concepts, technologies, and applications. Springer (2007)
18. Volz, J., Bizer, C., Gaedke, M., Kobilarov, G.: Discovering and maintaining links on the web of data. Springer (2009)
19. Zeginis, D., et al.: A collaborative methodology for developing a semantic model for interlinking Cancer Chemoprevention linked-data sources. Semantic Web (2013)

# Link Prediction in Linked Data of Interspecies Interactions Using Hybrid Recommendation Approach

Rathachai Chawuthai[1,2(✉)], Hideaki Takeda[2], and Tsuyoshi Hosoya[3]

[1] The Graduate University for Advanced Studies, Kanagawa, Japan
[2] National Institute of Informatics, Tokyo, Japan
{rathachai,takeda}@nii.ac.jp
[3] National Museum of Nature and Science, Tokyo, Japan
hosoya@kahaku.go.jp

**Abstract.** Linked Open Data for ACademia (LODAC) together with National Museum of Nature and Science have started collecting linked data of interspecies interaction and making link prediction for future observations. The initial data is very sparse and disconnected, making it very difficult to predict potential missing links using only one prediction model alone. In this paper, we introduce Link Prediction in Interspecies Interaction network (LPII) to solve this problem using hybrid recommendation approach. Our prediction model is a combination of three scoring functions, and takes into account collaborative filtering, community structure, and biological classification. We have found our approach, LPII, to be more accurate than other combinations of scoring functions. Using significance testing, we confirm that these three scoring functions are significant for LPII and they play different roles depending on the conditions of linked data. This shows that LPII can be applied to deal with other real-world situations of link prediction.

**Keywords:** Biological classification · Collaborative filtering · Community structure · Hybrid recommendation approach · Interspecies interaction · Linked data · Link prediction

## 1 Introduction

The technologies of semantic web and linked data have begun connecting world's data through the Internet [1]. Biodiversity observational data is one of timely issues of linked open data. It has been gathered and shared for several years by some research communities such as GBIF[1], LODAC[2], TDWG[3], etc., so information of living things throughout the world has started linking. In Japan, Linked Open Data for ACademia (LODAC) has collected taxon concepts in Resource Description Framework (RDF)[4]

---

[1] Linked Open Data for ACademia (http://lod.ac)
[2] Global Biodiversity Information Facility (http://www.gbif.org)
[3] Biodiversity Information Standard (http://www.tdwg.org)
[4] http://www.w3.org/TR/rdf11-concepts

© Springer International Publishing Switzerland 2015
T. Supnithi et al. (Eds.): JIST 2014, LNCS 8943, pp. 113–128, 2015.
DOI: 10.1007/978-3-319-15615-6_9

and provided Simple Protocol and RDF Query Language (SPARQL)[5] endpoint for public access [2]. It results in a great benefit for scientists and people who are interested in the interconnected information of biodiversity.

Moreover, LODAC and National Museum of Nature and Science[6] have started collecting links of fungus-host interactions from a list of fungi recorded in Japan [3] and transforming them into RDF. This project is initiated not only for offering an online dataset of interspecies interaction but also preforming a recommendation system for predicting potential links of species. It becomes a navigator for biologists to make further biological observation and help them to reduce observation's cost including time and budget. For this reason, this study aims to predict potential missing interspecies interactions in the network of linked data, and give them rankings.

This paper, which is one part of the according project, introduces an approach to the Link Prediction in Interspecies Interaction network (LPII). In this study, we analyze the interaction between fungi and their hosts, which can be either animals or plants, in order to give predictive scores to missing links of fungus-host. The difficulty of this research is to deal with a very sparse and rarely connected dataset, so a large number of missing links are waiting to be discovered. This situation is similar to the beginning phase of linked data that pieces of data are linked in small scale; so much effort to discover more interconnections is required. In this case, a well-known scoring function for link prediction such as corroborative filtering [4] alone is not enough; hence more perspectives such as community detection [5] and prior knowledge of biological classification [6] are considered to be scoring functions for our model.

The accuracy of this prediction model is verified by the Area Under the receiver operating characteristic Curve (AUC) [7] that is generally used in the link prediction problem [8], and the suitability of the model is evaluated by domain experts. Roles of all scoring functions based on different conditions of the dataset are discussed as well.

Therefore, the contribution of our research is to introduce LPII, a hybrid approach that combines three concepts of collaborative filtering, community structure, and biological classification. To express our approach more clearly, a diagram in Fig. 1 presents a workflow as a big picture of our work including four main steps. First, existent links of interspecies interactions are queried, and then transformed into a bipartite graph. Second, the graph is analyzed, nonexistent links are introduced, and then their predictive scores are given using the combination of three scoring functions, which indicate the possibility to discover interactions between species based on the three mentioned concepts. Next, the missing links between species are ranked by the given scores. At last, biologists use the predicted result to be a guideline for making observations and feed their observed results back to the database again. These steps are described in more detail in the following sections.

This paper is organized as follows. The background and related work are reviewed in Section 2. Our approach, LPII, is introduced in Section 3. Our experiments are described in Section 4. Then, the result is discussed in Section 5. Finally, we conclude and suggest future work in Section 6.

---

[5]    http://www.w3.org/TR/rdf-sparql-query
[6]    http://www.kahaku.go.jp/english

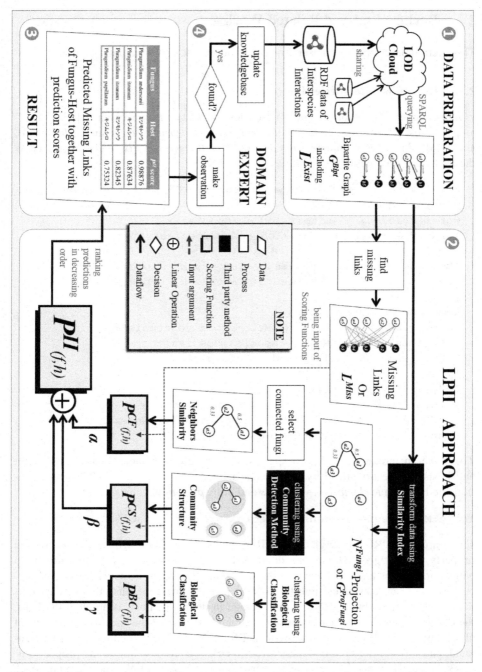

**Fig. 1.** Workflow diagram describing the summary of this research is presented in four steps: (1) data preparation, (2) LPII approach, (3) predicted result, and (4) activity of domain experts

## 2     Background and Related Work

This section describes the dataset, link prediction model, and evaluation model.

### 2.1     Dataset of Interspecies Interaction

LODAC developed ontology and dataset describing taxon concept in RDF. For example, a rust fungus species named *Melampsora yezoensis* from genus *Melampsora*, and a host plant species named *Salix pierotii* from the genus *Salix* can be expressed as the following statements.

```
@prefix rdf:     <http://www.w3.org/1999/02/22-rdf-syntax-ns#> .
@prefix rdfs:    <http://www.w3.org/2000/01/rdf-schema#> .
@prefix lodac:   <http://lod.ac/species/> .
@prefix species: <http://lod.ac/ns/species#> .


lodac:Melampsora    species:hasTaxonRank    species:Genus.
lodac:Salix         species:hasTaxonRank    species:Genus.

lodac:Melampsora_yezoensis
   rdfs:label                "Melampsora yezoensis"@la ;
   species:hasTaxonRank    species:Species ;
   species:hasSuperTaxonlodac:Melampsora .

lodac:Salix_pierotii
   rdfs:label                "Salix pierotii"@la ;
   rdf:type                  species:ScientificName ;
   species:hasSuperTaxonlodac:Salix .
```

To define an interaction between species, in this case, commensalism of rust fungi and hosts can be organized by a predicate named *species:growsOn*. Thus, the relationship between *Melampsora yezoensis* and *Salix pierotii* becomes

```
lodac:Melampsora_yezoensis species:growsOn lodac:Salix_pierotii.
```

There are 903 URIs of fungi, 2001 URIs of hosts, and 2,966 triples formed by the property *species:growsOn* in our dataset. This data also acts as a bipartite graph whose one side is a set of fungi and the other side is a set of hosts. However, the occurrence appears to suggest that the dataset is very sparse because the average out-degree of fungi is about 3.28 and the average in-degree of hosts is about 1.48. Thus, the density of actual interactions is approximately 0.15% of the maximum number of all possible links. It means that more than 1.8 millions of links are theoretically calculated, and further more relationships await to be discovered. Without any proper strategies, this activity may lead to high experimental cost, so the link prediction approach becomes a solution to handle this situation.

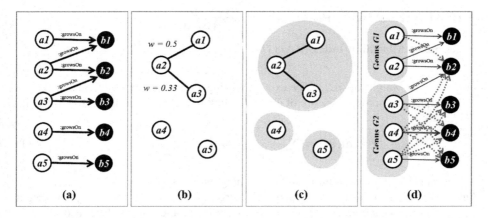

**Fig. 2.** (a) Bipartite graph indicates fungi from A-side nodes grow on hosts from B-side nodes. (b) $N^{Fungi}$-Projection is built from a bipartite graph using Jaccard index. (c) Community structure is detected from $N^{Fungi}$-Projection and represented by grey clusters. (d) Red dash lines are predicted links according to biological classification of A-side nodes.

## 2.2    Link Prediction Model

For link prediction in a bipartite graph, the collaborative filtering method [4], which gives predictive scores for missing fungus-host pairs based on number of common hosts among fungi, is used. It transforms one side of a bipartite graph into a classical graph using a similarity index, find some close neighbors, and scores all predicted links. In general, a bipartite graph $G^{Bipt}=(N^A, N^B, L^{Exist})$ is firstly defined by a set $N^A$ of A-side nodes, a set $N^B$ of B-side nodes, and a set $L^{Exist} \subseteq N^A \times N^B$ of existent links. For example, the graph containing $N^A=\{a1, a2, a3, a4, a5\}$, $N^B=\{b1, b2, b3, b4, b5\}$, and $L^{Exist} =\{(a1,b1), (a2,b1), (a2,b2), (a3,b2), (a3,b3), (a4,b4), (a5,b5)\}$ are demonstrated in Fig. 2(a). A set of missing links ($L^{Miss}$), which does not exist in the $G^{Bipt}$, is defined by $L^{Miss} = N^A \times N^B - L^{Exist}$. Hence, $L^{Miss}$ is $\{(a1,b2), (a1,b3),(a1,b4), (a1,b5), (a2,b3), (a2,b4), (a2,b5), (a3,b1), (a3,b4), (a3,b5), (a4,b1), (a4,b2), (a4,b3), (a4,b5), (a5,b1), (a5,b2), (a5,b3), (a5,b4)\}$. Next, a predictive score of each missing link is calculated using collaborative filtering through the similarity of either A-side or B-side nodes. In the research, A-side nodes are considered, so it needs to produce a similarity network of A-side nodes called $N^A$-projection. The $N^A$-projection of $G^{Bipt}$ is $G^{ProjA} = (N^A, E^{ProjA})$ that is defined by as set of edges $E^{ProjA}=\{(x,y)|x,y \in N^A$ and $\Gamma(x) \cap \Gamma(y) \neq \emptyset\}$ where $\emptyset$ is an empty set, and $\Gamma(n)$ returns a set of nodes that have direct interactions with the node $n$, such as $\Gamma(a2)=\{b1, b2\}$. Therefore, the $G^{ProjA}$ becomes $(\{a1, a2, a3, a4, a5\}, \{(a1,a2), (a2,a3)\})$ that is presented in Fig. 2(b). After that, weight of each edge in $N^A$-projection is measured using a proper similarity index, such as

— **Common Neighbors** (CN) [8]:          $|\Gamma(x) \cap \Gamma(y)|$          ,

— **Jaccard Index** [9]:          $\dfrac{|\Gamma(x) \cap \Gamma(y)|}{|\Gamma(x) \cup \Gamma(y)|}$          ,

— **Sørensen index** [10]:          $\dfrac{|\Gamma(x) \cap \Gamma(y)|}{|\Gamma(x)|+|\Gamma(y)|}$          ,

— **Hub Depressed Index** (HDI) [8]:    $\dfrac{|\Gamma(x)\cap\Gamma(y)|}{\max\,(\Gamma(x),\Gamma(y))}$    , and

— **Resource Allocation Index** (RA) [11]:    $\sum_{z\in\Gamma(x)\cap\Gamma(y)}\dfrac{1}{|z|}$ .

For example, weight of *(a1,a2)* or *w(a1,a2)* using Jaccard index is 0.5, and *w(a2,a3)* is 0.33. Last, the predictive score of a missing link *(ai,bi)* using collaborative filtering, $P^{CF}(ai,bi)$, is

$$P^{CF}(ai, bi) = \sum_{\substack{aj\,\in\,neighborsOf(ai)\\ \cap\,linksTo(bi)}} w(ai, aj) \tag{1}$$

where *neighborsOf(ai)* returns a set of neighbors of the node *ai* under $G^{ProjA}$; *linksTo(bi)* gives a set of nodes that have direct links to *bi*; and *w(ai,aj)* is weight between nodes *ai* and *aj*. For example, $P^{CF}(a1,b2)$ computed by *w(a1,a2)* is *0.5*.

A closer look at the projection of fungi ($N^{Fungi}$-projection or $G^{ProjFungi}$) from the interspecies interaction dataset shows that the degrees of most fungi are very low and some of them do not have neighbors, so the collaborative filtering alone does not provide enough prediction for this situation. The authors of [12] suggested recommending user-item pairs based on probability of item in a cluster of users. In the case of our dataset, there are two perspectives of clustering of fungi, one is host-based similarity, and the other one is clustering based on background information.

The former perspective is a clustering based on fungi that individuals' relationship similarity among fungi is not explicitly given. Fore example, it cannot find $P^{CF}(a1,b3)$ because *w(a1,a3)* is not presented in the collaborative filtering. This situation indicates that a community of fungi becomes an appropriate solution, so community structure of fungi in $N^{Fungi}$-projection needs to be detected. In the community detection problem, a cost function is a key player. A well-known cost function is the Modularity, which measures the quality of the division of a graph into communities [13]. The modularity function $Q$ is defined by

$$Q = \frac{1}{2m}\sum_{ai,aj\in C}\left(w(ai, aj) - \frac{d(ai)\,d(aj)}{2m}\right) \tag{2}$$

where node *ai* and node *aj* are in a same community, *m* is half of number of edges appearing in an undirected graph, the function *w(ai,aj)* is the appropriate weight function, and functions *d(aj)* and *d(aj)* returns degrees of node *ai* and node *aj* in $N^{Fungi}$-projection respectively. Many researches about graph mining proposed community detection methods such as,

— **Walktrap** that employed random walk and estimated total Modularity in every step before merging clusters [14],
— **Fast Greedy** that used the same random walk as Walktrap but calculated only local Modularity [14],
— **Edge Betweenness** that is a top-down clustering method where edges are removed in the decreasing order of their edge-betweenness scores [15], and
— **InfoMap** that used an information theoretic clustering on a graph to be a map of random walks representing information flow on a network [16].

The output of a community detection method is a set of sub graphs of $N^{Fungi}$-projection having dense connections between the nodes within a cluster but sparse connections between nodes in different clusters as demonstrated in Fig. 2(c).

Besides the community detection, the latter perspective is a clustering based on static knowledge of fungi. Some reviews indicated that some groups of fungi are mostly found at some particular plants, for instance, fungi from genus *Cyttaria* always grow on plants from genus *Nothofagus* [6]. This fact seems to suggest that clustering based on biological classification is meaningful for finding missing associations between fungi and hosts. For example, if *{a1, a2}* is under the same genus, and *{a3, a4, a5}* is from another genus, the possible missing links is displayed are Fig. 2(d).

### 2.3     Evaluation Model

In the evaluation phase, the $L^{Exist}$ is split into training links ($L^{Train}$) and test links ($L^{Test}$). The $L^{Train}$ is an input of prediction model for preparing predicted links ($L^{Predict}$) together with the predictive score of missing interspecies interactions called $P^{II}$. Thus, $L^{Predict}$ is a subset of $L^{Test} \cup L^{Miss}$. In this case, to the best of our knowledge, Precision is not suitable for our case, because the number of $L^{Predict}$ is much larger than $L^{Test}$. Moreover, it exactly relies on positive and negative results rather than ranking of predictive scores. Area Under the receiver operating characteristic Curve (AUC) is always used to evaluate prediction algorithms because the comparison of the ranking of both $L^{Miss}$ and $L^{Test}$ is calculated [7]. The AUC becomes higher when elements of $L^{Test}$ have high rankings among $L^{Predict}$ descending order by $P^{II}$. The AUC is

$$AUC = \frac{n' + 0.5n''}{n} \tag{3}$$

where there are $n$ comparisons, $n'$ is times of $P^{II}$ score of $L^{Test}$ being higher than $L^{Miss}$, and $n''$ is times they have the same score. For example, let $L^{Predict} = \{(a1,b1), (a2, b2), (a3,b4), (a4,b5)\}$, $L^{Test} = \{(a1,b1), (a2, b2)\}$, $P^{II}(a1,b1)=0.5$, $P^{II}(a2,b2)=0.3$, $P^{II}(a3,b4)=0.3$, and $P^{II}(a4,b5)=0.4$; the comparisons are $P^{II}(a1,b1)>P^{II}(a3,b4)$, $P^{II}(a1,b1)>P^{II}(a4,b5)$, $P^{II}(a2,b2)<P^{II}(a4,b5)$, and $P^{II}(a2,b2)=P^{II}(a3,b4)$; so AUC values equals $(2 \times 1 + 1 \times 0.5) \div 4 = 0.625$.

## 3     LPII Approach

According to previous sections, the link prediction approach to collaborative filtering alone does not well address the issue of our dataset. A hybrid approach consisting of collaborative filtering and other suitable methods is regularly presented by some studies such as [12] and [18]. For this reason, the hybrid approach to Linked Predication on Interspecies Interaction (LPII) is introduced. To make the purpose more clearly, we presented a workflow in the second step of Fig. 1. as a big picture of this section.

Our approach is initiated based on the views that (1) a fungus should be found at a host shared by its host-based neighbors, (2) fungi from the same community should be found at hosts shared by most members, and (3) fungi in the same biological classification should have similar fungus-host intersections. The hybrid approach of our work combines three prediction models, also known as scoring functions for link prediction, that are collaborative filtering ($P^{CF}$), community structure ($P^{CS}$), and biological classification ($P^{BC}$). The total scoring function, which combines all functions, for predicting interspecies interaction ($P^{II}$) of a fungus $f$ and a host $h$ is represented by

$$P^{II}(f,h) = \alpha \cdot P^{CF}(f,h) + \beta \cdot P^{CS}(f,h) + \gamma \cdot P^{BC}(f,h) \tag{4}$$

where $\alpha$, $\beta$, and $\gamma$ are constant variables indicating importance of scoring functions respectively.

In our research, $P^{CF}$ introduced by [4] is proposed to be a baseline that is calculated using the equation (1). To prepare data for the prediction process, $G^{Bipt}=(N^{Fungi}, N^{Hosts}, L^{Exist})$ is constructed by transforming RDF dataset into a bipartite graph; then $G^{ProjFungi} = (N^{Fungi}, E^{ProjFungi})$ is built by

$$E^{ProjFungi} = \{(f_i, f_j) | f_i, f_j \in N^{Fungi} \ and \ \Gamma(f_i) \cap \Gamma(f_j) \neq \emptyset\} \tag{5}$$

where
$$\Gamma(x) = \{y | (x,y) \in L^{Exist}\} \tag{6}$$

Next, $P^{CS}(f,h)$ is derived from the idea of [12], if a host $h$ is frequently shared by most fungi under the same community as fungus $f$, the fungus $f$ is more likely found at the host $h$. This research considered Naïve Bayes [17], which is a likelihood classification function, to evaluate the probability of finding the fungus $f$ at the host $h$. Because cluster is known, the probability to find $h$ in the community of $f$ is simply defined by

$$P^{CS}(f,h) = \frac{\sum_{f_i \in MC(f)} 1\{(f_i,h) \in L^{Exist}\}}{\sum_{f_i \in MC(f), h_j \in N^{Host}} 1\{(f_i,h_j) \in L^{Exist}\}} \tag{7}$$

where $MC(f)$, which stands for "*Members of Community of*", returns all members of the community of the fungus $f$; and $1\{(f_i,h) \in L^{Exist}\}$ returns 1 if a link $(f_i,h)$ is existent, otherwise 0; and $1\{(f_i,h_j) \in L^{Exist}\}$ returns 1 for every existent link given by $f_i$.

Last, $P^{BC}(f,h)$ is computed based on probability of a host $h$ comparing to all hosts shared by all fungi under the same biological classification as the fungus $f$. This function is expressed by

$$P^{BC}(f,h) = \frac{\sum_{f_i \in MB(f)} 1\{(f_i,h) \in L^{Exist}\}}{\sum_{f_i \in MB(f), h_j \in N^{Host}} 1\{(f_i,h_j) \in L^{Exist}\}} \tag{8}$$

The definition is similar to equation (7), whereas the term $MB(f)$, which stands for "*Members of Biological classification of*", returns the set of all fungi under the same biological classification than a fungus $f$.

## 4    Experiment

In order to demonstrate the feasibility and suitability of the LPII approach, the experiment was designed to achieve the following objectives.

1. To prove that our model is suitable for RDF data of interspecies interaction.
2. To prove that the combination of collaborative filtering, community structure, and biological classification provides better prediction result than other ones.
3. To find out the relative importance of roles that collaborative filtering, community structure, and biological classification play in the prediction model.

## 4.1   Preparing Data

The initial input of this model is the bipartite graph of interspecies interaction, $G^{Bipt}=(N^{Fungi}, N^{Hosts}, L^{Exist})$, that is generated from RDF data as presented in the first step of Fig. 1. Let a variable *?f* represent a fungus and a variable *?h* represent a host, the following SPARQL statements are as suggested.

— $N^{Fungi}$ is initiated by "`select ?f where {?f species:growsOn ?h .}`",
— $N^{Host}$ is initiated by "`select ?h where {?f species:growsOn ?h .}`", and
— $L^{Exist}$ is initiated by "`select ?f, ?h where {?f species:growsOn ?h .}`".

Besides $G^{Bipt}$, $G^{ProjFungi} = (N^{Fungi}, E^{ProjFungi})$ together with the weight function $w$ are built and executed based on $L^{Exist}$. Since our data is very sparse, links in $E^{ProjFungi}$ are about 1,500 and it is not enough for performing community detection. Thus, biological classification of hosts species is utilized. In this case, $L^{Exist*}$ is introduced for collecting relationships between fungus species and host genus, so $L^{Exist*}$ is retrieved from the SPARQL statement "`select ?f, ?hGe where {?f species:growsOn ?h. ?h species:hasSuperTaxon ?hGe. ?hGe species:hasTaxonRank species:Genus. }`". Then, in our experiment, the term $L^{Exist}$ of the equations (6), (7), and (8) was replaced by the bag $L^{Exist*}$; and the value of argument $h$ of $P^{CS}(f,h)$ and $P^{BC}(f,h)$ was altered to the genus of that host $h$. Therefore, $E^{ProjFungi}$ contains interactions up to 5,530 links that are enough for performing network analysis.

In addition, the function $MB(f)$ is evaluated by a query "`select ?member where {?member species:hasSuperTaxon ?fGe. :f species:hasSuperTaxon ?hGe. ?hGe species:hasTaxonRank species:Genus. }`", where *:f* is a URI of the focusing fungus.

## 4.2   Executing Link Prediction

Before testing the second objective, a well-suited similarity index for $P^{CF}$ and a community detection method for $P^{CS}$ have to be selected. The first experiment is to evaluate AUC from stated similarity indices: Common Neighbors (CN), Jaccard Index, Sørensen Index, Hub Depressed Index (HDI), and Resource Allocation Index (RA). Next, the similarity index providing highest AUC is used to build $E^{ProjFungi}$. Then, comparison among community detection methods is performed. According to Section 2.2, Walktrap, Fast Greedy, Edge Betweeness, and InfoMap are candidates. After that, the similarity index and community detection method providing highest AUC result are selected to be key players in the next experiment.

To prove the second objective, our approach is evaluated on the dataset. There are several steps of testing in each interaction as follows:

— 20% of $L^{Exist}$ is split into $L^{Test}$.
— The remaining 80% of $L^{Exist}$ becomes $L^{Train}$.
— $L^{Exist*}$ is initiated according to the $L^{Train}$.
— $E^{ProjFungi}$ is built according to the the $L^{Exist*}$ using the selected similarity index.
— Community structure of fungi is detected based on $E^{ProjFungi}$ using the selected community detection method.
— $L^{Predict}$ that is a subset of $L^{Test} \cup L^{Miss}$ is generated.
— The predictive score of each predicted link in $L^{Predict}$ is computed by $P^{CF}(f,h)$, $P^{CS}(f,h)$, and $P^{BC}(f,h)$, and then the score is normalized to 0-1 range.
— AUC is calculated based on combinations of the scoring functions:
  • Stand-alone function:  $(P^{CF})$, $(P^{CS})$, and $(P^{BC})$
  • Summation:  $(P^{CF} + P^{CS})$, $(P^{CF} + P^{BC})$, $(P^{CS} + P^{BC})$, and $(P^{CF} + P^{CS} + P^{BC})$
  • Multiplication: $(P^{CF} \times P^{CS})$, $(P^{CF} \times P^{BC})$, $(P^{CS} \times P^{BC})$, and $(P^{CF} \times P^{CS} \times P^{BC})$

### 4.3    Significance Testing

In this section, we run significance testing [19] to prove the importance of scoring functions: $P^{CF}$, $P^{CS}$, and $P^{BC}$. An experiment is proposed based on the view that the variation of scoring functions results in the change of AUC. In this case, we specify varied weights of scoring functions and validate AUC in every iteration. After that, the statistical significance testing is made against the assumption that the transition of each factor does not have an impact on the change of AUC. This experiment is done with the whole dataset and its samples divided by conditions of each fungus node. When every node located in $G^{ProjFungi}$ was thoroughly investigated, we found that the degree distribution is skewed and can be categorized into the following groups.

— **High degree** means the degree is more than 50.
— **Normal degree** means the degree is between 5 and 50.
— **Low degree** means the degree is less than 5.

Besides degree distribution, community size has a high variance as presented in Fig. 3. The distribution of communities' size is summarized by the following list.

— **Big community** means the number of members is more than 50.
— **Normal community** means the number of members is between 5 and 50.
— **Small community** means the number of members is less than 5.

These characteristics of fungus node are combined into several conditions such as, *(Degree=High, Community=Big)*, *(Degree=Normal, Community=Big)*, etc. Then, the variation of the significance of $P^{CF}$, $P^{CS}$, and $P^{BC}$ is tested on each condition.

In addition, this experiment intends to find out the suitable value of $\alpha$, $\beta$, and $\gamma$ of equation (4). In this case, patterns of $\alpha$, $\beta$, and $\gamma$ that are meaningful for high value of AUC are monitored. Compromised values of all coefficients according to conditions of dataset are studied as well.

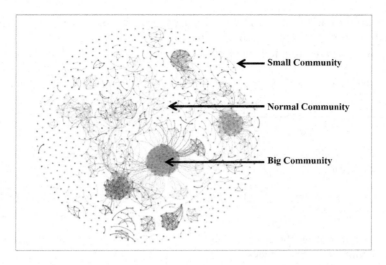

**Fig. 3.** Visualization of community structure of fungi shows various sizes of clusters

# 5     Result and Discussion

Section 4 has already introduced three experiments according to the objectives. The Section 4.1 demonstrated that $G^{Bipt}$ and $G^{ProjFungi}$ could be prepared from RDF data of interspecies interaction, so our approach was well compatible with RDF data. Next, the other two remaining experiments are discussed.

## 5.1     Evaluating the Prediction Model

The accurate comparison among all similarity indices was logged in Table 1. Each index resulted in not much different AUC. In this case, Jaccard index was selected because it gave the highest AUC, and it was used to construct $G^{ProjFungi}$. After that, the result of comparison among community detection methods was displayed in the Table 2. Based on this result, Walktrap was selected for the next experiment.

Next, our experiment provided about 400,000 predicted links, and AUC was evaluated based on every combination of scoring functions of our prediction model. There were three types of combinations: stand-alone function, summation, and multiplication, as mentioned in the Section 4.2. The result of this experiment is shown in Table 3. It has been found that among all stand-alone scoring functions, the collaborative filtering approach, $P^{CF}$, provided the best AUC. The $P^{CF}$ was also considered as a baseline in our experiment, so no stand-alone function was better than the collaborative filtering method. However, the summation of them, $P^{CF} + P^{CS} + P^{BC}$, offered the best result among all combinations. The result also showed that the multiplication of them did not give good scores. The available result seemed to suggest that the linear combination of $P^{CF}$, $P^{CS}$, and $P^{BC}$ was the most meaningful. Although the AUC value was slightly better than the baseline, just 0.01 increasing denoted that some test links were raised up to 4,000 higher ranks. In our experiment, many zero-score test links computed by only $P^{CF}$ were moved up to the top 10% rankings after $P^{CS}$ and $P^{BC}$ were added. Thus, the prediction model expressed by the equation (4) has been confirmed.

**Table 1.** AUC calculated by scoring funtions from different similarity indices

|       | CN    | Jaccard   | Sørensen | HDI   | RA    |
|-------|-------|-----------|----------|-------|-------|
| AUC   | 0.837 | **0.859** * | 0.841    | 0.854 | 0.854 |

**Table 2.** AUC calculated by scoring function from different communinty detection methods together with Jaccard index

|       | Walktrap   | Fast Greedy | Edge Betweeness | InfoMap |
|-------|------------|-------------|-----------------|---------|
| AUC   | **0.823** * | 0.747       | 0.652           | 0.749   |

**Table 3.** AUC calculated by combinations of scoring functions for link prediction using Walktrap togehter with Jaccard index

| Combination | Scoring Function(s) | AUC |
|-------------|---------------------|-----|
| Stand-alone function | $P^{CF}$ | 0.859 |
|                      | $P^{CS}$ | 0.823 |
|                      | $P^{BC}$ | 0.680 |
| Summation of functions | $P^{CF} + P^{CS}$ | 0.867 |
|                        | $P^{CF} + P^{BC}$ | 0.876 |
|                        | $P^{CS} + P^{BC}$ | 0.865 |
|                        | $P^{CF} + P^{CS} + P^{BC}$ | **0.892** * |
| Multiplication of functions | $P^{CF} \times P^{CS}$ | 0.817 |
|                             | $P^{CF} \times P^{BC}$ | 0.862 |
|                             | $P^{CS} \times P^{BC}$ | 0.827 |
|                             | $P^{CF} \times P^{CS} \times P^{BC}$ | 0.818 |

## 5.2    Evaluating the Significance of Each Scoring Function

Although the summation of all scoring functions was confirmed, it did not mean that all scoring functions played equal role in the LPII. After significant testing was done, result on the whole dataset presented in the last row of Table 4 showed that P-Values of both $P^{CF}$ and $P^{CS}$ were lower than 0.05, whereas P-Value of $P^{BC}$ was higher than 0.05. It could be interpreted that both $P^{CF}$ and $P^{CS}$ were significant for the prediction model but not $P^{BC}$, so the scoring function $P^{BC}$ should be removed from the prediction model generally. However, the Table 3 confirmed that the combination of three functions was better than two functions. When we analyzed the behavior of each coefficient thoroughly, as presented in Fig. 4, we found that the increase of each coefficient had an effect on AUC. Fig. 4(a), (d), and (e) demonstrated that $\alpha$ and $\beta$ were consistent with each other and also with AUC. Because the increasing sequence values of $\alpha$ and $\beta$ was always accompanied by the increasing sequence of AUC as expressed by the dark color in the heat map. However, Fig. 4(b) and (c) showed that the pattern of $\gamma$ was not coherent with other ones especially AUC. A closer look at the behavior of $\gamma$ and AUC in Fig. 4(f) showed that $\gamma$ should be very small value otherwise AUC would be dropped. Thus, it can be interpreted that $P^{BC}$ is still necessary for our approach.

**Table 4.** Statistical significance of each scoring function were detemined by characteristics of nodes having different degree and community sizes under $G^{projFungi}$. *(Note: Number of * indicates significance level, while P-Value≤0.05 is not siginificant)*

| Characteristic of Node | | P-Value | | |
|---|---|---|---|---|
| Degree | Community Size | $P^{CF}$ | $P^{CS}$ | $P^{BC}$ |
| High | Big | < 0.001 *** | < 0.001 *** | 0.094 |
| Normal | Big | < 0.001 *** | < 0.001 *** | < 0.001 *** |
| Normal | Normal | < 0.001 *** | < 0.001 *** | < 0.001 *** |
| Normal | Small | < 0.001 *** | < 0.001 *** | < 0.001 *** |
| Low | Big | < 0.001 *** | < 0.001 *** | 0.011 * |
| Low | Normal | 0.058 | 0.003 ** | 0.032 * |
| Low | Small | 0.056 | 0.084 | < 0.001 *** |
| Whole dataset | | < 0.001 *** | < 0.001 *** | 0.086 |

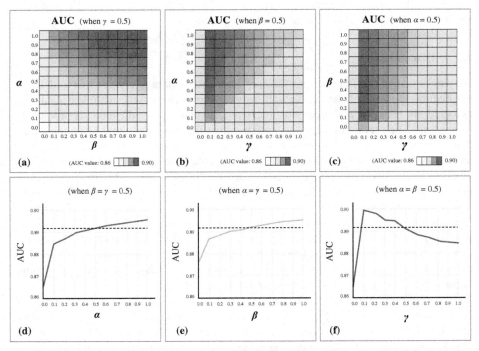

**Fig. 4.** Heat maps (a), (b), and (c) display AUC resulted from two coefficients while the other one is 0.5. Charts (d), (e), and (f) demonstrates the AUC impacted by each individual coefficient when others are 0.5, and a dash line shows AUC value when $\alpha = \beta = \gamma$.

Moreover, when data was investigated based on a condition of each fungus node mentioned in Section 4.3, it was found that $P^{CF}$, $P^{CS}$, and $P^{BC}$ have different roles in different conditions. One can note that there is no high degree fungus in either normal or small communities, so these two conditions were excluded. Table 4 demonstrated the significance of $P^{CF}$, $P^{CS}$, and $P^{BC}$ according to the conditions of nodes in the graph $G^{ProjFungi}$.

— $P^{CF}$ is not significant if nodes have low degree and are not in a big community.
— $P^{CS}$ is not significant if nodes have low degree and are in a small community.
— $P^{BC}$ is not significant if nodes have high degree and are in a big community.

The result satisfied the condition that when there were not enough fungus-host interactions, fungi have low degree and stay in small community, so the background knowledge of fungi such as biological classification, became important criterion for the prediction model. According to the Fig. 3, there were a lot of fungi corresponding to this condition, so $P^{BC}$ needs to be used.

Besides, appropriate values for $\alpha$, $\beta$, and $\gamma$ appearing in the equation (4) were determined. As we selected and analyzed top list of $\alpha$, $\beta$, and $\gamma$ that provided high AUC in every condition of the dataset, there was no pattern among these values and there was no clear implication on the significance of scoring functions. Although the golden values of all coefficients cannot be claimed, a closer look at the data and Fig. 4 indicated that $\gamma$ should be smaller than $\alpha$ and $\beta$ in general.

### 5.3    Discussion

As mentioned in the first section, dealing with sparse data is always unavoidable in the beginning phase of data collection, especially the early stage of linked data. The associations between resources are not dense, so it is really difficult to predict the future link. The prediction based on collaborative filtering is suitable when the data is dense enough. Then, prediction based on community structure becomes important for data that is less dense but high clustered. When data is less connected, projection of a bipartite graph and community detection are hardly possible to be implemented, so the link prediction based on clustering defined by background information of resources becomes a key player. In case of normal linked data, taxonomy of resources identified by well-known predicates such as *rdf:type*, *rdf:subClassOf*, *skos:broader*, *skos:narrower*, etc., are evaluated. Therefore, the link prediction based on the dimensions of similarity between collaborators, probability to find links among community members, and probability to find links among nodes having similar background knowledge are always necessary for early stage of linked data.

In addition, domain experts such as biologists are to evaluate the predicted results. A fungus-host link having high score is highly possible to be discovered, because the focusing host is frequently interacted by fungi that are adjacent neighbors, members of the same community, and under the same biological classification. For example:

— `lodac:Phragmidium_mucronatum   species:growsOn lodac:ハマナス` .
— `lodac:Phragmidium_fusiforme    species:growsOn lodac:ハマナス` .
— `lodac:Phragmidium_potentillae  species:growsOn lodac:イワキンバイ` .

which are not presented in the dataset [3], have been discovered. Then, the result of observation is preserved in an RDF repository and shared to Linked Open Data (LOD) cloud as presented in the third, fourth, and first steps of the workflow in Fig. 1. Therefore, due to our investigation and outcome, we can assure that this research is suitable and feasible to be carried out in the real-world situations.

# 6    Conclusion and Future Work

This study is an attempt to address the issue of link prediction in sparse network of linked data. This paper introduces LPII, a hybrid recommendation approach for link prediction. The research can be summarized into four steps as demonstrated in the Fig. 1. In the first step, a bipartite graph of fungus-host associations is generated from linked data of interspecies interaction using SPARQL query. In the second step, this bipartite graph is executed by our prediction model that combines three scoring functions based on different perspectives: collaborative filtering, community structure, and biological classification. Collaborative filtering prediction is evaluated based on the view that fungi will be found at hosts where similar fungi are already found. The other perspectives are based on possibility of links between fungi and hosts that are always found in the same community and in the same biological classification of fungi. When our approach was evaluated by AUC, it has been found that the linear combination of three scoring functions is more accurate than other combinations. Moreover, all perspectives are statistically significant and play different roles in different characteristics of data. Collaborative filtering and community structure are highly significant when fungus degree and its community size are not low, whereas biological classification becomes highly important when node degree and community size are not high. The experiment also suggested that the coefficient of the biological classification should be lower than the other ones in order to improve AUC. After the calculation of LPII, in the third step, the list of missing links together with predictive scores is provided. At last, biologists make an observation over the prediction result of fungi and hosts as presented in the fourth step. An observed result will be preserved in an RDF repository and published to LOD cloud in order to be knowledgebase for a next prediction.

However, when more data is observed and collected, the data becomes denser. The link prediction inside a cluster may not be appropriate anymore. While using link prediction across clusters [20] together with semantic distance among resources [21] would be considered in our future work.

**Acknowledgements.** We would like to thank Assoc. Prof. Ryutaro Ichise, Dr. Rémy Cazabet, Dr. Nimit Pattanasri, Dr. Vorapong Suppakitpaisarn, and Mr. Pannawit Samatthiyadikun for their assistance and guidance in machine learning, data analysis, social network analysis, and technical support.

# References

1. Christian, B., Heath, T., Lee, B.T.: Linked data-the story so far. International Journal on Semantic Web and Information Systems, 1–22 (2009)
2. Minami, Y., et al.: Towards a data hub for biodiversity with LOD. In: Takeda, H., Qu, Y., Mizoguchi, R., Kitamura, Y. (eds.) JIST 2012. LNCS, vol. 7774, pp. 356–361. Springer, Heidelberg (2013)
3. Katumoto, K.: List of fungi recorded in Japan. Kanto Branch of the Mycological Society of Japan (2010)

4. Huang, Z., Li, X., Chen, H.: Link prediction approach to collaborative filtering. In: The 5th ACM/IEEE-CS Joint Conference on Digital Libraries. ACM (2005)
5. Feng, X., Zhao, J.C., Xu, K.: Link prediction in complex networks: a clustering perspective. Eur. Phys. J. B **85**(1–3) (2012)
6. Peterson, K.R., et al.: Cophylogeny and biogeography of the fungal parasite Cyttaria and its host Nothofagus, southern beech. Mycologia **102**(6), 1417–1425 (2010)
7. Hanley, J.A., McNeil, B.J.: The meaning and use of the area under a receiver operating characteristic (ROC) curve. Radiology **143**(1), 29–36 (1982)
8. Lu, L., Zhou, T.: Link prediction in complex networks: A survey. Physica A (Elsevier) **390**(6), 1150–1170 (2011)
9. Hamers, L., Hemeryck, Y., et al.: Similarity measures in scientometric research: the Jaccard index versus Salton's cosine formula. Information Processing & Management **25**(3), 315–318 (1989)
10. Sørensen, T.: A method of establishing groups of equal amplitude in plant sociology based on similarity of species and its application to analyses of the vegetation on Danish commons. Biologiske **5**, 1–34 (1948)
11. Zhou, T., et al.: Predicting missing links via local information. The European Physical Journal B **71**(4), 623–630 (2009)
12. Huang, C.L., Lin, C.W.: Collaborative and content-based recommender system for social bookmarking website. World Academy of Science, Engineering and Technology **68**, 748–753 (2010)
13. Newman, M.E.: Modularity and community structure in networks. PNAS **103**(23), 8577–8582 (2006)
14. Pons, P., Latapy, M.: Computing communities in large networks using random walks. J. Graph Algorithms Appl. **10**(2), 191–218 (2006)
15. Newman, M.E.: Detecting community structure in networks. The European Physical Journal B-Condensed Matter and Complex Systems **38**(2), 321–330 (2004)
16. Rosvall, M., Bergstrom, C.T.: Maps of random walks on complex networks reveal community structure. Proceedings of the National Academy of Sciences **105**(4), 1118–1123 (2008)
17. Lowd, D., Domingos, P.: Naive Bayes models for probability estimation. In: Proceedings of the 22nd International Conference on Machine Learning, pp. 529–536. ACM (2005)
18. Rojsattarat, E., Soonthornphisaj, N.: Hybrid recommendation: combining content-based prediction and collaborative filtering. In: Liu, J., Cheung, Y., Yin, H. (eds.) IDEAL 2003. LNCS, vol. 2690, pp. 337–344. Springer, Heidelberg (2003)
19. Hastie, T., Tibshirani, R., Friedman, J.: The Elements of statistical learning. Springer series in statistics. Springer, New York (2001)
20. Kim, J., Choy, M., Kim, D., Kang, U.: Link prediction based on generalized cluster information. In: WWW 2014 Companion, pp. 317–318 (2014)
21. Roddick, J.F., Hornsby, K., Vries, D.: A unifying semantic distance model for determining the similarity of attribute values. In: Proceedings of the 26th Australasian Computer Science Conference, vol. 16, pp. 111–118 (2003)

# CURIOS Mobile: Linked Data Exploitation for Tourist Mobile Apps in Rural Areas

Hai H. Nguyen$^{(\boxtimes)}$, David Beel, Gemma Webster,
Chris Mellish, Jeff Z. Pan, and Claire Wallace

dot.rural Digital Economy Hub, University of Aberdeen, Aberdeen, UK
{hai.nguyen,d.e.beel,gwebster,c.mellish,
jeff.z.pan,claire.wallace}@abdn.ac.uk

**Abstract.** As mobile devices proliferate and their computational power has increased rapidly over recent years, mobile applications have become a popular choice for visitors to enhance their travelling experience. However, most tourist mobile apps currently use narratives generated specifically for the app and often require a reliable Internet connection to download data from the cloud. These requirements are difficult to achieve in rural settings where many interesting cultural heritage sites are located. Although Linked Data has become a very popular format to preserve historical and cultural archives, it has not been applied to a great extent in tourist sector. In this paper we describe an approach to using Linked Data technology for enhancing visitors' experience in rural settings. In particular, we present CURIOS Mobile, the implementation of our approach and an initial evaluation from a case study conducted in the Western Isles of Scotland.

## 1 Introduction

From the data perspective, most heritage based mobile applications (apps) to date have used content specifically tailored for the apps. This content usually follows a pre-set geographical route and story-lines while requiring a large amount of human effort to generate. Even though this approach can provide user-friendly and concise content to visitors, it is not very practical for small local community groups. Local community groups often develop their cultural heritage collections in an archive instead of a collection of stories that can be presented to visitors. Moreover, if this data covers a large, non-linear geographical area, it is not realistic to use pre-set geographical routes and expect visitors to follow them.

Linked Open Data is a set of best practices to publish structural data on the web, as introduced by Tim Berners-Lee [1]. It has several important advantages over traditional relational databases such as integrability and reusability, which has made it become increasingly popular within the cultural heritage sector. There have been several efforts to bring cultural heritage archives into Linked Data formats such as in the CultureSampo project [2] and the OpenART [3] project. The ultimate goal of such projects is to allow data to be able to contextualised, reused, and integrated further.

© Springer International Publishing Switzerland 2015
T. Supnithi et al.(Eds.): JIST 2014, LNCS 8943, pp. 129–145, 2015.
DOI: 10.1007/978-3-319-15615-6_10

The work described in this paper was carried out as parts of the CURIOS Mobile project at the University of Aberdeen, a follow-up of the CURIOS project.[1] The CURIOS project aims to produce a set of software tools to allow small local community groups to produce and consume their cultural heritage archives in Linked Open Data formats. The main software is a Linked Data Content Management System (CMS) [4,5].This Linked Data CMS provides a platform for novice users to produce more Linked Data by providing a friendly user interface as in the traditional CMSs such as Wordpress or Mediawiki.

Given the rich content generated in Linked Data formats, CURIOS Mobile explored the ways to exploit Linked Data in order to provide visitors with an enjoyable user experience while exploring rural areas. The objective of CURIOS Mobile was to deliver a tourist mobile application which can 1) exploit the current linked dataset generated by the CURIOS CMS and 2) work reasonably well with unreliable mobile Internet connection in rural areas. The Hebridean Connections' dataset has been used as the main case study for the CURIOS Mobile project.

There have been several attempts to bring Linked Data and the Semantic Web closer to mobile devices: DBPedia Mobile [6], mSpace Mobile [7], *Who's Who* [8], etc. However, previous work either focuses on only specific problems such as context discovery and visualisation, or assumes that data connection is always available and reliable. In contrast, in this paper we present a generic framework to exploit Linked Data archives for tourist activities, especially in a rural context where limited or no data connection is available to mobile devices. Below are the summary of the paper's main contributions.

1. A generic framework to use Linked Data archives for tourist activities via a mobile application.
2. Linked Data-based caching solutions to the unreliable data connection issue in rural areas.
3. A recommendation mechanism to choose which information to present to visitors on the site based on data characteristics.
4. A simple mechanism to generate text-based descriptions directly from RDF triples based on the techniques introduced in [9].

This paper is organised as follows. In Section 2 we briefly introduce the context of this paper; in particular the technologies and dataset we are using, as well as the broader framework as to where the system described in this paper fits. In Section 3 we introduce the main challenges for developing and deploying tourist mobile applications in rural areas. In Section 4 we introduce our approach, including a brief overview of the system and our solutions to the challenges mentioned in Section 3. In Section 5 we present some preliminary results of our implementation. Section 6 is the discussion of related work. Section 7 includes the conclusion and some pointers to future work.

---

[1] http://curiosproject.abdn.ac.uk

## 2    Background

### 2.1    Linked Open Data

A 4-star Linked Open Data as described in Tim Berners-Lee's note [1] would use HTTP URIs[2] to denote things (i.e., individuals) and W3C standards such as RDF or OWL[3] to describe such individuals' information or to relate one individual to another. The Resource Description Framework (RDF) is used as a standard format to describe things and their relationships within a linked dataset as RDF triples. The RDF triples are then stored in a type of database system, namely a triplestore and can be retrieved or maintained via a specific query language, SPARQL[4]. The vocabularies used to describe things and their relationships using RDF are usually defined in an OWL ontology.

### 2.2    The CURIOS Project

The CURIOS project aims to provide a sustainable and extensible software system for historical societies to produce and consume cultural heritage data in the form of Linked Open Data. By combining Linked Data standards and software with Drupal, a popular open source CMS, CURIOS provides users with limited knowledge on semantic technology a friendly front-end in order to produce linked data without noticing the underlying technologies (e.g., SPARQL, RDF). In CURIOS, the data entered by users are stored in a triplestore while the configuration of how data are presented to users is stored in Drupal's traditional SQL database. This approach allows the linked dataset maintained by CURIOS to be loosely coupled to Drupal meaning it can be reused in different applications or by other software. There have been two main case studies conducted to evaluate the CURIOS system: one involving historical societies based in the Western Isles of Scotland (Hebridean Connections) and another one with a local historical group at Portsoy, a fishing village located in the North East of Scotland. These two case studies are very different in terms of dataset's scale and the organisation structure. However, CURIOS has been well-received from both communities.

### 2.3    The Hebridean Connections Case Study

Hebridean Connections is a project connecting local historical societies across the Western Isles of Scotland (a.k.a. Outer Hebrides). Thousands of records about the genealogy, places, traditions, cultural and history of the islands have been generated by local historical societies and their contributors. Before the release of CURIOS, the data had been preserved in multiple physical archives by each local historical society and in a relational database maintained via proprietary

---

[2] Uniform Resource Identifier
[3] Web Ontology Language
[4] Simple Protocol And RDF Query Language

software. This dataset has been digitised and preserved in a Linked Open Data standard format (RDF). More and more linked data has been added into the archive using the CURIOS Linked Data Content Management System since its first release (February 2014). The CURIOS system has been deployed on the Hebridean Connections website recently (available at http://www.hebrideanconn ections.com).

As of 28/07/2014, the dataset in the Hebridean Connections case study consists of 864,429 RDF triples before inference, incorporated within a relatively simple OWL ontology. These triples form a total of 44,358 records. Basically, a CURIOS record is the set of triples, usually presented together, describing a particular subject (identified by a URI). Formally, a CURIOS record is defined as follows.

**Definition 1 (CURIOS Record).** *A CURIOS record $r_s$ of a subject $s$ is a set of RDF triples of the form $< s, p, o >$ where $s$ is the URI identifying the record by its numeric identifier, $p$ is either a datatype property or an object property, and $o$ is either a literal value or a URI.*

Figure 1 shows an example of a person record within the Herbidean Connections' dataset. Each record in the dataset will have at least a subject ID(e.g., 23160), a title (e.g., "Angus Macleod") and a record type associated with it (e.g., hc:Person). Moreover, depending on its type (e.g., hc:Person, hc:Location, etc.), a record can have different datatype properties. For example, only hc:Person records can have hc:occupation datatype property while hc:Location records can have geographical information (i.e., hc:easting and hc:northing). A record might also contain "links" (i.e., object properties) to other records such as hc:childOf.

```
hc:23160   hc:title "Angus Macleod"
hc:23160   hc:subjectID 23160
hc:23160   rdf:type hc:Person
hc:23160   hc:sex "Male"
hc:23160   hc:description "Angus Macleod was born in 1916 to 8 Calbost..."
hc:23160   hc:approvedForPub "yes"
hc:23160   hc:bkReference "CEP 2335"
hc:23160   hc:isChildOf hc:23112
hc:23160   hc:isBornAt hc:369
```

**Fig. 1.** Part of a record of a person in Hebridean Connections' dataset

Figure 2 shows some statistics of the Hebridean Connections dataset (as of 28/07/2014), including the total number of records grouped by categories, published records, records with geographical information and records with a description. Note that even though the records have already been digitised and stored in a triplestore, not all of them are available to the public for browsing because parts of the datasets are still under revision. The records are grouped into 17 categories and records about people and places make up a large proportion of the

| Record types | #records | #published | #geog info | #with description |
|---|---|---|---|---|
| Vehicles | 30 | 6 | | 29 |
| Sound files | 39 | 39 | | 39 |
| Gaelic verses | 55 | 51 | | 55 |
| Historical events | 81 | 71 | | 79 |
| Businesses | 82 | 65 | 4 | 69 |
| Natural landscape features | 92 | 64 | 62 | 77 |
| Organisations | 123 | 103 | | 111 |
| Resources | 190 | 88 | | 146 |
| Buildings and public amenity | 206 | 122 | 75 | 142 |
| Objects and artefacts | 216 | 215 | | 215 |
| Stories, reports and traditions | 448 | 435 | | 446 |
| Boats | 487 | 412 | | 450 |
| Locations | 1,137 | 700 | 204 | 757 |
| Landmarks and archaeological sites | 2,255 | 2,221 | 2,198 | 2,243 |
| Croft and Residences | 2,818 | 1,527 | 498 | 1,368 |
| Image Files | 3,226 | 3,185 | | 3,065 |
| People | 32,873 | 17,398 | | 27,883 |
| **Total** | 44,358 | 26,702 | 3,041 | 37,174 |

**Fig. 2.** The Hebridean Connections dataset

dataset. Column **#geog info** shows the number of records with geographical information (e.g., having `hc:easting` and `hc:northing` properties to represent an Ordnance Survey grid reference). Most records also have a description, a human-generated text giving more information about the record. Some of the description might have annotations, i.e., links to other records. However, not all records have a description, as shown in the last column of Figure 2.

# 3   Challenges to Tourist Mobile Apps in Rural Areas

## 3.1   Unreliable Connectivity and Expensive Download Costs

Tourist mobile applications can be grouped into two main groups: *on-the-fly download* and *one-off download* approaches. The on-the-fly download approach provides the most up-to-date information and requires much less initial download. A typical category of applications following this approach is the so-called mobile web-apps. Every time a user requests some information, the application pulls the response onto the device and displays it to the user. This information might or might not be stored on the device. An advantage of this approach is that users generally only download (pay for) what they browse. This approach is best suited for central places of interests such as museums and galleries where mobile Internet connection is generally good or where Wi-Fi connection is available. Unfortunately, in rural areas the apps using this approach will become very unresponsive, or in the worst case, will not work at all. For example, when there is no Internet connection, the Brighton Museum app[5] only prompts a message that users need to have Internet connection to simply open the app.

---

[5] https://play.google.com/store/apps/details?id=com.surfaceimpression.
brightonmuseums

The one-off download approach, in contrast, relies on an assumption that the visitors intentionally use the mobile app for their exploration activities. A typical scenario is that the visitors download the app at home or at places with good Wi-Fi connection and then bring the app with them on-site for offline usages. This approach therefore requires a heavy initial download in order to ensure good user experience in terms of rich contents (more audio, images and videos) and a responsive user interface (all data is kept on the device). Some examples using this approach are the Timespan - Museum Without Walls iOS app[6] and Great Escape Moray published by the National Library of Scotland[7]. These apps require an initial download from 350MB to more than 450MB. Despite bringing rich and layered content to users, the one-off download approach suffers several drawbacks, especially in a rural context. Firstly, visitors need to know about the app before coming to the site. This is suitable for popular tourist sites but not for the small-scaled, remote areas less known for tourist activities. Secondly, as the data is downloaded as a whole, it is not as pertinent for tourists who only have certain interests in the information. For example, the user might only be interested in certain places or people only, and in this instance it would be best for the app to only pre-download such information. Thirdly, applications using this approach cannot cope with frequently updated data. For this approach, updates are done much less often and usually require another heavy data download.

### 3.2   Manual Recommendation in a Large Archive

Given the large total number of records in the archive (about 45 thousands records in which over 3000 are places), it is challenging to choose which ones should be presented to the users. A simple solution is to ask for recommendations from local people. Hebridean Connections, our local partner, helped to recommend 325 records related to the Pairc area of Lewis, Scotland. Of these 325, 64 records are about places of interests but only 55 of these are available to the public due to on-going revisions. Clearly, recommending 325 records (64 places) over 45 thousand records (3000 places) is not a trivial task and requires much time and effort.

In addition to this, even when there is a list of suggested records from local people, it is also not easy to choose which records should be given to the users and/or should be cached in the users mobiles. For example, some records are very general and can be related to many other records such as "World War I", "Pairc Historical Society". For example, a major event such as "World War I" can be associated with hundreds of other records ranging from people to places and stories. However, a user viewing the "World War I" record is not necessarily interested in all several hundreds of related records. In fact, presenting all records would be not only expensive in terms of downloading cost and time, especially in the rural context, but also confusing to users, as there would be too many records to browse.

---

[6] https://itunes.apple.com/gb/app/museum-without-walls-scotlands/id556429487
[7] http://www.nls.uk/learning-zone/great-escapes

### 3.3   Presenting Records without a Description

For such a large, crowd-sourced archive like Hebridean Connections, it is often the case that not all records have a detailed description. Recall from Section 2.3, Figure 2 shows that only 83.8% (37,174 out of 44,358) of the records have a description. This might not be a very important issue if the data are only browsed on the web or shared in a linked data cloud. However, for tourist applications, the lack of detailed, human-readable description will significantly affect the experience of visitors. Fortunately, despite the lack of a human-generated description, these records still have a set of triples with datatype properties and object properties, which can be used to generate a very simple description. Therefore, instead of presenting to visitors a set of raw RDF triples and no description, the system should be able to generate a simple description based on the RDF data of that record.

## 4   The CURIOS Mobile System

### 4.1   System Architecture

The CURIOS Mobile System adopts the client-server software model, as shown in Figure 3. As can be seen, CURIOS Mobile is a component of the whole CURIOS

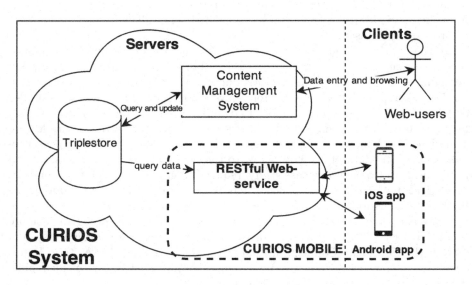

**Fig. 3.** CURIOS Mobile System

System, in which it reuses the database (the triple store) maintained via the CURIOS CMS. This database is wrapped by a RESTful[8] web service, which

---

[8] REpresentational State Transfer (REST) is a web architecture style commonly used for the implementation of web-based APIs [10].

provide data to the mobile devices via an API (Application Programming Interface). By using this software model, the same API/Web-services can be used for multiple mobile platforms: Android, iOS, Windows Phone, etc. However, unlike the traditional client-server software model, where the client side is very thin and only shows results retrieved from the server, the client side in CURIOS Mobile system also caches and stores data in a database. Below we describe briefly the components of the CURIOS Mobile system as well as the main CURIOS system and how they interact to each other.

**The CMS** is not a component of the CURIOS Mobile system in particular, but a key component for the whole CURIOS system, where it allows users (members of historical societies) to enter and curate data. It is also the central place for the public to browse and explore the archive as well as to contribute to the dataset in some way via social media tools such as the commenting system. The CMS uses both a relational database as a back-end for Drupal's internal data (e.g., nodes, entities) and a triplestore to store data (records) generated from users.

**The triplestore** is used for both the CMS and the CURIOS Mobile system. The RDF triples are stored in the Jena TDB persistent storage system. As the same triplestore serves multiple applications (e.g., the CMS and the RESTful web service), Jena Fuseki SPARQL server [11] is used for data retrieval and maintenance. Data held in the triplestore are generated and maintained via updates sent from the CMS. In return, the triplestore answers the requests sent from the CMS and the RESTful web service and hence allow users or client apps to browse the dataset.

In theory, the advantage of Linked Data technologies is to enable open data, and hence the triplestore can be accessed (read-only) by web-users directly without going through the middle-services such as the CMS or the RESTful web service. Therefore, the CURIOS Mobile system or the CMS are just examples of how the linked datasets can be used and reused. In the CURIOS Mobile project, there are two important reasons why direct access to the triplestore should be avoided: access control and data pre-processing. Recall from Section 2.3, parts of the dataset are not visible to the public and some datasets, e.g., in the Hebridean Connections case study, require a complex hierarchy of user-roles and associated permissions, which needs to be implemented in a middleware like the CMS. Another reason is that some of the data are not yet ready to be presented in their current form and hence require some pre-processing before sending back to the client apps/end-users. For example, the coordinate system used in the Hebridean Connections' dataset is the Ordnance Survey Grid Reference (e.g., using easting/northing) while most mapping API/services for mobile devices nowadays use the latitude/longitude system.

**The RESTful web service** provides an easy, flexible way for mobile devices to request and retrieve data from the server.[9] Beside the main role as an API for

---

[9] The RESTful web-service is currently hosted at http://curiosmobile.abdn.ac.uk:8080/CuriosMobile and the API Documentation is given at http://docs.curiosmobile.apiary.io.

mobile clients to access the data stored in the triplestore, this sub-component of CURIOS Mobile also has three more tasks: caching recommended data from the triplestore; generating narratives for records without a description; and pre-processing data before sending to the client apps under JSON format. The web service is also responsible for validating/updating cached entries, e.g., to see if a record has been update recently, and send corresponding updates to the client apps so that the second layer of data caching (see the next section) is kept up-to-date.

**The client apps** include a GUI (Graphical User Interface) for end-users to browse the dataset and a second layer of data caching. More specifically, data retrieved from the RESTful web service will be stored in a SQLite database of the device. This caching layer would optimise the performance of data browsing using the mobile devices, especially with unreliable Internet connectivity. Three main tasks of the client apps are:

- to allow users to view the archived records using the mobile devices,
- to cache records in advance based on users' location and the previous browsing history, and
- to send notifications about places of interest when they approach them.

## 4.2 Data Caching for Rural Settings

As mentioned above, CURIOS Mobile maintains two layer of data caching, one in the RESTful web service and one in the client apps. The first layer is to avoid overhead while querying data against the triplestore as well as to keep track of which records have been out of date. In this section we focus on the second layer of data caching since this is used to overcome the problem of unreliable Internet connection in rural areas.

The caching services can be used to download relevant data when the visitors have good Internet connection such as WIFI or 3G. Relevant data will already be downloaded and stored in client apps even before the visitor arrives at the places of interest. When the visitor goes to rural areas with limited or even no access to the Internet, the app will still be able to use seamlessly. Below we present two caching services implemented in CURIOS Mobile: *Location-based Caching* and *Semantics-based Caching*.

**Location-based Caching** is to cache the records with geographical information (i.e., a pair of easting/northing or latitude/longitude) based on the current user's location. Therefore, only the records located around the geographical area the user is visiting are cached. To adapt different case studies and different geographical areas, we use a parameter, the *euclidean distance* (denoted by $distance_e$) between the user's current location and the record's location, to adjust the level of of caching. Records within the radius of $distance_e$ from the user's current location are cached. However, one should be careful while adjusting $distance_e$ as this parameter should be proportional to the average distance between places of interest to be most effective. For example, if $distance_e$ is much greater than the average distance between places, most places will be cached, and the client apps end up downloading all the dataset even when the visitors are not keen on viewing all the places. In contrast, if $distance_e$ is smaller compared to the range where mobile data connection (signal) is available, it is likely that no record can be downloaded and cached due to no Internet connection.

**Semantics-based Caching.** As the data used within CURIOS Mobile is represented as linked data, it is also possible to perform ahead caching for the related records with respect to what the users have been viewing. We call this style of caching *Semantics-based Caching*. Similar to Location-based Caching, we use a parameterisable distance to control the level of caching needed. We refer to this distance as the *semantic distance*, as defined in Definition 2.

**Definition 2 (Semantic Distance).** *The semantic distance of two CURIOS records $r_{s1}$ and $r_{s2}$ in a triplestore $R$, denoted by $distance_s(r_{s1}, r_{s2})$, is the length of the shortest path in the RDF graph connecting $s1$ and $s2$.*

If two records are directly linked, i.e., $< s1, p, s2 > \in R$ then $distance_s(r_{s1}, r_{s2}) = 1$. However, like location-based caching, it is pertinent that this distance should be chosen carefully to avoid downloading too much data, which might not be relevant enough to users. For example, CURIOS Mobile only downloads and caches records one link away from the viewing record, i.e., $distance_s(r_{s1}, r_{s2}) = 1$. These records will be downloaded in the background process and hence cannot interfere with users current activities. Subject to successful downloads, these records are stored in the app's SQLite database. In reality, some records are general and linked to many other records, and it is not practical to download and store all of them. Therefore, it is necessary to have a mechanism to sort the records based on the level of interest, and then only pick the top of them (i.e., the most interesting ones). This mechanism will be described in detail in the following section.

## 4.3   Auto-Recommendation of Things of Interest

As presented in Section 3.2, it is not trivial to manually choose which records should be presented to users in a large archive like the Hebridean Connections's dataset. To tackle this challenge, we have constructed a *utility function* to assess the level of interest in each record. Firstly, if a user is interested in some records or contents by viewing or searching for them, not only these records but also the related records (selected via the semantic distance in Definition 2) can be the potential candidates for caching. Secondly, to avoid over-caching uninteresting records, related records will be sorted, picked, downloaded and cached based on their level of interest (the result of the utility function).

While assessing the level of interest of a record, there are three factors taken into account: 1) the recommendations from local people, 2) the quality of the record description (based on text length) and 3) the number of links to and from that record, as described in the below equation:

$$Utility(r) = p_1 \times is\_suggested(r) + p_2 \times description\_quality(r) + p_3 \times links\_quality(r)$$

where $r$ is a CURIOS record, $is\_suggested(r)$ is a binary bit representing whether the record has been suggested by locals, $description\_quality(r)$ ranging from 0 to 1 represents the quality of $r$ in terms of its description and $links\_quality(r)$ ranging from 0 to 1 represents the quality of the record $r$ in terms of the links from/to it, and $p_1, p_2, p_3$ the preferences given to each factor of the utility function.

Note that the utility function can be easily modified to adapt different datasets/case studies. For example, in CURIOS Mobile we set $p_1$, $p_2$ and $p_3$ the values of 0.5, 0.3 and 0.2 respectively so that the priority is then given to suggested records and records with higher quality description. For example, a record without a description and which is not suggested by locals will unlikely be selected and cached. Links to and from that record are also taken into account, but with a lower preference. This factor only counts 20% towards the final utility value, because this might add noise to the utility function, particularly in cases where the record is a generic term such as a book, a historical society or an important historical event (e.g., "World War I").

The utility value of a record can be used in a couple of ways. Firstly, the app can have a threshold value to allow some records to be downloaded and cached in the clients apps based on the utility values. This initial set of cached records will be used to give basic information to the user in the beginning. After that, as soon as the user starts to browse this set of records, other records related to this set will also be selected and cached using the semantic distance. Eventually the set of cached records will be growing until there is no Internet connectivity or no user activity. Secondly, when a record is linked to many other records, it is possible to compare the utility values of these linked records to decide which ones should be downloaded and presented to users. In CURIOS Mobile, these tasks are done on the server side, before sending back the record to mobile clients.

### 4.4  Description Generation from RDFs

To present records without a description to users (see Section 3.3), we adopted the *free generation* approach similar to the one in the Triple-Text system [9]. An example of a CURIOS record which only contain RDF triples and no description is shown in Figure 4. Now the general ideas are that each text sentence is generated from an RDF triple and the verbs in these sentences are constructed directly from RDF property names. An advantage of this approach is that it is domain-independent, meaning that this can be used as a generic approach for CURIOS Mobile to exploit any archived linked data generated via the CURIOS system. However, this approach requires the ontology designers to think ahead of properties and concept names so that they can be used later to generate correct verbs and nouns. Narrative generation is done on the server side of CURIOS Mobile (i.e., in the RESTful web service) using the SimpleNLG library [12].

The description generation process for a record has three steps. Firstly, all RDF triples within the record containing irrelevant data such as metadata (e.g., record owner, publication status) are filtered out. Secondly, we build an abstract model of a record from the RDF triples. A record has a title, a type, a set of datatype properties and their values (e.g., `hc:occupation:Teacher`) and a set

```
hc:10738   hc:ownedBySociety   "CEBL"
hc:10738   hc:approvedForPub   "yes"
hc:10738   rdf:type hc:Person
hc:10738   hc:subjectID 10738
hc:10738   hc:title "Margaret Smith"
// datatype properties
hc:10738   hc:sex "Female"
hc:10738   hc:BKReference "CEBL 73"
hc:10738   hc:alsoKnownAs "Mairead Iain Mhoireach"
hc:10738   hc:dateOfBirth hc:Dr6411
hc:10738   hc:dateOfDeath hc:Dr641
// object properties
hc:10738   hc:married hc:10708
hc:10738   hc:livedAt hc:792
hc:10738   hc:livedAt hc:780
hc:10738   hc:isChildOf hc:861
hc:10738   hc:isChildOf  hc:863
hc:10738   hc:isParentOf  hc:5521
hc:10738   hc:isParentOf  hc:6015
hc:10738   hc:isParentOf  hc:4219
hc:10738   hc:isParentOf  hc:17394
hc:10738   hc:informationObtainedFrom  hc:5181
```

**Fig. 4.** RDF triples of the record `hc:10738`

of object properties (e.g., `hc:isChildOf`) and their values' titles. The names of datatype properties and object properties are usually nouns and verbs respectively. In Figure 4, although `hc:dateOfBirth` and `hc:dateOfDeath` are listed as datatype properties, they are in fact object properties linking to a `hc:DateRange` individual. The values of the `hc:DateRange` individual is computed from a a pair of time points. A pair of time points can cover different periods, e.g., a date, a year, a decade, a century, etc. This approach is to deal with inexact dates that occur very frequently in the cultural heritage domain. This also explains why in the generated text there are different formats of "date" (see Figure 5). Thirdly, we generate a description for the record based on this abstract model, sentence by sentence. The generated description of a record is a paragraph consisting of: a sentence describing the record's title and type, a set of sentences generated from the datatype properties, and a set of sentences generated from the object properties. Because datatype properties are usually named as nouns, we used possessive adjectives (e.g., his, her, its) for sentence construction. Similarly, we use pronouns (e.g., he, she, it) for sentences constructed from object properties. All generated sentences are added into a paragraph and realised by SimpleNLG.

Figure 5 shows the description generated from the RDF triples in Figure 4. Datatype properties which are not recognised as a noun such as `hc:alsoKnownAs` are used as complements for the subject in the first sentence. For object properties with multiple values such as `hc:isParentOf` and `hc:isChildOf`, instead of producing multiple similar sentences which cause redundancy and disinterest,

Margaret Smith also known as Mairead Iain Mhoireach is a person.
Her date of death is Thu, 18 Dec 1873. Her date of birth is 1801.
Her bk reference is CEBL 73. Her sex is Female. Margaret Smith is
parent of John Gillies, Henrietta Gillies, Christina Gillies and
Peter Gillies. She is child of John Smith and Catherine Maciver.
She marries Angus Gillies. She lives at Bosta and 1 Earshader. She
informations obtained from Register of Births, Marriages and Death.

**Fig. 5.** The generated description for `hc:10738` from the triples in Figure 4

the RDF triples are aggregated into one sentence. If there are too many values for an object properties (e.g., more than 5), the system only generates the total count and some records, sorted by the level of interest using the utility function mentioned in Section 4.3. It can be seen that this approach cannot deal with object properties whose names are not verbs. For example, `hc:informationObtainedFrom` is translated as a verb "informations obtained from". Also, the articles such as "a/an/the" are also missing.

## 5 Preliminary Results

In this section we present some preliminary evaluation of our implementation of the data caching services. The testing were done at the Pairc area in the Isle of Lewis, Scotland. As shown in Figure 6, three Android-based devices with different settings are used to evaluate the caching services.

| #Device | Network | Device | OS | Caching distance | Test |
|---------|---------|--------|-----|------------------|------|
| Device 1 | 3 UK | Moto G (3G) | Android 4.4.2 | 12km (3G available) | Both |
| Device 2 | O2 UK | Moto G (3G) | Android 4.4.2 | 7km (only GPRS available) | Location-based caching |
| Device 3 | No 3G | Nexus 7 | Android 4.4.3 | Not applicable | Semantics-based caching |

**Fig. 6.** Devices used for testing

We aimed to use Device 1 for testing the combination of both caching services, Device 2 for testing location-based caching only, and Device 3 for testing semantics-based caching only. Device 1 and 2 were able to use data connection (e.g., 3G or GPRS) while Device 3 could only access to WIFI connection and hence Device 3 could not use the location-based caching during the journey. Because 3G data connection is only available within 10-12km of the site and only GPRS is available within 5-7km of the sites, to test the *location-based caching* service, we adjusted the caching distance ($distance_e$ in Section 4.2) to 12km and 7km for Device 1 and 2 respectively. By doing so, Device 1 could use the 3G network for its location-based caching service while Device 2 could only use the slower GPRS for its caching service. Information about how strong the signal would be in different areas along the route can be obtained easily from any operator's online coverage map.

To test the *semantics-based caching* service, before going to the site, users for Device 1 and Device 3 were asked to randomly browse and read records for 15 to 20 minutes. After browsing, the number of cached records for Device 1 and Device 3 were 230 and 128 respectively. The number of cached records in Device 2 remained 0.

During the journey towards the site, Device 1, with a 12km caching distance setting, had increased the number of cached records from 230 to 285 and settled at this number (all 55 recommended places had been cached) at about 10km away from the Pairc area. The user with Device 1, with 285 cached records, could browse most records about places and related things such as people, stories, etc. on site even without an Internet connection. Only 15% of the clicked record could not be rendered, meaning that 85% of clicked records are cached by both caching services.

Device 2 got some messages showing that some of the to-be-cached records had not been downloaded successfully while driving at a speed of approximately 80-90km per hour. When stopped, however, after remaining in a location for several minutes the caching resumed correctly. This is because Device 2 could only use GPRS data connection which is much slower than 3G meaning that the downloads were interrupted while moving fast between places with and without a data connection. At 6km away from the sites, 16/55 records had been cached and when arriving at Ravenspoint, a place within the Pairc area and about 4-5km away from the nearest suggested place, all 55/55 records had been cached in Device 2. This suggests that the location-based caching service worked well, even with a slow data connection such as GPRS.

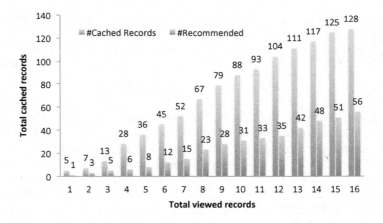

**Fig. 7.** The cache's growth using the semantics-based caching service

For Device 3 (only using *semantics-based caching*), the user was asked to browse the places which they would like to visit on the map as well as the records related to these places randomly. We measured the number of records which the user clicked to view, the number of cached records after viewing the record, and how many of the cached records were recommended by locals. Figure 7 shows the growth of the cache while the records were browsed randomly. Firstly, the cache size (blue column) almost grows linearly to the number of viewed records. This means that the system did not over-cache records and hence could keep the cost of data download low for visitors. Obviously, as a visitor approaches to browsing all records then the cache will settle at some point because there would be nothing left to cache. Secondly, in average, recommended records (red column) count about 35% of the cache, meaning that the semantics-based caching service downloaded a good amount of high quality records, which would be very likely to be viewed by visitors when they arrived at the site. Like location-based caching, records cached using semantics-based caching can be used to send pushed notifications to users while they are passing the places related to the cached records. This process can be done completely offline as only user's current location is required. Therefore, users would be able to read records along the way, even in the areas with no connectivity.

The initial results suggested that although the combination of both caching services worked best, it would still be possible for devices without 3G service like Device 3 to use semantics-based caching to give users information about interesting places (and related records) as they are passing them. As for location-based caching, the results showed that not only 3G but also GPRS data service was sufficient for downloading the records. However, the coverage pattern of Internet availability and the user's moving speed should be taken into account while performing location-based caching.

## 6    Related Work

There have been a significant amount of work on the Semantic Mobile Web. DBPedia Mobile [6] is a location-aware mobile app which can explore and visualise DBPedia resources around the user's location. DBPedia Mobile also maintains a cache of resources on the server side similar to the first layer of caching in CURIOS Mobile. mSpace Mobile [7] is a framework that can retrieve Semantic Web information (RDF) from different sources and visualise the retrieved data to users. However, unlike our approach, in both approaches mentioned above there is no or only minor caching on the client side. These approaches require a reliable Internet connection to operate and hence would be difficult to work in a rural context. In terms of caching services, the most relevant work to our approach is *Who's Who* [8] althought this work does not allows location-based caching. In *Who's Who*, data retrieved from the server side are kept in a local RDF store in the devices and SPARQL queries are used to retrieve requested information from the device's RDF store. When needed, the device will ask for more triples from the server. The main difference between our caching service and *Who's who* is that in the mobile device's caching layer, we use the mobile SQLite database instead of an RDF store. This solution is much simpler for deployment as no RDF store needs to be installed and operated in the device while a good performance can still be achieved. Another difference is that our approaches does use the semantics-based caching service for caching related data in addition to the requested data.

Previous works on natural language generation from linked data mostly focus on using the ontology statements to generate text, such as NaturalOWL [13] or SWAT [14]. These approaches therefore require taking the ontology as an input to generate text. However, this is not possible in CURIOS Mobile, where the server side has access to only the triple store via a SPARQL server but not the OWL ontology. As a result, it is necessary that the description of a record must be generated directly from RDF triples. Therefore we adopted the approach introduced in [9]. Although it was not perfect (e.g., grammatical errors and ontology design commitments), we believe that it is a simple yet acceptable solution to the problem of presenting records without a description.

# 7   Conclusion and Future Work

In this paper we described CURIOS Mobile, a framework to exploit linked data-based archive for tourist activities in rural areas. We showed how our approach can overcome the connectivity issues in rural areas by using different caching services, especially the semantics-based caching which takes advantage of the Linked Data format. In addition, we introduced the mechanisms to recommend things of interest in a large archive such as the Hebridean Connection dataset. Finally, we presented a simple mechanism to generate text from RDF triples for records without a description.

There are two important directions of future work. Firstly, the current CURIOS Mobile system only focuses on the consumption of linked data and hence it would be interesting to explore how to extend the system to allow the production of linked data from visitors. Secondly, we would like to investigate how to improve the utility function and the narrative generation service so that visitors' preferences and activity history can be taken into account.

**Acknowledgements.** The research described here is funded by the SICSA Smart Tourism programme and the award made by the RCUK Digital Economy programme to the dot.rural Digital Economy Hub (award reference: EP/G066051/1). We would like to thank our project partners, Hebridean Connections and Bluemungus, for their comments and support.

# References

1. Berners-Lee, T.: Linked-data design issues. W3C design issue document, June 2009. http://www.w3.org/DesignIssue/LinkedData.html
2. Mäkelä, E., Hyvönen, E., Ruotsalo, T.: How to deal with massively heterogeneous cultural heritage data - lessons learned in CultureSampo. Semantic Web **3**(1), 85–109 (2012)

3. Allinson, J.: Openart: Open metadata for art research at the tate. Bulletin of the American Society for Information Science and Technology **38**(3), 43–48 (2012)
4. Taylor, S., Jekjantuk, N., Mellish, C., Pan, J.Z.: Reasoning driven configuration of linked data content management systems. In: Kim, W., Ding, Y., Kim, H.-G. (eds.) JIST 2013. LNCS, vol. 8388, pp. 429–444. Springer, Heidelberg (2014)
5. Nguyen, H.H., Taylor, S., Webster, G., Jekjantuk, N., Mellish, C., Pan, J.Z., Rheinallt, T.A.: CURIOS: Web-based presentation and management of linked datasets. In: Proceedings of the ISWC 2014 Posters & Demos Track (2014)
6. Becker, C., Bizer, C.: Exploring the geospatial semantic web with dbpedia mobile. Journal of Web Semantics **7**(4), 278–286 (2009)
7. Wilson, M.L., Russell, A., Smith, D.A., Owens, A., Schraefel, M.C.: mSpace mobile: a mobile application for the semantic web. In: End User Semantic Web Workshop (2005)
8. Cano, A.E., Dadzie, A.-S., Hartmann, M.: Who's who – a linked data visualisation tool for mobile environments. In: Antoniou, G., Grobelnik, M., Simperl, E., Parsia, B., Plexousakis, D., De Leenheer, P., Pan, J. (eds.) ESWC 2011, Part II. LNCS, vol. 6644, pp. 451–455. Springer, Heidelberg (2011)
9. Sun, X., Mellish, C.: An experiment on "free generation" from single RDF triples. In: Proceedings of the 11th European Workshop on Natural Language Generation, ENLG 2007, pp. 105–108 (2007)
10. Fielding, R.T.: Architectural Styles and the Design of Network-based Software Architectures. PhD thesis (2000)
11. Seaborne, A.: Fuseki: serving RDF data over HTTP (2011). http://jena.apache. org/documentation/serving_data/ (accessed October 27, 2012)
12. Gatt, A., Reiter, E.: Simplenlg: a realisation engine for practical applications. In: Proceedings of the 12th European Workshop on Natural Language Generation, pp. 90–93 (2009)
13. Androutsopoulos, I., Lampouras, G., Galanis, D.: Generating Natural Language Descriptions from OWL Ontologies: the NaturalOWL System. J. Artif. Intell. Res. (JAIR) **48**, 671–715 (2013)
14. Stevens, R., Malone, J., Williams, S., Power, R., Third, A.: Automating generation of textual class definitions from OWL to English. J. Biomedical Semantics **2** (2011)

# Building of Industrial Parts LOD for EDI
# - A Case Study -

Shusaku Egami[1]($\boxtimes$), Takahiro Kawamura[1], Akihiro Fujii[2],
and Akihiko Ohsuga[1]

[1] Graduate School of Information Systems,
University of Electro-Communications, Tokyo, Japan
{egami.shusaku,kawamura,ohsuga}@ohsuga.is.uec.ac.jp
[2] Faculty of Science and Engineering, Hosei University, Tokyo, Japan
fujii@hosei.ac.jp

**Abstract.** A wide variety of mechanical parts are used as products in the area of manufacturing. The code systems of product information are necessary for realizing Electronic Data Interchange (EDI) of business-to-business. However, each code systems are designed and maintained by different industry associations. Thus, we built an industrial parts Linked Open Data (LOD), which we called "N-ken LOD" based on a screw product code system (N-ken Code) maintained by Osaka fasteners cooperative association (Daibyokyo). In this paper, we first describe building of N-ken LOD, then how we linked it to external datasets like DBpedia, and built product supplier relations in order to support the EDI.

**Keywords:** Linked Data · Linked Open Data · Semantic Web · Manufacturing

## 1 Introduction

Japanese LOD is increasing every year; for example, regional information, disaster information, sub-culture, and more domains, whereas LOD is still not familiar in the manufacturing area. The spread of LOD will contribute to development of information technology in the manufacturing area. For example, developers can build applications utilizing LOD, such as publishing data about productive relationship and product information. By these applications, the relation between products and external data are visualized and they may lead to sales support that links to Electric Commerce (EC) site, EDI, and the creation of new business. In our previous work [1] [2], we focused on the mechanical parts distribution in the manufacturing area, and built a fastener LOD (N-ken LOD). Specifically, we built LOD based on a screw product code system (N-ken code) for EDI maintained by Osaka fasteners cooperative association (Daibyokyo[1]). Furthermore, we developed an application that links CAD to think of utilization of N-ken LOD.

---

[1] http://www.daibyokyo.com/

© Springer International Publishing Switzerland 2015
T. Supnithi et al.(Eds.): JIST 2014, LNCS 8943, pp. 146–161, 2015.
DOI: 10.1007/978-3-319-15615-6_11

In order to populate N-ken LOD, however, it is necessary to raise utility; for example, increasing links as additional information and product information. Thus, we describe a method to improve the utility of the LOD in addition to our building method of N-ken LOD in this paper. Specifically, we describe a method for linking DBpedia Japanese [3] based on similarity of labels, and building of product supplier relations. Especially, we describe a procedure to resolve contradiction between the structure of N-ken LOD and external datasets.

The rest of the paper is organized as follows. In section 2, an overview of Japanese LOD is described. In section 3, N-ken LOD is explained. In section 4, a method for linking to DBpedia Japanese is described. In section 5, a method for building of product supplier relations is described. In section 6, evaluation results and discussion are presented. Finally, in section 7, we conclude this work with future works.

## 2  Related Work

In recent years, several kinds of existing data have been converted to Resource Description Framework (RDF) and published on the Web as "Linked" Open Data in the world. In Japan, there are also several projects to publish dataset as LOD; for example, Yokohama Art LOD, Linked Open Data for ACademia (LODAC) Museum [4], DBpedia Japanese, Japanese Wikipedia Ontology [5], and many more. Also, Linked Open Data Challenge Japan has been held every year since 2011. This challenge increases datasets for LOD, applications using LOD and developer tools. Some applications that support for application development using LOD are also published, such as LinkData.org [6] provided by RIKEN and SparqlEPCU [7] provided by Chubu University. However, building of LOD is not progressing in the industrial area. Graube et al. [8] have proposed an approach that integrating the industrial data using Linked Data, but have not published LOD. Our N-ken LOD is only LOD in the industrial area.

An approach of generating RDF from tabular data is Direct Mapping. This approach generates RDF from rows as subject, columns as property, and values of cells as object. D2R Server [9] can output RDF from relational database such as MySQL. W3C recommended RDB to RDF Mapping Language (R2RML)[2] for mapping relational databases to RDF in more detail. In addition, there are tools to generate RDF from tabular data such as RDF Refine[3]. RDF Refine is an extension of Google Refine that is a data cleaning tool. However, if we used these approaches for N-ken Code, we cannot provide intended RDF, because N-ken Code consists of multiple multidimensional tabular data. Hence, we generated RDF using Apache POI[4] and Apache Jena[5].

Exact matching and measuring of string similarity are commonly-used techniques for ontology alignment (e.g. TF-IDF, cosine similarity, edit distance, Jaro-Winkler, and more) [10]. Moreover, there is an approach utilizing synonymous

---

[2] http://www.w3.org/TR/r2rml/

[3] http://refine.deri.ie/

[4] http://poi.apache.org/

[5] https://jena.apache.org/

relations such as 'owl:sameAs' and 'skos:exactMatch' [11]. These approaches are used when ontology and LOD are dynamically changing and their internal links or texts are rich. However, N-ken LOD has less text and internal links, and thus it was difficult to increase links between heterogeneous LOD using these approaches alone. In this paper, therefore, we propose an approach of instance matching with high precision, in the case that LOD has little text and internal links, and matching instances are little in other LOD.

## 3    N-Ken LOD

N-ken LOD is LOD based on a screw product code system for EDI called "N-ken Code" which is owned by Daibyokyo. The project of "N-ken Code" started aiming to develop a common code by a volunteer research group of screw Business-to-Business information processing in Higashi-Osaka, Japan. The group was organized by about 30 screw trading companies gathered from Osaka, Kyoto and Nara in order to develop a common EDI system. Since 1997, this group has worked with coding unified screw names and related matters, enactment of the protocol, determination of communication means, and implementation of the core system concept for the introduction of EDI. In 2004, Web-EDI was built in ASP; it has been used in the group member companies. In 2011, this group was dissolved and N-ken Code is now maintained by Daibyokyo.

### 3.1    Building of N-Ken LOD

N-ken Code consists of structure of product classifications, classification code, trade name code, form notation name, nominal designation, and others in tabular data. Figure 1 shows the structure of product classifications. According to product types, they are classified into four classes, large classification, middle classification, small classification, and subtyping. As indicated by the arrows, however, there is the case that they are classified into three classes of large classification, medium classification, and subtyping, or large classification, small classification, and subtyping. Furthermore, they may be classified into two classes of large classification and subtyping in some cases. Finally, the products are concluded as the trade name and the abbreviation of the form name. We converted this data to RDF based on the basic principles of Linked Data, and we are publishing it on the Web as N-ken LOD[6].

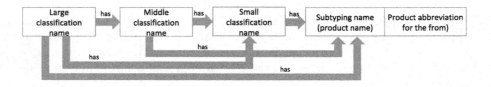

**Fig. 1.** Structure of product classifications

---

[6] http://monodzukurilod.org/neji/

**Table 1.** General kind master (Cross-recessed round head machine screw H-type JIS)

| Trade name code | NMCPA | | | |
|---|---|---|---|---|
| Large classification name | Machine screw | | | |
| Middle classification name | Cross-recessed machine screw | | | |
| Small classification name | Cross-recessed round head machine screw | | | |
| Subtyping name | Cross-recessed round head machine screw H-type JIS | | | |
| Screen notation name | (+) round head machine screw H-type/J | | | |
| Form notation name | (+) round head machine screw H-type/J | | | |
| Product classification code | 01 | 01 | 01 | 01 |

**Table 2.** General product master(Cross-recessed round head machine screw H-type JIS)

| Subtyping name | Cross-recessed round head machine screw H-type JIS | | | |
|---|---|---|---|---|
| Form notation name | (+) round head machine screw H-type/J | | | |
| Trade name code | NMCPA | | | |
| Product classification code | 01 | 01 | 01 | 01 |
| Nominal designation | M1.6X888 | | | |
| | ... | | | |
| | M10X888 | | | |

We explain our approach of generating RDF from N-ken Code based on "cross-recessed round head machine screw H-type JIS" as an example. Table 1 shows the general kind master. Table 2 shows the general product master. Due to space limitation, some rows and columns of the original data are omitted.

First, trade name code, large classification, middle classification, small classification, subtyping, screen notation name, form notation name, and product classification code are extracted from the general kind master. Also, nominal designations are extracted from the general product master. The product classification is composed of the numeric string of up to eight digits by 1-4 frames of two-digit numbers. Large classification code is entered in the first frame, middle classification code is entered in the second frame, small classification code is entered in the third frame, and subtyping code is entered in the fourth frame. The Uniform Resource Indicators (URIs) of resources are created based on the product classification code, because all entities should be presented by a unique URI in RDF. The URI of the resource of each classification is defined as follows.

> http://monodzukurilod.org/neji/resource/{classification code}

The prefix of this URI is "neji", and that means "screw" in Japanese. Table 3 shows examples of URI and name of resources. Figure 2 shows an RDF graph of N-ken LOD representing "cross-recessed round head machine screw H-type JIS".

**Table 3.** Examples of URI and name of resources

| URI(QName) | Name |
|---|---|
| neji:01 | Machine screw |
| neji:0101 | Cross-recessed machine screw |
| neji:010101 | Cross-recessed round head machine screw |
| neji01010101 | Cross-recessed round head machine screw H-type JIS |

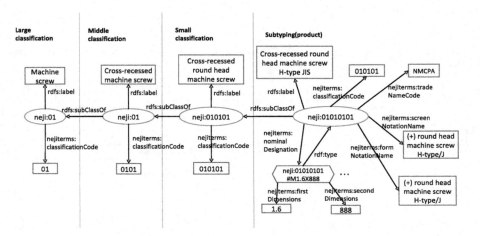

**Fig. 2.** RDF graph (Cross-recessed round head machine screw H-type JIS)

## 4   Linking to DBpedia Japanese

### 4.1   Linking to Resources of Additional Data

N-ken LOD includes links to resources of DBpedia Japanese as an external LOD. However, the linking method was a simple exact matching of label strings, and as a result, the number of linked resources were very little. Thus, we employ a similarity measure to increase the links.

First of all, we collect resources that are classified as "manufacturing" and "machine element" category from a SPARQL endpoint of DBpedia Japanese. Regarding each resource of N-ken LOD, we calculate similarities with all resources that we collected from DBpedia Japanese. The similarity is calculated by the Equation 1, that have some weight parameters on classical string-distance measure.

$$sim1(l_a, l_b) = 1 - \left( \frac{EditDist(l_a, l_b)}{max(|l_a|, |l_b|)} \cdot \alpha \cdot \beta \right) \tag{1}$$

$l_a$ is a 'rdfs:label' of the resource of N-ken LOD, $l_b$ is a 'rdfs:label' of the resource of DBpedia Japanese. $EditDist(l_a, l_b)$ is the edit distance of $l_a$ and $l_b$, and it is the minimum number of editing operations needed to make $l_a$ and $l_b$ the same string. There are two types of the edit distance. One is the type, where each cost of deletion, insertion, and substitution is 1. The other is the type, where the cost of the substitution is 2. We use the latter in this method. $\alpha$ and $\beta$ are weights to increase precision. When we caluculate the edit distance between $l_a$ and $l_b$, coincidental characters can appear at multiple points. In this case, Longest Common Subsequence (LCS) between $l_a$ and $l_b$ is divided into multiple sets. Therefore, in the case that LCS between $l_a$ and $l_b$ is divided into more than three sets, we set the value of $\alpha$ to 1.5. For the other case, the value of $\alpha$ is 1. Even if the similarity based on the edit distance is high, in fact, there is the case that the candidate resource of DBpedia Japanese has no relation with the screw. Therefore, in the case that the candidate resource of DBpedia Japanese does not link to the resource that belongs to the "screw" category, we set the value of $\beta$ to 1.5. For the other case, the value of $\beta$ is 1. N-ken LOD has a variety of industrial parts, but we selected "screw" category because most resources are screw-related in N-ken LOD. We altered 'owl:sameAs' to 'rdfs:seeAlso' as a link property. Linking by 'owl:sameAs' ensures consistency, even if all properties of two resources are replaced. Since the resources of N-ken LOD and linked candidate resources of DBpedia Japanese are not identical, we assumed that the use of 'owl:sameAs' is incorrect and used 'rdfs:seeAlso'. By using 'rdfs:seeAlso', we linked resources of N-ken LOD to resources of DBpedia Japanese, so that additional information can be obtained.

### 4.2 Selecting a Property in Consideration of Category

In the method of the previous section, there are the cases that N-ken LOD is incorrectly linked to the resources of DBpedia Japanese such as materials and processing methods. We show an example that is incorrectly linked as follows.

| Polyacetal_wing_bolt | rdfs:seeAlso | Polyacetal . |
| Knurled_nut | rdfs:seeAlso | Knurl . |

To avoid these incorrect links, we defined properties of materials and processing methods. First of all, we investigate two level deep categories of the linking candidate resource of DBpedia Japanese. If the resource is classified into the material category, we use 'nejiterms:material' as a link property. Also, if the resource is classified into the processing category, we use 'nejiterms:processing Method' as a link property. If the resource is classified into both the material category and the processing category, we select a lower category and use an appropriate property. If the resource is classified into a category of "screw", "machine element", "joining", or "fastener", we use 'rdfs:seeAlso' as linking property. This process is outlined in Algorithm 1.

---

**Procedure 1.** Calculate Similarity

---

**Input:** $NkenLODList, DBpediaList$
**Output:** $RDF$
1: $maxSim \Leftarrow 0$
2: $RDF \Leftarrow null$
3: **for** each label $l_a \in$ NkenLODList **do**
4:   **for** each label $l_b \in$ DBpediaList **do**
5:     $/ * get\ edit\ distance * /$
6:     $s1 \Leftarrow EditDist(l_a, l_b)$
7:     $/ * get\ the\ number\ of\ LCS\ divisions * /$
8:     $s2 \Leftarrow getNumCS(l_a, l_b)$
9:     **if** $s2 \geq 3$ **then**
10:       $alpha \Leftarrow 1.5$
11:     **else**
12:       $alpha \Leftarrow 1$
13:     **end if**
14:     $sim \Leftarrow 1 - (s1/max(l_a.length, l_b.length) * alpha)$
15:     **if** $sim > maxSim$ **then**
16:       $/ * get\ categories\ of\ l_b * /$
17:       $cat \Leftarrow getCategories(l_b)$
18:       **if** $cat$ contains $material$ **then**
19:         $property \Leftarrow$ "$nejiterms : material$"
20:         $maxSim \Leftarrow sim$
21:         $result \Leftarrow l_b$
22:       **else if** $cat$ contains $process$ **then**
23:         $property \Leftarrow$ "$nejiterms : processingMethod$"
24:         $maxSim \Leftarrow sim$
25:         $result \Leftarrow l_b$
26:       **else**
27:         $/ * investigate\ whether\ l_b\ links\ to$
28:           $the\ resource\ that\ belongs\ to$ "$screw$" $category * /$
29:         $w \Leftarrow getNumWikiPageWikiLink(l_b)$
30:         **if** $w = 0$ **then**
31:           $beta \Leftarrow 1.5$
32:         **else**
33:           $beta \Leftarrow 1$
34:         **end if**
35:         $sim \Leftarrow 1 - ((1 - sim) * beta)$
36:         **if** $sim > maxSim$ **then**
37:           $property \Leftarrow$ "$rdfs : seeAlso$"
38:           $maxSim \Leftarrow sim$
39:           $result \Leftarrow l_b$
40:         **end if**
41:       **end if**
42:     **end if**
43:   **end for**
44:   **if** $result\ != null$ **then**
45:     $/ * create\ triple * /$
46:     $triple \Leftarrow link(l_a, result, property)$
47:     $RDF.add(triple)$
48:   **end if**
49: **end for**

---

### 4.3   Linking to Broader Concepts in Consideration of Synonymous Relations

All articles in Japanese Wikipedia are classified into some categories. Index of the articles summarized by area is described in a category page. DBpedia Japanese defines the resource of the category page as 'skos:Concept'. The resource of DBpedia Japanese corresponding to the article in Japanese Wikipedia is classified into the category resource by 'dcterms:subject'. The large classification's resource is top-level concepts in N-ken LOD, and there are 27 resources. Each large classification has a tree. By selecting the category resource of DBpedia Japanese as the broad concept of the large classification, reasoning through the category of DBpedia Japanese is possible and it increases the utility of LOD. In the following, the method to link the category resource of DBpedia Japanese as the broader concept is described.

**Process 1.** We use the method to large classification resources that are linked to a resource of DBpedia Japanese by 'rdfs:seeAlso'. We assume that categories of the DBpedia Japanese resource that have rdfs:seeAlso relations with the large classification resource are synonymous with the broader concept of the large classification resource. Also, we investigate whether the DBpedia Japanese resource that has 'rdfs:seeAlso' relations with the large classification resource has been classified into "machine element", "manufacture", or these subcategory. If it is true, this category is linked as a broader concept of large classification resources of N-ken LOD. Figure 3 shows the process of linking to category resources of DBpedia Japanese in consideration of 'rdfs:seeAlso' relations using washer as an example.

**Process 2.** In the rest of large classification resources, the broader concepts of resources, where the end of the label is "screw", "bolt", or "nut" are defined "screw" category, and broader concepts of resources, where the end of the label is "parts" are defined "machine element" category. Also, the broader concepts of resources, where the end of the label is "rivet" are defined "joining" category based on the result of "rivet" in process 1. The other broader concepts of large classification are defined "fastener" category.

## 5   Building of Product Supplier Relations

It is considered that linking of N-ken LOD and product data of companies enables to apply N-ken LOD to EDI and Electronic Ordering System (EOS). However, most companies have not published product information as machine-readable format such as eXtensible Markup Language (XML) and Comma-Separated Values (CSV). In contrast, most companies have published product information as HTML. We, thus extracted product information from the website, and built product supplier relations.

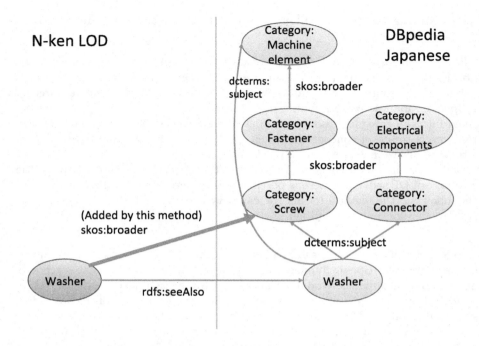

**Fig. 3.** The method that links to category resource of DBpedia Japanese

## 5.1    Getting product pages

First, we get a list of websites of fastener companies. The list consists of member companies' websites in Daibyokyo and websites classified into the following categories in "Yahoo! Category".

1. "nails, screws, and fasteners" category
2. All subcategories and descendants of (1)
3. All shortcut categories of (2)
4. All subcategories and descendants of (3)

Shortcut category is a category that does not exist in current category, and exists in another category and is related to the current category.

Second, we retrieve texts of pages that describe product information from websites in the list. Description of the product information is different in each website. There are websites that describe all product information on the top page, and websites that have a product catalog page. Thus, we retrieve text of the top page, product page, and child page of the product page in order to extract trade name from websites as much as possible. We assumed that the product page is the page linked from the top page by the anchor link that contains the string of "goods", "products", or "catalog".

## 5.2   Searching Trade Names

In many cases, products of screw and fastener have different names, even if they are the same meaning. Therefore, simple exact matching cannot extract the trade name from text of the product page. Using N-gram and the measuring of the string similarity, we search the trade name that is similar to a product resource (subtyping resource) of N-ken LOD from retrieved texts. N-gram splits the text into N-characters and makes the index of appearance position as heading of the split string. Also, N-gram splits the search term into N-characters. By the search based on the index, it is possible to search the string without omission. Since N-gram is the method of string search for an exact matching, it is not possible to search only product that exactly matches the name of the product resource of N-ken LOD. We use the measuring of string similarity when matching string, to search the trade name similar to the name of the product resource of N-ken LOD. However, since the target is not DBpedia Japanese, we cannot use Equation 1. Since the data is related to companies, we assumed that the high precision is required. Therefore, we measure string similarity using Equation 2.

$$sim2(l_a, l_b) = 1 - \frac{EditDist(l_a, l_b) - Trans(l_a, l_b)}{LCS(l_a, l_b)} \qquad (2)$$

$LCS(l_a, l_b)$ is LCS between $l_a$ and $l_b$. $Trans(l_a, l_b)$ is the cost of strings transposition. The string transposition is possible, when the same string is contained in the both strings that is $l_a$ and $l_b$. But the positions are different. In this method, we consider only the string consisting of two or more characters, and the cost is 1 per character. We assumed that use of LCS increases the precision, and the string transposition increases the recall.

## 5.3   Generate Triples

If the target string is found, we generate triples using the GoodRelations[7] that is the ontology for e-commerce, and add to the N-ken LOD.

```
@prefix rdf: <http://www.w3.org/1999/02/22-rdf-syntax-ns#> .
@prefix gr: <http://purl.org/goodrelations/v1#> .
@prefix rdfs: <http://www.w3.org/2000/01/rdf-schema#> .
@prefix foaf: <http://xmlns.com/foaf/0.1/> .
<http://monodzukurilod.org/neji/resource/{company      name}>      rdf:type
gr:BusinessEntity ;
rdfs:label "company name"@ja ;
gr:legalName "company name";
gr:offers <http://monodzukurilod.org/neji/resource/{classification code}> .
foaf:page <http://example.com/> .
```

---

[7] http://purl.org/goodrelations/v1

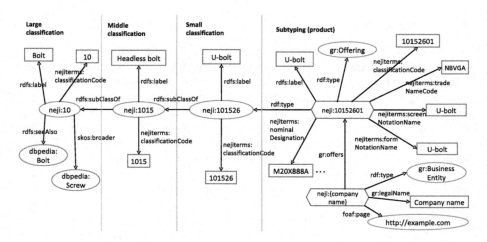

**Fig. 4.** RDF graph that is finally generated (U-bolt)

The 'gr:offers' property defines 'gr:BusinessEntitiy' class as domain, and defines 'gr:offering' class as range. Therefore, it is necessary to define 'gr:offering' as a 'rdf:type' of product resources that is range of 'gr:offers'. However, since a product resource is a subclass of a small classification, contradiction occurs if the type of product resource is defined as 'gr:offering'. Hence, we resolve contradiction by defining the type of a product resource as a small classification class and 'gr:offering'. Also, we delete instances of product resources and add the property of nominal designation to product resources. Figure 4 shows an RDF graph that was built based on "U-bolt" as an example.

# 6    Evaluation and Discussion

In this section, we evaluate the linking method described in section 4 and the building method of product supplier relations described in section 5.

## 6.1    Evaluation of Linking Accuracy to DBpedia Japanese

Table 4 shows the results of linking to DBpedia Japanese using the method described in section 4. Resources, where similarity is more than 0.2 are linked. In DBpedia Japanese, there is little information of screw and fastener. Therefore, it is unlikely that all resources of N-ken LOD link to the corresponding resources of DBpedia Japanese. Also, it is difficult to investigate all of the resources corresponding to the resource of N-ken LOD from DBpedia Japanese. Therefore, we randomly sampled 500 resources from N-ken LOD and manually checked these resources that have a corresponding to resource of DBpedia Japanese. We determined the number of resources that have a corresponding resource of DBpedia

**Table 4.** Results of linking to DBpedia Japanese

| Method | Correct triple | Precision | Recall | F-measure |
|---|---|---|---|---|
| EditDistance | 1502 | 0.57 | 0.46 | 0.51 |
| Method1 | 1426 | 0.80 | 0.44 | 0.56 |
| Method2 | 1456 | 0.76 | 0.45 | 0.56 |

**Table 5.** Examples of incorrect links

| N-ken LOD | DBpedia Japanese | Property |
|---|---|---|
| Bolt for chemical anchor | Chemical anchor | rdfs:seeAlso |
| Hexagon nut with flange | Flange | rdfs:seeAlso |
| Concrete drill | Concrete | material |
| TRF drill screw | Drill_(tools) | processingMethod |

Japanese as the number of correct answers. Method1 is the method that links to additional information. Method2 is the method that links to material category and process category. Precisions of Method1 and Method2 are higher than that of classic approach, and F-measures of Method1 and Method2 are higher than that of classic approach. Also, the recall of Method2 is higher than that of Method1, but the precision is lower than that of Method2. Therefore, the F-measure did not change. Recalls are low, even if any method is used. The reason is that there are many resources in N-ken LOD, which have the label including modifiers, such as standard name, application, and other parts name. Table 5 shows examples of incorrect links.

The modifier part of label such as "chemical anchor" and "flange" is longer than the part representing the classification of the resource, such as "bolt" and "nut". As a result, similarity with the modifier becomes higher and incorrect links increase. If the modifier is about material and processing method, we succeed to filter incorrect links. But, we could not filter with respect to such application and partial names, and it is the reason to lower the precision and recall. In addition, there are incorrect links in terms of material and processing method. For example, "concrete drill" (rotary hammer drill) is linked to "concrete" as material, but in fact, it is a drill to be used for concrete. In DBpedia Japanese, the concrete is classified into the concrete category and composite material category. Composite material category is subcategory of material category. Thus, since there are words used as an application in N-ken LOD, even if they have been classified into the material category in DBpedia Japanese, they are linked to incorrect entityAlso, "TRF drill screw" is linked to "drill_(tools)" as a processing method, but "drill_(tools)" is not a processing method. In DBpedia Japanese, process category includes many process-related resources and category such as processing method, tool, and work pieces. Therefore, there is a case of incorrectly linking to a resource that is not a processing method. It is considered to solve these problems, when it gets resources by selecting a category from DBpedia

Japanese and filtering, it is necessary to select as lower category as possible. In order to solve these problems, it is considered necessary to select as lower category as possible, when getting resources by selecting a category from DBpedia Japanese. Because there is a problem in the hierarchical structure of the DBpedia Japanese and Japanese Wikipedia, it is considered necessary to modify the hierarchy.

## 6.2    Evaluation of Building of Product Supplier Relations

We built product supplier relations using the method described in 5.2. We set the threshold to 0.66. In this method, we get 210 websites of fastener companies from Yahoo! Category and Daibyokyo. As a result, we were able to build product supplier relations from 66 websites. It is considered that one of the reasons that the number of websites largely decreased is that there are many companies which are not member of Daibyokyo in Yahoo! Category. It is considered that since N-ken LOD has been built based on N-ken Code, we could not extract trade names which are defined by the fasteners cooperative association in Tokyo, Kanagawa, and Aichi. The recall was low, whereas the precision is high. This reason is that there are many websites, whose product pages are not retrieved by our crawler. Structure of the websites vary from company to company. In this method, we assumed that a product page is linked from the top page by the anchor link that contains the string of "goods", "products", or "catalog", but, there are many websites, whose product pages are linked by images. Therefore, the crawler could not retrieve those product pages and extract trade names. This caused the low recall.

## 6.3    Discussion

In this paper, we described a method for expanding N-ken LOD in order to enhance utility of LOD. In N-ken LOD, since there is no description of resources and 'owl:sameAs' link, it was difficult to use the existing matching approach. Thus, we linked N-ken LOD to DBpedia Japanese by the method proposed in this paper. Since there are not enough resources related to screw and fastener in DBpedia Japanese, increasing the recall was difficult, but we could get the high precision. It is considered that a dictionary of screw and fastener is necessary to increase precision and recall furthermore.

N-ken LOD has been included in Japanese Linked Data cloud diagram[8] shown in Figure 5. Japanese Linked Data cloud diagram is a diagram that visualizes the connection of Japanese LOD to the following criteria:

1. Data publisher is a person on organization in Japanese
2. There is a label of Japanese in the dataset.
3. The dataset must contain at least 1000 triples.
4. The dataset must be connected via RDF links to a dataset that is already in the diagram, and require at least 10 links.

---

[8] http://linkedopendata.jp/?p=486

**Fig. 5.** Japanese Linked Data cloud diagram[12]

5. Access of the entire dataset must be possible via http (https), via an RDF dump, or via a SPARQL endpoint.

For criteria included in the original LOD cloud diagram[9], the dataset must be connected via RDF links to a dataset that is already in the diagram, and requires at least 50 links. N-ken LOD before this research did not apply to this criteria. By the result of this research, the number of links with external datasets was increased to 1456. Therefore, N-ken LOD was included in Japanese LOD cloud diagram criteria original LOD cloud shown in Figure 6. As can be seen from Figure 5,6, Japanese LOD in the industrial area is only N-ken LOD, and there is room for the spread of LOD in the industrial area. This is due to the fact that Open Data about enterprise information is not well developed in Japan. We believe that creating new services will be published by increasing industrial LOD

---

[9] http://lod-cloud.net/

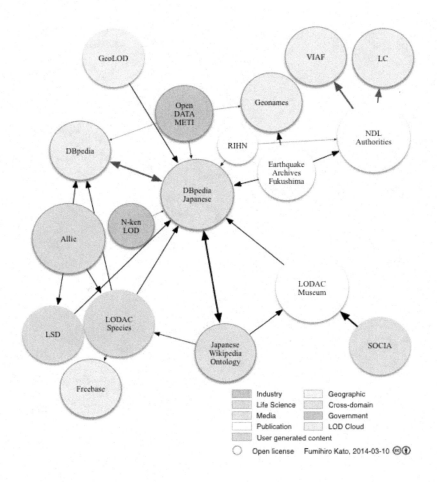

**Fig. 6.** Japanese LOD cloud diagram (original criteria)[12]

and linking them to N-ken LOD in the future. Then, understanding of usefulness of these services will promote Open Data in the industrial area.

## 7    Conclusion and Future Work

In this paper, we described a method for building and expanding the industrial parts LOD, called N-ken LOD. It is considered that N-ken LOD will enable the development of information systems to provide new services by linking to industry LOD in the future.

We extracted product information from several companies' websites, and we built product supplier relations. But it was difficult to extract all trade names, because the structures of websites related to a fastener are very different. In the future, we will improve the quality of the crawler and try to expand product supplier relations. Also, we plan to address the construction of other LOD in the industrial area.

**Acknowledgments.** This work supported by MEXT KAKENHI Grant Number 26330349. We are deeply grateful to Osaka fasteners cooperative association that has provided us with data.

# References

1. Egami, S., Shimizu, H., Taniguchi, S., Fujii, A.: Mashup Applications Based on Screw LOD. IEICE Technical Report **113**(178), 13–18 (2013)
2. Fujii, A., Egami, S., Shimizu, S.: EDI support with LOD. In: Proceedings of the Joint International Workshop: 2013 Linked Data in Practice Workshop (LDPW 2013) and the First Workshop on Practical Application of Ontology for Semantic Data Engineering (PAOS 2013). CEUR Workshop Proceedings (2013)
3. Kato, F., Takeda, H., Koide, S., Ohmukai, I.: Building DBpedia Japanese and linked data cloud in Japanese. In: Proceedings of the Joint International Workshop: 2013 Linked Data in Practice Workshop (LDPW 2013) and the First Workshop on Practical Application of Ontology for Semantic Data Engineering (PAOS 2013). CEUR Workshop Proceedings (2013)
4. Matsumura, F., Kobayashi, I., Kato, F., Kamura, T., Ohmukai, I., Takeda, H.: Producing and consuming linked open data on art with a local community. In: Proceedings of the Third International Workshop on Consuming Linked Data (COLD 2012). CEUR Workshop Proceedings (2012)
5. Morita, T., Sekimoto, Y., Tamagawa, S., Yamaguchi, T.: Building up a class hierarchy with properties from Japanese Wikipedia. In: IEEE/WIC/ACM International Conference on Web Intelligence (2012)
6. Shimoyama, S., Nishikata, K., Yoshida, Y., Toyoda, T.: Promotion and education of open data by creation and hosting RDF on LinkData.org repository. In: The 26th Annual Conference of the Japanese Society for Artificial Intelligence (2012)
7. Toshioka, K., Fujimoto, R., Yasue, S., Suzuki, Y.: A WEB Application Development Framework SparqlEPCU for RDF Store - Knowledge Warehousing and Interchange with XML/RDF/Ontology. Information Science Research Journal **20**, 3–16 (2013)
8. Graube, M., Pfeffer, J., Ziegler, J., Urbas, L.: Linked Data as integrating technology for industrial data. In: The 14th International Conference on Network-Based Information Systems (2011)
9. Bizer, C., Cyganiak, R.: D2R server - publishing relational databases on the semantic web. In: Poster at the 5th International Semantic Web Conference (2006)
10. Cheatham, M., Hitzler, P.: String similarity metrics for ontology alignment. In: Alani, H., et al. (eds.) ISWC 2013, Part II. LNCS, vol. 8219, pp. 294–309. Springer, Heidelberg (2013)
11. Gunaratna, K., Thirunarayan, K., Jain, P., Sheth, A., Wijeratne, S.: A statistical an schema independent approach to identify equivalent properties on Linked Data. In: 10th International Conference on Semantic Systems (2013)
12. Kato, F.: Japanese Linked Data Cloud diagram 2014-03-10 (2014). http://linkedopendata.jp/?p=486

# inteSearch: An Intelligent Linked Data Information Access Framework

Md-Mizanur Rahoman[1(✉)] and Ryutaro Ichise[1,2]

[1] Department of Informatics,
The Graduate University for Advanced Studies, Tokyo, Japan
mizan@nii.ac.jp
[2] Principles of Informatics Research Division,
National Institute of Informatics, Tokyo, Japan
ichise@nii.ac.jp

**Abstract.** Information access over linked data requires to determine subgraph(s), in linked data's underlying graph, that correspond to the required information need. Usually, an information access framework is able to retrieve richer information by checking of a large number of possible subgraphs. However, on the fly checking of a large number of possible subgraphs increases information access complexity. This makes an information access frameworks less effective. A large number of contemporary linked data information access frameworks reduce the complexity by introducing different heuristics but they suffer on retrieving richer information. Or, some frameworks do not care about the complexity. However, a practically usable framework should retrieve richer information with lower complexity. In linked data information access, we hypothesize that pre-processed data statistics of linked data can be used to efficiently check a large number of possible subgraphs. This will help to retrieve comparatively richer information with lower data access complexity. Preliminary evaluation of our proposed hypothesis shows promising performance.

**Keywords:** Linked data · Information access · Data access complexity · Data statistics

## 1 Introduction

Linked data has opened great potentiality of building large knowledge-base. According to the lod-2 linked open data statistics[1], currently linked open data hold 2122 datasets, among them 928 datasets are accessible which consist more than 61 billion RDF triples. Therefore, it is well understood that linked data presently contain a vast amount of knowledge that underscores the need for good and efficient data access options. Usually linked data adapt graph-based data structure. Therefore, like other graph-based information access systems, contemporary linked data information access systems try to determine subgraph that corresponds the required information need. However, on the fly checking of a large number of possible subgraphs increases the information access complexity.

---

[1] http://stats.lod2.eu/, accessed on April 9, 2014.

© Springer International Publishing Switzerland 2015
T. Supnithi et al.(Eds.): JIST 2014, LNCS 8943, pp. 162–177, 2015.
DOI: 10.1007/978-3-319-15615-6_12

This makes the information access frameworks less effective. A large number of contemporary linked data information access frameworks reduce the complexity by introducing different heuristics. However, some of them do not care about the complexity. Whatever the cases, a practically usable framework should retrieve richer information with lower complexity.

Over the contemporary linked data information access systems, the subgraph checking heuristics can be largely categorized into two categories. The first category systems (say the Language Tool-based Systems) use natural language tools to determine the possible subgraphs. The second category systems (say the Pivot Point-based Systems) track a Pivot Point (e.g., a particular keyword from the input query); then, from that Pivot Point, they try to explore the other points and thereby determine possible subgraphs. PowerAqua[11], TBSL [19], FREyA[3], SemSek[1], CASIA[18] etc. belong to the Language Tool-based Systems while Treo[7], NLP-Reduce[10] etc. belong to the Pivot Point-based systems. However, both category systems possess some drawbacks.

In the Language Tool-based Systems, input query usually is parsed by language tools such as language parsers. In such a case, the language parser outputs, which generally are parse trees, are used to predict the possible subgraph (over linked data graph). A parse tree is a tree with tagged[2] part of input query. Then by fitting the parse tree into linked data's <Subject, Predicate, Object> like structure, the Language Tool-based Systems try to predict possible subgraph over linked data graph. Over linked data, this <Subject, Predicate, Object> like structure is called an RDF triple. Therefore, an RDF triple could be considered as the basic part of linked data graph and is presented as <<Source Node, Edge, Destination Node>>.

However, the Language Tool-based Systems sometime predict inappropriate or misleading subgraphs. This is because, in most of the cases, language tools generate multiple parse trees. Or, they sometime tag input keywords with wrong tags. For example, consider an input query "Japanese national whose spouse born in Germany" which holds a Concept "Japanese national", two Predicates "spouse" and "born", and an Individual Entity "Germany". Over linked data graph, usually a Concept represents a Destination Node while an Individual Entity represents either a Source Node or a Destination Node[3]. On the other hand, a Predicate always represents an Edge. To fit the above input query in linked data graph, parse tree need to tag input query correctly. However, language tools sometime tag input query wrongly (e.g., whether "spouse" is a Predicate or Concept). Furthermore, even if language tool can tag input query correctly, because of multiple parse trees, correct parse tree choosing could be wrong which leads improper prediction of linked data subgraph. System with improper subgraph generates empty/wrong result. Tagging input query with all possible options and use of all parse trees, that were investigated in [16], could be a workaround. But it increases system's complexity.

On the other hand, in the Pivot Point-based Systems, since single point (e.g., a particular keyword from the input query) leads next point exploration,

---

[2] tagging could be POS tagging, NER tagging etc.
[3] http://www.linkeddatatools.com/introducing-rdf

inappropriate selection of pivot-point will affect system's performance. This approach also could miss contextual information attachments among the points, or could predict a subgraph that generates empty result. Dynamic programming based subgraph prediction, which was investigated in Treo, could be a possible workaround. However, in such a case, backtracking and picking of another point increases the data access complexity. This is because, if such approach is adapted, each of this backtracking needs to check for all the instances that correspond to the point. Moreover, in real world scenario, every point (i.e., keyword) corresponds to multiple instances. For example, with exact string matching, keyword "Germany" has at least 22 instances over DBpedia 3.8. This also increases system's complexity.

In this study, we focus above mentioned drawbacks - i) improper prediction of subgraphs and ii) missing of contextual information of input query - over the contemporary linked data information access systems. We propose a keyword-based linked data information access system which we call as **intelligent search** system (inteSearch). It does not generate empty results and does not miss contextual attachments among the keywords of input query.

The prime contributions of our proposed system is: although it explores a larger number of possible subgraphs, which helps in retrieving richer information, it can keep execution complexity low. We hypothesize that if we can analyze linked data structure and their data usage beforehand, we can exploit more number of subgraphs with lower complexity. Exploitation of more number of subgraphs will ensure richer information access capability. On the other hand, pre-processed data statistics will ensure lower information access complexity.

The remainder of this paper is divided as follows. Section 2 overviews our proposed information access framework, the inteSearch. In Section 3 we describe pre-possessing part of data and, in the following section, we describe proposed framework in details. In section 5 we show the results of implementing proposal through experimental results and discussion. In Section 6 we describe works related to this study. Finally, section 7 concludes our study.

## 2    Overview of Proposed Framework

Our proposed framework, the inteSearch, takes keywords as input and generates linked data information as output. For each keyword, we construct the Basic Keyword Graph (BG) that is the basic part of linked data's graph. Then by joining all Basic Keyword Graphs (BGs), we construct subgraph for input keywords called the Keyword Graph (KG) . It determines subgraph of linked data graph. We join BGs in a way called the Binary Progressive Approach [17] (details are described later) so that they can maintain contextual attachments among the keywords. We join two BGs for all joining options which generates multiple Keyword Graphs (KGs). We rank the KGs. Then for the ranked KGs, we iteratively construct and execute linked data query (i.e., SAPARQL query[4]) until we find non-empty output.

---

[4] http://www.w3.org/TR/rdf-sparql-query/

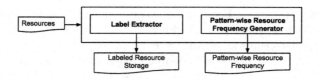

**Fig. 1.** The Pre-processor

In the inteSearch, each of its step relies on some kind of pre-processed data statistics. The contribution of this study, i.e., retrieving richer information with lower complexity, is achieved by this pre-processed data statistics. Therefore, before describing proposed framework in details, we describe about the construction of pre-processed data statistics. Following two sections describe them respectively.

## 3   Pre-Processed Data Statistics

The pre-processed data statistics helps in constructing the KG and ranking of the KG. As mentioned in Section 1, linked data store data in RDF triples. An RDF triple represents the basic part of the linked data graph. Therefore, if we analyze RDF triples and their usage, we could possibly predict the linked data graph.

The elements of RDF triples are linked data resources. For a linked dataset, we make statistics for all of the resources. A Pre-processor calculates these statistics. Below we describe the Pre-processor. We also present the data statistics for an exemplary dataset.

### 3.1   Pre-Processor

It takes individual resource and constructs two different statistics. Two sub-processors called the Label Extractor and the Pattern-wise Resource Frequency Generator construct them respectively. Figure 1 shows the Pre-processor. Below we describe it in details.

**Label Extractor.** It takes individual resource of dataset and store its corresponding literal values. We consider literal values are the primary means of data access. The resource representing predicates ($rrp$) e.g., label, name, prefLabel, etc., inform literal values of a resource. Linked data store data in RDF triples. So, for a resource $r$, $lv(r)$ generates the literal values of $r$ where

$$lv(r) = \{o \mid \exists < r, p, o > \in \text{RDF triples of dataset} \land p \in rrp\}$$

As the output of the Label Extractor process, we store pairs $< r, lv(r) >$ for all resources of the dataset. We mention this storage as the Labeled Resource Storage (LRS).

**Pattern-Wise Resource Frequency Generator.** It also takes individual resource of dataset and stores three different frequencies. In an RDF triple, a resource can appear either as a Subject element, or a Predicate element or an Object element. The Pattern-wise Resource Frequency of a resource is calculated by how many times resource appear as Subject element, Predicate element and Object element in RDF triples of dataset. It helps in predicting subgraphs of linked data graph.

If, for a resource $r$, $sf(r)$, $pf(r)$ and $of(r)$ are respectively considered as the Subject Pattern-wise Resource Frequency, the Predicate Pattern-wise Resource Frequency and the Object Pattern-wise Resource Frequency, they can be defined as:

$$sf(r) = | \{< r,p,o >| \exists < r,p,o >\in \text{RDF triples of dataset}\} |$$
$$pf(r) = | \{< s,r,o >| \exists < s,r,o >\in \text{RDF triples of dataset}\} |$$
$$of(r) = | \{< s,p,r >| \exists < s,p,r >\in \text{RDF triples of dataset}\} |$$

As the output of the Pattern-wise Resource Frequency Generator process, we store quad $< r, sf(r), pf(r), of(r) >$ for all resources of the dataset. We mention this storage as the Resource Pattern Frequency Storage (PFS).

### 3.2 An Exemplary Data Statistics

To show an exemplary data statistics, we introduce an exemplary dataset. Figure 2 shows the exemplary dataset. It describes simple profiling information for person named Amanda, Berlusconi, Cleyra and Donald by providing their relationships and birthplaces. Upper part of the dataset, which is separated by horizontal dotted bar, presents linked data's schema/ontology while the lower part holds instances, instance labels, and their linking. The schema/ontology maintains the Concept and Property by the types Class and Property respectively. On the other hand, instances are described by the upper part's Classes and Properties which are shown by the dotted arrows. For example, Class type of :dnld is :Person. However, for clarity of the figure, we do not show all of them. Linked data instances are presented with prefix ":" and labeled by literal values.

Table 1 shows some of the exemplary data statistics for Figure 2 dataset. Therefore as storage, the first and and second columns are stored in the LRS. and the first, third, fourth and fifth columns are stored in the PFS. Readers are requested to check the table construction. For example, the first row of the table shows data statistics for the resource :Country. In Figure 2, the literal value of :Country is described as Country, the $sf$(:Country) associated RDF triples is 2, the $pf$(:Country) associated RDF triples is 0 and the $of$(:Country) associated RDF triples is 4 (though it is shown 2 for the sake of clarity).

## 4    Detail Development of Proposed Framework: Intesearch

This section describes detail development of the inteSearch. In the inteSearch, we take keywords as input and generate required linked data information as output.

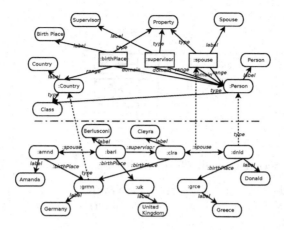

**Fig. 2.** An exemplary dataset

**Table 1.** Exemplary data statistics for the Figure 2 dataset

| $r$ | $lv(r)$ | $sf(r)$ | $pf(r)$ | $of(r)$ |
|---|---|---|---|---|
| :Country | Country | 2 | 0 | 4 |
| :birthplace | Birth Place | 4 | 4 | 0 |
| :grce | Greece | 2 | 0 | 1 |
| :dnld | Donald | 4 | 0 | 1 |
| :... | ... | ... | ... | ... |

Figure 3 shows process flow of the inteSearch. For each input keyword, the Basic Keyword Graph Generator process constructs the BG (Basic Keyword Graph) and calculates its weight. The Basic Keyword Graph Generator process is executed for all input keywords that constructs all the BGs (Basic Keyword Graphs). By joining the BGs, the Keyword Graph Generator process constructs the KGs (Keyword Graphs). The Ranker process ranks the KGs and generates ranked KGs. Then the SPARQL Query Generator process generates SPARQL query for the ranked KGs. It iteratively construct and execute linked data query (i.e., SAPARQL query) until it finds non-empty output.

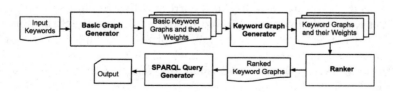

**Fig. 3.** Process flow of inteSearch

**Table 2.** The Basic Keyword Graph for keyword $k$

| | Basic Keyword Graph |
|---|---|
| $BG_s(k)$ | $k \xrightarrow{?p} ?o$ |
| $BG_p(k)$ | $?s \xrightarrow{k} ?o$ |
| $BG_o(k)$ | $?s \xrightarrow{pk} ?$ |

## 4.1   Basic Keyword Graph Generator

For each input keyword, it constructs the BG and calculates the weight of the BG. Below first we define the BG, then we describe how we select appropriate BG and calculate its weight.

**Basic Keyword Graph.** The BG is devised as a graph << Source Node, Edge, Destination Node >>. The intuition behind the BG construction is to replicate the basic part of linked data graph.

For keyword $k$, Table 2 shows three different types of BGs. They varied depending upon where the $k$ is placed in the BG. Here, the first type of BG (i.e, $BG_s(k)$) is graph where the Source Node is $k$ and the Edge and Destination Node are variables. In KG construction, variables are filled by the keywords. Or they will finally be used to retrieve required information (process will be described details later). Variables are presented by question mark following letters (e.g., $?s$, $?p$ and $?o$). The second type of BG (i.e, $BG_p(k)$) is graph where the Edge is $k$, and the Source Node and Destination Node are variables. In the same way, the third type of BG (i.e, $BG_o(k)$) is graph where Destination Node is $k$, and the Source Node and the Edge are variables. Variable holding Source Nodes and Destination Nodes are used to connects two BGs and called as the Connectors (shown with bold font question mark follow letters).

**Selection and Weight Calculation of BG.** For a keyword, we select the BG by the pre-processed data statistics. The LRS (Labeled Resource Storage) and the PFS (Resource Pattern Frequency Storage) are used in this selection. For the LRS and PFS, see Section 3.

The LRS is used to retrieve keyword corresponding resources. Two parameters the Per Keyword Maximum Allowed Resources ($\eta$) and the String Similarity Threshold ($\zeta$) is used in this retrieval. We capture maximum $\eta$ number of top similar resources i.e., $\forall r \in \{r_1, r_2, ..., r_i\}$ which have $sim(k, lv(r)) \geq \zeta$ where $i \leq \eta$. Here $sim(k, lv(r))$ is a string similarity function that takes a keyword $k$ and literal values of $r$ $lv(r)$ from the LRS and gives similarity value which ranges between 0 (zero) and 1 (one) - 0 means "no match" while 1 means "exact match".

Considering the value of $sim(k, lv(r))$, for each resource $r \in \{r_1, r_2, ..., r_i\}$, we measure how frequent $r$ is in its Pattern-wise Resource Frequency. It gives insight about the selection of BG. The PFS is used to find the Pattern-wise Resource Frequency. We hypothesize that if more similar resource holds larger Predicate Pattern-wise Resource Frequency, we should select the $BG_p$. Otherwise, we should select the $BG_s$ and the $BG_o$.

To select the BG of $k$, we calculate three scores (i.e., SScore($r$), PScore($r$) and OScore($r$)) for each $r \in \{r_1, r_2, ..., r_i\}$ of $k$ as below.

$$SSocre(r) = sim(k, lv(r)) * sf(r)$$
$$PScore(r) = sim(k, lv(r)) * pf(r)$$
$$OScore(r) = sim(k, lv(r)) * of(r)$$

Then, for all $r \in \{r_1, r_2, ..., r_i\}$ of $k$, we find each score's maximum values as SMaxScore(k), PMaxScore(k) and OMaxScore(k) respectively. They are calculated as

$$SMaxSocre(k) = \max_{j:r_j \in \{r_1, r_2, ..., r_i\}} SScore(r_j)$$
$$PMaxSocre(k) = \max_{j:r_j \in \{r_1, r_2, ..., r_i\}} PScore(r_j)$$
$$OMaxSocre(k) = \max_{j:r_j \in \{r_1, r_2, ..., r_i\}} OScore(r_j)$$

Now, if the value of PMaxScore(k) is bigger than both the value of SMaxScore(k) and OMaxScore(k), we select the $BG_p(k)$ and set its weight as SMaxScore(k). Otherwise, we select both the $BG_s(k)$ and the $BG_o(k)$ and set their weights with the larger value between SMaxScore(k) and OMaxScore(k).

## 4.2  Keyword Graph Generator

It starts after the construction of all input keyword corresponding BGs. Here we join input keyword corresponding BGs to construct the KG. We also calculate the weight of the KG.

We join BGs so that joining can maintain contextual attachments among the keywords. We join BGs in their Connectors. In the BGs, Variable holding Source Nodes and Destination Nodes are the Connectors. The contextual attachments among the keywords are achieved by a joining approach which is called the Progressive Joining Approach and was investigated in [17]. Below we define about the approach and then we describe how we adapt it in the inteSearch.

**Progressive Joining Approach.** To maintain contextual attachments among the keywords, we join the BGs with the Progressive Joining Approach. In the Progressive Joining Approach, if $m$ number of input keywords are given as $\{k_1, k_2, k_3, ..., k_m\}$, we start joining BGs for first (i.e, $k_1$) and second (i.e. $k_2$) keywords. It construct an Intermediate-version KG. Then this Intermediate-version KG gets joined with the next keyword (i.e., $k_3$) corresponding BG which is continued till the last keyword (i.e, $k_m$) corresponding BG. In this joining, keyword order has great role. We only join next keyword corresponding BG to the Connectors of immediate prior keyword corresponding BG. This ensures contextual attachments among the keywords.

The rationale behind the Progressive Joining Approach is it assumes input keywords are given in an order; and to find keyword order, it follows order of keywords as they are given in their corresponding natural language sentence. In other words, the Progressive Joining Approach assumes that instead of putting keywords randomly, users will put keywords in an order that conforms the natural language structure of a sentence. This assumption is supported by Michael A. Covington's work [2] where he showed that major language such as Chinese, English French almost never allow variations in the word orders that make up sentences. The word order-based information access is also common in contemporary linked data information access [16,17,19].

**Adaptation of Progressive Joining Approach.** In joining, the Connectors get replaced either by the keywords or by new Connectors. Here, two criteria are used to join the BGs. They are:

Criterion 1  = joining will always retain at least one Connector
            or a variable holding Edge
Criterion 2  = if last keyword corresponding BG is still not joined,
            joining will always retain at least one Connector

Criterion 1 ensures information access. This is because, the Connectors or variable holding Edges will attach required information form the linked data graph. Criterion 2 ensures next BG joining. That is, if last keyword corresponding BG is yet to join, it will have Connector to join.

Above criteria based joining generates two types of KG.

Adhesive Type KG  = when a BG get merged into another BG
            or an Intermediate-version KG
Explosive Type KG = when BG get expanded with another BG
            or an Intermediate-version KG

Table 3 shows the KGs for an exemplary joining scenario. For keyword $k_1$ and $k_2$, it joins a $BG_p(k_1)$ and a $BG_o(k_2)$. First column shows the Adhesive Type KG while second column shows the Explosive Type KGs. Here, the Connectors are shown with bold font question mark following letters and variable holding Edges are shown with normal font question mark following letters.

Since joining between two BGs increases number of KGs, Table 4 shows possible number of KG increase in joining of two BGs. Such as gray shaded cell shows number of KG increase for joining of Table 3 BGs.

Table 5 shows an exemplary KG construction and extension scenario for keyword $k_1$, $k_2$ and $k_3$. It follows the Progressive Joining Approach. At the first column an Intermediate-version KG is shown which was constructed for keywords $k_1$ and $k_2$. This Intermediate-version KG is going to join with one of the $k_3$ corresponding BG (shown at the third column). At the second column, it retrieves the last joined BG (shown in the rectangle) and join between the immediate prior BG and the next BG (shown in the fourth column). Therefore KG is expanded and generates three KGs (shown in the fifth column).

The KG is constructed from its constituent BGs. Therefore, weight of the KG is calculated from the constituent BGs. The minimum weight of constituent BGs is the weight of the KG.

**Table 3.** The KGs in an exemplary $BG_p(k_1)$ and $BG_o(k_2)$ joining

| | Adhesive Type KG | Explosive Type KG |
|---|---|---|
| $BG_p(k_1)$ [?s]—$k_1$→[?o] | | [?o_1]←$k_1$—[?s_1]—$^{?p_2}$→[$k_2$] |
| $BG_o(k_2)$ [?s]—$^{?p}$→[$k_2$] | [?s_1]—$k_1$→[$k_2$] | [?s_1]—$k_1$→[?o_1]—$^{?p_2}$→[$k_2$] |

**Table 4.** Possible number of KG increase (i.e., Adhesive Type KG and Explosive Type KG) for two BGs joining

| | $BG_s(k_k)$ | $BG_p(k_k)$ | $BG_o(k_k)$ |
|---|---|---|---|
| $BG_s(k_l)$ | 2 | 3 | 3 |
| $BG_p(k_l)$ | 3 | 4 | 3 |
| $BG_o(k_l)$ | 3 | 3 | 2 |

**Table 5.** Exemplary construction and extension scenario of KG

| Intermediate-version KG | Immediate prior BG | Next BG | Joining between last joined BG and next BG | Increase of KG |
|---|---|---|---|---|
| | | | | [$k_1$]←$?s_1$—$k_2$→[$k_3$] |
| | | | [?s_1]—$k_2$→[$k_3$] | [$k_1$] |
| | | | [?o_2]←$k_2$—[?s_1]—$k_3$→[$k_3$] | [?o_2]←$k_2$—[?s_1]—$^{?p_3}$→[$k_3$] |
| [$k_1$]←$^{?p_1}$—[?s_1]—$k_2$→[?o_2] | [$k_1$]—$^{?p_1}$—[?s_1]—$k_2$→[?o_2] | [?s]—$^{?p}$→[$k_3$] | [?s_1]—$k_2$→[?o_2]—$^{?p_3}$→[$k_3$] | [$k_1$]←[?s_1]—$k_2$→[?o_2]—$^{?p_3}$→[$k_3$] |

As it is seen, joining of BGs generates multiple KGs, we need to rank them to find the best KG. Weight of the KG is used to rank the KG.

## 4.3 Ranker

For all KGs, it ranks each of them by its weight and depth-level. We already talk about the weight. Now we will talk about the depth-level. The depth-level of a KG is how many Edges a KG holds. For example, depth-level of BG is 1 (one).

Higher ranked KGs are with lower depth-level and higher weight. In ranking, the priority of lower depth-level supersedes the priority of higher weight. That is, we firstly order KGs from lower depth-level to higher depth-level and then we order them for their respective weights. So, if there is a KG that holds lower-depth level with lower weight, it will still be ranked higher than a KG that holds higher-depth level with higher weight.

## 4.4 SPARQL Query Generator

We construct SPARQL query for the ranked KGs. We iteratively construct and executed linked data query (i.e., SAPARQL query[5]) until we find non-empty output. We consider this output as our intended output.

To construct SPARQL query, we replace keywords of KG by the keyword corresponding resources. For same keyword corresponding resources, we accommodate them with "UNION" relation.

---

[5] http://www.w3.org/TR/rdf-sparql-query/

## 5   Experiment

In the experiment, we use question answering over linked data 3 (i.e., QALD-3[6]) open challenge test question set. It includes ninety nine natural language questions from DBpedia. "Give me all movies directed by Francis Ford Coppola" (Q.#28) is one these questions . Since our proposed framework is a keyword-based framework, we manually devise keywords. For example, for the above question, we construct keywords as {Film, director, Francis Ford Coppola}. The order of the keywords is as the keywords are appeared in the question.

We discarded a few questions, such as questions that are mentioned "OUT OF SCOPE" (by the QALD-3 organizer), require Boolean type answers, aggregation functions, temporal precedence (such as *latest, past five years*, etc.) understanding, and questions for which resources are not found in the dataset. They are out of the scope of our research.

We use parameters value for $\eta$ (Per Keyword Maximum Allowed Resources) and $\zeta$ (String Similarity Threshold) (See Section 4.1) as 20 and 1.00 respectively. We index data statistics storage (i.e., LRS and RFS) with Lucene 4.3.0. Indexing helps in BG construction with fixed complexity. For DBpedia we manually define *rrp* (resource representing predicates) (See Section 3) as {[7]label, [8]title}. We then implement inteSearch using the Java Jena (version 2.6.4) framework. The inteSearch hardware specifications are as follows:

Intel®Core™i7-4770K central processing unit (CPU) @ 3.50 GHz based system with 16 GB memory.

We loaded DBpedia dataset in Virtuoso (version 06.01.3127) triple-store, which was maintained in a network server. To evaluate inteSearch, we performed three experiments and analyzed their results. These experiments will be described in detail below.

### 5.1   Experiment 1

The first experiment was performed to evaluate how the inteSearch performs. We check whether the inteSearch can retrieve richer information. We will report inteSearch performance according to the keyword group, i.e., number of keywords each question holds. It will give more understanding about the performance of inteSearch. Keyword groups are separated by the number of keywords each question can hold. For example, the exemplary question shown in Section 5 falls into a "Three Keyword Group" question, because it holds three keywords. We executed the inteSearch for each group of questions and evaluated their results in terms of Average Recall, Average Precision, and Average F1 Measure. Based on the given answers of QALD-3 test questions, Recall is the fraction of relevant answers that the inteSearch can retrieve. Precision is the fraction of retrieved answers that are relevant, and the F1 Measure is the harmonic mean of precision

---

[6] http://greententacle.techfak.uni-bielefeld.de/~cunger/qald/index.php?x=home& q=3

[7] http://www.w3.org/2000/01/rdf-schema#

[8] http://purl.org/dc/elements/1.1/

**Table 6.** inteSearch Recall, Precision, and F1 Measure grouped by number of keywords for the DBpedia questions

|  | No of Qs | Recall (Avg) | Precision (Avg) | F1 Measure (Avg) |
|---|---|---|---|---|
| One Keyword Group | 1 | 1.00 | 1.00 | 1.00 |
| Two Keyword Group | 45 | 0.90 | 0.96 | 0.92 |
| Three Keyword Group | 13 | 0.77 | 0.77 | 0.77 |
| Four Keyword Group | 8 | 0.75 | 0.75 | 0.75 |
| Five Keyword Group | 3 | 1.000 | 1.000 | 1.000 |
|  |  | 0.87 | 0.90 | **0.88** |

and recall. For each of the keyword-group questions, we calculated the Average Precision and Average Recall by summing up the individual Recall and individual Precision, and then dividing them by the number of questions for each group. However, it was impossible to calculate the Average F1 Measure using the same method because the individual F1 Measure cannot be calculated if the Recall of that individual question is zero. In such cases, we put the individual question's F1 Measure at zero as well, and then calculated the average F1 Measure for each group of questions.

Table 6 shows our keyword-group-wise result analysis for Recall, Precision, and F1 Measure. The bottom of the table shows overall average performance for the questions that inteSearch can execute. As you can see, the performance of the "One/Two/Five Keyword Group" questions indicates (because of their F1 Measure > 0.90) that our BG (basic keyword graph) selection proposal works well. We investigated reason of comparatively lower performance of "/Three/Four Keyword Group" questions. We found that we sometime calculate wrong weight of BG which eventually gives incorrect KG ranking. This is because, for a keyword $k$, we select two BGs (i.e, $BG_s(k)$ and $BG_o(k)$) if it is not $BG_p(k)$ and set maximum weight of SMaxScore and OMaxScore (See Section 4.1) for both $BG_s(k)$ and $BG_o(k)$. It leads incorrect weighting of BG.

The performance of the questions for more than one keywords also validates our BG joining policy. Therefore, we conclude that internal structure of the linked data and their statistics have more significant impact on subgraph determining over linked data graph, which can be used potentially over keyword-based linked data information retrieval. Finally, from the overall performance of inteSearch (F1 Measure (Avg) 0.88), we conclude the inteSearch can retrieve richer information.

## 5.2    Experiment 2

The second experiment was performed to evaluate the data access complexity. The data access complexity is measured for the actual execution cost.

Here we will report execution cost for each keyword group questions. We want to see whether the inteSearch can cope with number of keyword increase. Table 7 shows the execution cost for One/Two/Three/Four/Five Keyword Group questions. We execute each question for three times, then for each keyword group,

**Table 7.** Execution cost (in milliseconds) for QALD-3 DBpedia test questions

| One Keyword Group | Two Keyword Group | Three Keyword Group | Four Keyword Group | Five Keyword Group |
|---|---|---|---|---|
| 710 | 2441 | 2774 | 3585 | 3720 |

**Table 8.** Average Recall, Precision and F-1 Measure for DBpedia QALD-3 test questions

|  | # of Questions | Processed | Right | Partially | Recall | Precision | F1-Measure |
|---|---|---|---|---|---|---|---|
| squall2sparql[6] | 99 | 99 | 80 | 13 | 0.88 | 0.93 | 0.90 |
| CASIA[18] | 99 | 52 | 29 | 8 | 0.36 | 0.35 | 0.36 |
| Scalewelis[9] | 99 | 70 | 32 | 1 | 0.33 | 0.33 | 0.33 |
| inteSearch | 99 | 70 | 60 | 1 | 0.87 | 0.90 | 0.88 |

we calculate average execution time. Execution costs are shown in milliseconds. It is seen that execution cost mainly increases for two factors: i) number of keywords ii) construction of SPARQL query. Execution cost linearly increases according to the number of keyword in a question. This is because, if number of keyword increases, number of KG also gets more. Our detail observation shows that more than half of the execution cost requires in SPARQL query generation. This is because, in SPARQL query generation, we need to go to triple-store and iteratively need to check whether replacement of keyword related resource is compatible. This rationalize our pre-processed data statistics generation. Because, if system needs to interact dataset on the fly it will increase execution complexity.

Execution cost could be reduced if number of KG is less or SPARQL could maintain big sized query. However, we consider that current execution cost is still reasonable (3.72 Seconds for Five Keyword Group questions). We conclude that greedy utilization of data statistics can check larger number of subgraphs in reasonable complexity.

## 5.3    Experiment 3

The third experiment was performed to evaluate the performance with other systems. We evaluated the performance comparison between inteSearch and QALD-3 challenge participant systems, specifically squall2sparql[6], CASIA[18] and Scalewelis[9]. For the inteSearch, the answered questions were 70 DBpedia questions that had also been used in Experiment 1. Table 8 shows a performance comparison between the QALD-3 challenge participant systems and inteSearch. In the evaluation report[9], the challenge participant systems reported on how many questions each system processed. Next, for the processed questions, each system reported its right number of questions, partially right number questions, Average Recall, Average Precision, and Average F1 Measure. Table 8 columns two, three, four, and five, respectively, show these performance levels.

---

[9] http://greententacle.techfak.uni-bielefeld.de/~cunger/qald/index.php?x=home\&q=3

The inteSearch can not able to outperform QALD-3 challenge participant systems. But it performs almost like the squall2sparql performs. In the inteSearch, we generate output for the first non-empty output generating KG which is very restrictive. Since, over the linked data graph, focus of this study is to check larger number of subgraphs and determine the best subgraph with lower complexity, we consider the inteSearch can access richer information. However, it is necessary to mention that the systems were not fully comparable, because inteSearch is a keyword-based system, while the others are natural language based systems. However, we present this performance comparison based on the assumption that if the required keywords are given to inteSearch, inteSearch will work in a very sophisticated manner. We also acknowledge that automatic identification of such keywords will further increase complexities. This point will need to be investigated in our future work.

On the other hand, to compare execution cost, we do not find any report which shows execution cost of squall2sparql and Scalewelis. Therefore, we compare execution cost between inteSearch and CASIA. We find that for each question, inteSearch takes average 2.66 Seconds while CASIA takes average 83.28 Seconds - which is a huge boost in execution cost comparison.

We conclude that the pre-processed data statistics facilitates in checking the larger number of subgraph and determining required subgraph over linked data graph.

## 6   Related Work

Information access over linked data information access requires the subgraph determining because linked data is also graph-based data[5, 12, 13].

Over the contemporary linked data information access systems, the subgraph checking have been investigated through different heuristics. They can be largely categorized into two categories. The first category systems (say the Language Tool-based Systems) use natural language tools to determine possible subgraphs. The second category systems (say the Pivot Point-based Systems") track a Pivot Point (e.g., a particular keyword from the input query), then, from that Pivot Point, they try to explore the other points and thereby determine possible subgraphs. PowerAqua[11], TBSL[19], FREyA[3], SemSek[1] CASIA[18] etc. belong to the Language Tool-based Systems while Treo[7], NLP-Reduce[10] etc. belong to the Pivot Point-based systems. However, both category systems possess some drawbacks.

Apart from the above category systems, we find some studies that try to find subgraph to subsume the data graph automatically[5, 13, 20]. Zenz et al., tries to find possible subgraphs on the fly[20]. But on the fly subgraph finding is costly. On the other hand, Elbassuoni et al., and Niu et al., try to construct keyword graph like us, but none of them try with pre-processed data statistics[5, 13].

On the other hand, pattern based semantic data consumption has recently got good attentions [8, 14]. Such as, Zhang et al., uses statistical knowledge patterns to identify synonymous resources[21]. We also analyze data for frequent pattern analysis.

Delbru et al., has proposed node based linked data indexing to get boost in resource retrieval efficiency[4], however, mere retrieval of node can not adapt required semantics. Therefore we indexed resources for their Pattern-wise Resource Frequency so that they can give some semantics. Picalausa et al., introduces structural indexing, which we believe a good initiative in linked data indexing, but they focus indexing for basic graph pattern where they do not present multiple input keywords adaptation[15].

## 7    Conclusion

Over linked data information access, we analyze linked data to access richer quality information with lower complexity. We investigate linked data RDF triples and prepare a pre-processed data statistics. Our assumption is that resource and Pattern-wise Resource Frequency have influences over linked data information access. We determine the best subgraph from more number of possible subgraphs which supports us in retrieving richer information. Although we investigate more number of possible subgraphs, we can keep system with lower complexity by the pre-processed data statistics. Experiment shows effectiveness our subgraph (i.e, the BGs and KGs) construction. We also ensure the contextual information attachment among the input keywords. The proposed framework is evaluated for standard question set. Current implementation of inteSearch uses exact resource matching, though framework supports any types of matching, which can be considered a weak point. However resource matching is not focus of our study rather in the inteSearch, we efficiently try to predict the best subgraph (over linked data graph) that can generate richer information. Currently we construct data statistics for single resource. However, we assume that connection between resources can be predicted by some kind of features. In future, we want to investigate such features that can predict multiple resource attachments. Therefore we want to construct data statistics for multiple resources. In such a case, framework will require lower number of subgraph checking which will further improve information access efficiency.

## References

1. Aggarwal, N., Buitelaar, P.: A system description of natural language query over dbpedia. In: Proceedings of Interacting with Linked Data, pp. 96–99 (2012)
2. Covington, M.A.: A dependency parser for variable-word-order languages. In: Derohanes (eds.) Computer Assisted Modeling on the IBM 3090, pp.799–845 (1992)
3. Damljanovic, D., Agatonovic, M., Cunningham, H.: FREyA: An interactive way of querying linked data using natural language. In: Proceedings of the 1st Workshop on Question Answering over Linked Data, pp. 125–138 (2011)
4. Delbru, R., Toupikov, N., Catasta, M., Tummarello, G.: A node indexing scheme for web entity retrieval. In: Aroyo, L., Antoniou, G., Hyvönen, E., ten Teije, A., Stuckenschmidt, H., Cabral, L., Tudorache, T. (eds.) ESWC 2010, Part II. LNCS, vol. 6089, pp. 240–256. Springer, Heidelberg (2010)
5. Elbassuoni, S., Blanco, R.: Keyword search over rdf graphs. In: Proceedings of the 20th ACM International Conference on Information and Knowledge Management, pp. 237–242 (2011)

6. Ferr, S.: squall2sparql: a Translator from Controlled English to Full SPARQL 1.1. Working Notes for CLEF 2013 Conference (2013)
7. Freitas, A., Oliveira, J., O'Riain, S., Curry, E., Pereira da Silva, J.: Treo: best-effort natural language queries over linked data. In: Proceedings of the 16th International Conference on Applications of Natural Language to Information Systems, pp. 286–289 (2011)
8. Gangemi, A., Presutti, V.: Towards a pattern science for the semantic web. Semantic Web 1(1–2), 61–68 (2010)
9. Guyonvarch, J., Ferr, S., Ducass, M.: Scalable Query-based Faceted Search on top of SPARQL Endpoints for Guided and Expressive Semantic Search. Research report PI-2009, LIS - IRISA, October 2013
10. Kaufmann, E., Bernstein, A., Fischer, L.: NLP-Reduce: A nave but domain-independent natural language interface for querying ontologies. In: Proceedings of the 4th European Semantic Web Conference (2007)
11. Lopez, V., Motta, E., Uren, V.S.: PowerAqua: Fishing the semantic web. In: Sure, Y., Domingue, J. (eds.) ESWC 2006. LNCS, vol. 4011, pp. 393–410. Springer, Heidelberg (2006)
12. Manning, C.D., Raghavan, P., Schütze, H.: An Introduction to Information Retrieval. Cambridge University Press (2009)
13. Niu, Z., Zheng, H.-T., Jiang, Y., Xia, S.-T., Li, H.-Q.: Keyword proximity search over large and complex rdf database. In: Proceedings of IEEE/WIC/ACM International Conferences on Web Intelligence and Intelligent Agent Technology, pp. 467–471 (2012)
14. Nuzzolese, A.G., Gangemi, A., Presutti, V., Ciancarini, P.: Encyclopedic knowledge patterns from wikipedia links. In: Aroyo, L., Welty, C., Alani, H., Taylor, J., Bernstein, A., Kagal, L., Noy, N., Blomqvist, E. (eds.) ISWC 2011, Part I. LNCS, vol. 7031, pp. 520–536. Springer, Heidelberg (2011)
15. Picalausa, F., Luo, Y., Fletcher, G.H.L., Hidders, J., Vansummeren, S.: A structural approach to indexing triples. In: Simperl, E., Cimiano, P., Polleres, A., Corcho, O., Presutti, V. (eds.) ESWC 2012. LNCS, vol. 7295, pp. 406–421. Springer, Heidelberg (2012)
16. Rahoman, M.-M., Ichise, R.: An automated template selection framework for keyword query over linked data. In: Takeda, H., Qu, Y., Mizoguchi, R., Kitamura, Y. (eds.) JIST 2012. LNCS, vol. 7774, pp. 175–190. Springer, Heidelberg (2013)
17. Rahoman, M.-M., Ichise, R.: Automatic inclusion of semantics over keyword-based linked data retrieval. IEICE Transactions of Information and Systems **E97-D**(11) (2014)
18. He, S., Liu, S., Chen, Y., Zhou, G., Liu, K., Zhao, J.: CASIA@QALD-3: A Question Answering System over Linked Data. Working Notes for CLEF 2013 Conference (2013)
19. Unger, C., Bühmann, L., Lehmann, J., Ngomo, A.-C. N., Gerber, D., Cimiano, P.: Template-based question answering over RDF data. In Proceedings of the 21st World Wide Web Conference, pp. 639–648 (2012)
20. Zenz, G., Zhou, X., Minack, E., Siberski, W., Nejdl, W.: From keywords to semantic queries-incremental query construction on the semantic web. Journal of Web Semantics **7**(3), 166–176 (2009)
21. Zhang, Z., Gentile, A.L., Blomqvist, E., Augenstein, I., Ciravegna, F.: Statistical knowledge patterns: Identifying synonymous relations in large linked datasets. In: Alani, H., Kagal, L., Fokoue, A., Groth, P., Biemann, C., Parreira, J.X., Aroyo, L., Noy, N., Welty, C., Janowicz, K. (eds.) ISWC 2013, Part I. LNCS, vol. 8218, pp. 703–719. Springer, Heidelberg (2013)

# Publishing Danish Agricultural Government Data as Semantic Web Data

Alex B. Andersen, Nurefşan Gür(✉), Katja Hose,
Kim A. Jakobsen, and Torben Bach Pedersen

Department of Computer Science, Aalborg University, Aalborg, Denmark
alex@bondoandersen.dk, {nurefsan,khose,kah,tbp}@cs.aau.dk

**Abstract.** Recent advances in Semantic Web technologies have led to a growing popularity of the Linked Open Data movement. Only recently, the Danish government has joined the movement and published several datasets as Open Data. These raw datasets are difficult to process automatically and combine with other data sources on the Web. Hence, our goal is to convert such data into RDF and make it available to a broader range of users and applications as Linked Open Data. In this paper, we discuss our experiences based on the particularly interesting use case of agricultural data as agriculture is one of the most important industries in Denmark. We describe the process of converting the data and discuss the particular problems that we encountered with respect to the considered datasets. We additionally evaluate our result based on several queries that could not be answered based on existing sources before.

## 1 Introduction

In recent years, more and more structured data has become available on the Web, driven by the increasing popularity of both the Semantic Web and Open Data movement, which aim at making data publicly available and free of charge. Several governments have been driving forces of the Open Data movement, most prominently `data.gov.uk` (UK) and `data.gov` (USA), which publish Open Data from departments and agencies in the areas of agriculture, health, education, employment, transport, etc.

The goal is to enable collaboration, advanced technologies, and applications that would otherwise be impossible or very expensive, thus inspiring new services and companies. Especially for governments, it is important to inspire novel applications, which will eventually increase the wealth and prosperity of the country. While publication of raw data is a substantial progress, the difficulty in interpreting the data as well as the heterogeneity of publication formats, such as spreadsheets, relational database dumps, and XML files, represent major obstacles that need to be overcome [9,12,15] – especially because the schema is rarely well documented and explained for non-experts. Furthermore, it is not possible to evaluate queries over one or multiple of these datasets.

The Linked (Open) Data movement (http://linkeddata.org/) encourages the publication of data following the Web standards along with *links* to other data

© Springer International Publishing Switzerland 2015
T. Supnithi et al.(Eds.): JIST 2014, LNCS 8943, pp. 178–186, 2015.
DOI: 10.1007/978-3-319-15615-6_13

sources providing semantic context to enable easy access and interpretation of structured data on the Web. Hence, publishing data as Linked Data (LD) [7,8] entails the usage of certain standards such as HTTP, RDF, and SPARQL as well as HTTP URIs as entity identifiers that can be dereferenced, making LD easily accessible on the Web. RDF allows formulating statements about resources, each statement consists of subject, predicate, and object – referred to as a triple. Extending the dataset and adding new data is very convenient due to the self-describing nature of RDF and its flexibility. In late 2012, the Danish government joined the Open Data movement by making several raw digital datasets [3] freely available. Among others, these datasets cover transport, tourism, fishery, companies, forestry, and agriculture. To the best of our knowledge, they are currently only available in their raw formats and have not yet been converted to LD. We choose agriculture as a use case, as it is one of the main sectors in Denmark, with 66% of Denmark's land surface being farmland[1]. Thus, there is significant potential in providing free access to such data and enabling efficient answering of *sophisticated* queries over it.

In this paper, we show how we made Danish governmental Open Data available as LD and evaluate the challenges in doing so. Our approach is to transform the agricultural datasets into RDF and add explicit relationships among them using links. Furthermore, we integrate the agricultural data with company information, thus enabling queries on new relationships not contained in the original data. This paper presents the process to transform and link the data as well as the challenges encountered and how they were met. It further discusses how these experiences can provide guidelines for similar projects. We developed our own ontology while still making use of existing ontologies whenever possible. A particular challenge is deriving spatial containment relationships not encoded in the original datasets. For a detailed discussion about the whole process, we refer the reader to the extended version of this paper [2]. The resulting LOD datasets are accessible via a SPARQL endpoint (http://extbi.lab.aau.dk/sparql) as well as for download (http://extbi.cs.aau.dk/).

The remainder of this paper is structured as follows; Section 2 describes our use case datasets and discusses the main challenges. Then, Section 3 describes the process and its application to the use case. Section 4 evaluates alternative design choices, while Section 5 concludes and summarizes the paper.

## 2    Use Case

We have found the agricultural domain to be particularly interesting as it represents a non-trivial use case that covers spatial attributes and can be extended with temporal information. By combining the agricultural data wih company data, we can process and answer queries that were not possible before as the original data was neither linked nor in a queryable format.

Late 2012, the Ministry of Food, Agriculture, and Fisheries of Denmark (FVM) (http://en.fvm.dk/) made geospatial data of all fields in Denmark freely

---

[1] http://www.dst.dk/en/%20Statistik/emner/areal/arealanvendelse.aspx

available – henceforth we refer to this collection of data as *agricultural data*. This dataset combined with the *Central Company Registry (CVR) data* (http://cvr.dk/) about all Danish companies allows for evaluating queries about fields and the companies owning them. In total, we have converted 5 datasets provided by FVM and CVR into Linked Open Data. We downloaded the data on October 1, 2013 from FVM [10] and from CVR.

**Agricultural Data.**    The agricultural data collection is available in Shape format [6], this means that each *Field*, *Field Block*, and *Organic Field* is described by several coordinate points forming a polygon.

*Field.*    The Field dataset has 9 attributes and contains all registered fields in Denmark. In total, this dataset contains information about 641,081 fields.

*Organic Field.*    This dataset has 12 attributes and contains information about 52,060 organic fields. The dataset has attributes that we can relate to the company data, i.e., the CVR attribute is unique for the owner of the field and references the CVR dataset that we explain below. The fieldBlockId attribute describes to which "Field Block" a field belongs to.

*Field Block.*    The Field Block dataset has 12 attributes for 314,648 field blocks and contains a number of fields [11]. Field Blocks are used to calculate the funds the farmers receive in EU area support scheme.

**Central Company Registry (CVR) Data.**    The CVR is the central registry of all Danish companies and provides its data in CSV format. There are two datasets available that we refer to as *Company* and *Participant*.

*Company.*    This dataset has 59 attributes [5] and contains information, such as a company's name, contact details, business format, and activity for about more than 600,000 companies and 650,000 production units.

*Participant.*    This dataset describes the relations that exist between a participant and a legal unit. A participant is a person or legal unit that is responsible for a legal unit in the company dataset, i.e., a participant is an owner of a company. The Participant dataset describes more than 350,000 participants with 7 attributes.

The use case data comes in different formats and contains only a few foreign keys. Further, there is little cross-reference and links between the datasets and no links to Web sources in general. Spatial relationships are even more difficult to represent in the data and querying data based on the available polygons is a complex problem. In particular, to enable queries that have not been possible before, we cleanse and link the (Organic) Field datasets to the Company dataset so that we can query fields and crops of companies related to agriculture. The particular challenges that we address are:

- Disparate data sources without common format
- Lack of unique identifiers to link different but related data sources
- Language (Danish)
- Lack of ontologies and their use

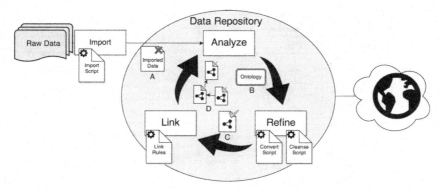

**Fig. 1.** Process overview

# 3 Data Annotation and Reconciliation

In this section, we outline the process that we followed to publish the datasets described in Section 2. The complete procedure with its main activities is depicted in Fig. 1.

All data in the data repository undergoes an iterative integration process consisting of several main activities:

**Import:** Extract the data from the original sources
**Analyze:** Gain an understanding of the data and create an ontology
**Refine:** Refine the source data by cleansing it and converting it to RDF
**Link:** Link the data to internal and external data

Data that has been through the integration process at least once may be published and thus become Linked Open Data that others can use and link to. In the remainder of this section, we will discuss these steps in more detail.

**Import.** The raw data is extracted from its original source into the repository and stored in a common format such that it is available for the later activities. The concrete method used for importing a dataset depends on the format of the raw data. The agriculture datasets and CVR datasets introduced in Section 2 are available in Shape and CSV formats. Shape files are processed in ArcGIS[2] to compute the spatial joins of the fields and organic fields, thus creating foreign keys between the datasets. As the common format we use a relational database.

**Analyze.** The goal of this step is to acquire a deeper understanding of the data and formalize it as an ontology. As a result of our analysis we constructed a URI scheme for our use case data based on Linked Data Principles [8]. We strive to use existing ontologies as a base of our own ontologies. To do this, we make use of predicates such as `rdfs:subClassOf`, `rdfs:subPropertyOf`, and `owl:equivalentClass`, which can link our classes and properties to known ontologies. Fig. 2 provides an overview of the ontology that we developed for our use case with all classes and properties. All arrows are annotated with predicates.

---

[2] http://www.esri.com/software/arcgis

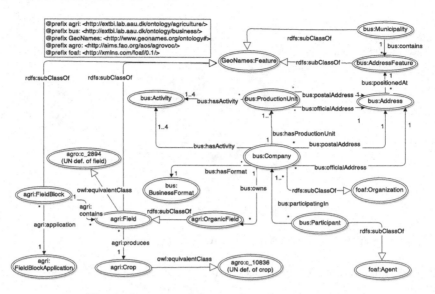

**Fig. 2.** Overview of the ontology for our use case

The arrows with black tips represent relations between the data instances. The arrows with white tips represent relations between the classes.

In short, we designed the ontology such that a Field is contained within a Field Block, which is expressed with the property `agri:contains` and is determined by a spatial join of the data. Organic Field is a subclass of Field and therefore transitively connects Field to Company. Field is also defined as being equivalent to the UN's definition of European fields from the AGROVOC [14] vocabulary. In addition we make use of other external ontologies and vocabularies, such as *GeoNames* [16], *WGS84* [4], and *FOAF (Friend of a Friend)* [1].

**Refine.** The Refine activity is based on the understanding gained in the Analyze activity and consists of data cleansing and conversion. Fig. 1 illustrates the data cleansing process where imported data and ontologies are used to produce cleansed data.

In our use case, we implemented data cleansing by using views that filter out inconsistent data as well as correct invalid attribute values, inconsistent strings, and invalid coordinates. Then we use Virtuoso Opensource [13] mappings to generate RDF data.

**Link.** The Link activity consists of two steps: internal linking and external linking, which converts the refined data into integrated data. The Link activity materializes the relationships between concepts and classes identified in the Analyze activity as triples. The example below shows our internal linking of the Field and the Field Block classes using the `geonames:contains` predicate.

```
agri:contains rdf:type owl:ObjectProperty ;
    rdfs:domain agri:FieldBlock ;
    rdfs:range agri:Field ;
    rdfs:subPropertyOf geonames:contains .
```

External linking involves linking to remote sources on instance and ontology level. On the ontology level, this means inserting triples using predicates such as `rdfs:subClassOf`, `rdfs:subPropertyOf`, and `owl:equivalentClass` that link URIs from our local ontology to URIs from remote sources. On instance level, we link places mentioned in the CVR data to equivalent places in GeoNames [16] using triples with the `owl:sameAs` predicate as illustrated in Fig. 3.

**Fig. 3.** External linking on instance level

The overall process has provided us with analyzed, refined, and linked data; in total *32,457,657* triples were created. The result of completing this process is published and registered on **datahub.io**[3]. In case we wish to integrate additional sources, we simply have to reiterate through the process.

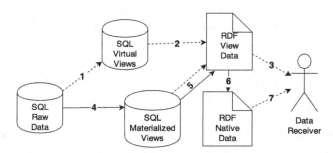

**Fig. 4.** Data flow for the materialization strategies

## 4   Experiments

In the following, we first describe three alternative design choices in the materialization of the data. They represent trade-offs between data load time and query time. We then discuss the results of our experimental evaluation, for which we ran an OpenLink Virtuoso 07.00.3203 server on a 3.4 GHz Intel Core i7-2600 processor with 8 GB RAM operated by Ubuntu 13.10, Saucy. The materialization strategies that we have considered are: *Virtual, relational materialization,* and *native*. Fig. 4 shows the different paths that data is traveling on; starting as raw data and ending at the user who issued a query. The solid lines represent data flow during the integration process whereas dashed lines represent data flow at query time.

---

[3] http://datahub.io/dataset/govagribus-denmark

**Table 1.** Load times in seconds

| Step | Virtual | Materialized | Native |
|---|---|---|---|
| Data Cleansing | 74.92 | 603.35 | 603.35 |
| Load Ontology | 1.01 | 1.01 | 1.01 |
| Load Mappings | 8.76 | 12.35 | 12.35 |
| Dump RDF | 0.00 | 0.00 | 4684.82 |
| Load RDF | 0.00 | 0.00 | 840.04 |
| **Total** | 84.68 | 616.70 | 6141.56 |

**Table 2.** Runtimes in seconds

| Query | Virtual | Materialized | Native |
|---|---|---|---|
| AQT 1 | 5.92 | 3.39 | 1.04 |
| AQT 2 | 13.32 | 7.00 | 0.23 |
| AQT 3 | 10.81 | 7.70 | 0.05 |
| AQT 4 | – | – | 0.14 |
| AQT 5 | – | 20.37 | 0.86 |
| SQT 1 | – | – | 2.35 |
| SQT 2 | 0.09 | 0.12 | 0.10 |
| SQT 3 | 2188.85 | 1.81 | 0.40 |
| SQT 4 | 6.57 | 2.35 | 1.63 |
| SQT 5 | – | 23.79 | 3.29 |
| **Average** | 370.93 | 8.31 | 1.01 |

**Virtual.** In the virtual strategy we perform data cleansing based on SQL views in the relational database. RDF mappings are formulated on top of these cleansing views to make the data accessible as RDF. To increase performance, we create a number of indexes on primary keys, foreign keys, and spatial attributes. In Fig. 4, using this strategy data flows through the arrows marked with 1, 2, and 3 at query time.

**Relational Materialization.** Here we materialize the above mentioned SQL views as relational tables. We create similar indexes as above but on the obtained tables. In Fig. 4, data flows through arrows 4, 5, and 3 – with 4 during load time and 3 and 5 during query time.

**Native RDF.** In this strategy, we extract all RDF triples from the materialized views and mappings and load them into a triple store. In Fig. 4, data flows through arrows 4, 5, and 6 during load time and arrow 7 during query time.

To test our setup, we created a number of query templates that we can instantiate with different entities and that are based on insights in agricultural contracting gained from field experts. Some of them contained aggregation and grouping (Aggregate Query Templates, AQT) others only standard SPARQL 1.0 constructs (Standard Query Templates, SQT). For the virtual and relational materialization strategies we measured the load times for each step during loading – the results are shown in Table 1. Table 2 shows the execution times for our query templates on the three materialization strategies. Queries that run into a timeout are marked by a dash. As we can see, the native RDF strategy is faster than the two others, and relational materialized is generally faster than virtual. There is obviously a notable overhead when using views and mappings. On the other hand, the virtual strategy has very fast load time compared to the other strategies since no data has to be moved or extracted – in fact, the cleansing is delayed until query time. The relational materialized strategy is one order of magnitude faster in load time than the native strategy as it has less overhead during loading.

We can therefore conclude that the virtual strategy is well suited for rapidly changing data as it has minimum load time, the materialized strategy represents a trade-off between load time and query time and is suitable for data with low update rates, and the native strategy decouples RDF data from the relational data and is very suitable for static data.

## 5 Conclusion

Motivated by the increasing popularity of both the Semantic Web and the Open (Government) Data movement as well as the recent availability of interesting open government data in Denmark, this paper investigated how to make Danish agricultural data available as Linked Open Data. We chose the most interesting agricultural datasets among a range of options, transformed them into RDF format, and created explicit links between those datasets by matching them on a spatial level. Furthermore, the agricultural data was integrated with data from the central company registry. All these additional links enable queries that were not possible directly on the original data. . The paper presents best practices and a process for transforming and linking the data. It also discusses the challenges encountered and how they were met. As a result, we not only obtained an RDF dataset but also a new ontology that also makes use of existing ontologies. A particularly interesting challenge was how to derive spatial containment relationships not contained in the original datasets because existing standards and tools do not provide sufficient support. The resulting LOD datasets were made available for download and as a SPARQL endpoint.

**Acknowledgments.** This research was partially funded by "The Erasmus Mundus Joint Doctorate in Information Technologies for Business Intelligence – Doctoral College (IT4BI-DC)".

## References

1. The Friend of a Friend (FOAF) Project. http://www.foaf-project.org/
2. Andersen, A.B., Gür, N., Hose, K., Jakobsen, K.A., Pedersen, T.B.: Publishing Danish Agricultural Government Data as Semantic Web Data. Technical Report DBTR-35, Aalborg University (2014). http://dbtr.cs.aau.dk/DBPublications/DBTR-35.pdf
3. Arendt, J.B.: Denmark releases its digital raw material. Ministry of Finance of Denmark, October 2012. http://uk.fm.dk/news/press-releases/2012/10/denmark-releases-its-digital-raw-material/
4. W3C-Dan Brickley. W3C Semantic Web Interest Group: Geo. http://www.w3.org/2003/01/geo/wgs84_pos, www.wgs84.com
5. Erhvervsstyrelsen. Record layout: Juridiske enheder og P-enheder. http://www.cvr.dk/Site/Resources/Files/Media/RecordlayoutABO110.pdf
6. ESRI. Shapefile technical description. An ESRI White Paper (1998). http://www.esri.com/library/whitepapers/pdfs/shapefile.pdf
7. Heath, T., Bizer, C.: Linked Data: Evolving the Web into a Global Data Space. Synthesis Lectures on the Semantic Web. Morgan & Claypool Publishers (2011)
8. Berners Lee, T.: Design issues, July 2006. http://www.w3.org/DesignIssues/LinkedData.html

9. Maali, F., Cyganiak, R., Peristeras, V.: A publishing pipeline for linked government data. In: Simperl, E., Cimiano, P., Polleres, A., Corcho, O., Presutti, V. (eds.) ESWC 2012. LNCS, vol. 7295, pp. 778–792. Springer, Heidelberg (2012)
10. Agriculture Ministry of Food and Fisheries of Denmark. FVM Geodata Download. https://kortdata.fvm.dk/download/index.html
11. Danish Ministry of the Environment. Markblokkort (datasæt). http://www.geodata-info.dk/Portal/ShowMetadata.aspx?id=1eb89ebb-f674-4ad1-9e53-d1e252226596
12. Skjæveland, M.G., Lian, E.H., Horrocks, I.: Publishing the norwegian petroleum directorate's FactPages as semantic web data. In: Alani, H., Kagal, L., Fokoue, A., Groth, P., Biemann, C., Parreira, J.X., Aroyo, L., Noy, N., Welty, C., Janowicz, K. (eds.) ISWC 2013, Part II. LNCS, vol. 8219, pp. 162–177. Springer, Heidelberg (2013)
13. OpenLink Software. Virtuoso RDF Views - Getting Started Guide, June 2007. http://www.openlinksw.co.uk/virtuoso/Whitepapers/pdf/Virtuoso_SQL_to_RDF_Mapping.pdf
14. Agricultural Information Management Standards. AGROVOC Linked Open Data. http://aims.fao.org/aos/agrovoc/
15. Villazón-Terrazas, B., Vilches-Blázquez, L.M., Corcho, O., Gómez-Pérez, A.: Methodological guidelines for publishing government linked data. Linking Government Data, pp. 27–49. Springer, New York (2011)
16. Wick, M.: GeoNames Ontology. http://www.geonames.org/ontology/documentation.html

# A Lightweight Treatment of Inexact Dates

Hai H. Nguyen[1]([⊠]), Stuart Taylor[1], Gemma Webster[1], Nophadol Jekjantuk[1],
Chris Mellish[1], Jeff Z. Pan[1], Tristan ap Rheinallt[2], and Kate Byrne[3]

[1] Dot.rural Digital Economy Hub, University of Aberdeen, Aberdeen AB24 5UA, UK
hai.nguyen@abdn.ac.uk
[2] Hebridean Connections, Ravenspoint, Kershader, Isle of Lewis HS2 9QA, UK
[3] School of Informatics, University of Edinburgh, Edinburgh EH8 9AB, UK

**Abstract.** This paper presents a *lightweight* approach to representing
inexact dates on the semantic web, in that it imposes minimal ontological commitments on the ontology author and provides data that can
be queried using standard approaches. The approach is presented in the
context of a significant need to represent inexact dates but the heavy-
weight nature of existing proposals which can handle such information.
Approaches to querying the represented information and an example user
interface for creating such information are presented.

## 1 Introduction

There is as yet no standard approach for representing uncertain information in
the semantic web. Existing proposals [1] suggest rather radical changes in representation, and the associated reasoning algorithms, in order to capture uncertainty. In this work, we seek solutions that can be combined more easily with
standard practices but which can handle some common special cases. We focus
in particular on the expression of inexact *dates*. Dates are very important in the
semantic web, which is reflected in the existence of standard data representations
(`xsd:date` and `xsd:dateTime`) for exact dates. However in many situations only
partial information about a date is available. This arises particularly in applications of the semantic web to cultural heritage. When information about the
past is represented, however, there is frequently uncertainty about when events
happened, artefacts were made, people were born, etc.

In our own work, the University of Aberdeen and Hebridean Connections are
working with historical societies based in the Western Isles of Scotland to produce a linked data resource documenting the people, places, events, boats, businesses etc. of their past. The current data consists of over 850,000 RDF triples,
incorporated within a relatively simple OWL ontology. There are 13 different
OWL properties that introduce dates for specific events (e.g. date of birth, date
of origin for a photograph, date demolished for a building). Our ontology is the

The research described here is supported by the award made by the RCUK Digital Economy programme to the dot.rural Digital Economy Hub; award reference:
EP/G066051/1. Many thanks to Panos Alexopoulos for useful conversations.

© Springer International Publishing Switzerland 2015
T. Supnithi et al.(Eds.): JIST 2014, LNCS 8943, pp. 187–193, 2015.
DOI: 10.1007/978-3-319-15615-6_14

| General Class | Pattern | Example | Frequency | Subtotal | Covered |
|---|---|---|---|---|---|
| Exact to the day | y-m-d | 1780-06-13 | 12949 | | |
| | d-m-y | 10/6/45 | 725 | | |
| | d-M-y | 12 MAY 1780 | 272 | | |
| | M-d-y | May 12 1780 | 8 | | |
| | | | | 13954 | yes |
| Exact to the month | y-m | 1780-12 | 274 | | |
| | M-y | Aug 1780 | 443 | | |
| | m-y | 03/1780 | 2 | | |
| | | | | 719 | yes |
| Exact to the year | y | 1978 | 10825 | 10825 | yes |
| Exact to the decade | dec | IN 1860'S | 1415 | 1415 | yes |
| Exact to a range of years | y-y | 1939-45 | 242 | | |
| | beforey | pre 1918 | 2 | | |
| | aftery | AFT 1890 | 3 | | |
| | | | | 247 | yes |
| Exact to the century | cent | 20th Century | 4 | 4 | yes |
| Vague within less than a month | mend | Aug/Sept 1972 | 26 | 26 | yes (using a date range) |
| Vague within more than a month but less than a year | yend | 1978/79 | 7 | 7 | yes (using a date range) |
| Vague year | cy | C. 1932 | 566 | | yes |
| | moddec | early 1950s | 86 | | (using a |
| | | | | 652 | date range) |
| Vague around a decade | cdec | c 1950s | 2 | | yes |
| | modcent | LATE 1600S | 3 | | (using a |
| | | | | 5 | date range) |
| Not directly interpretable as a date | unk | D.I.I. | 3069 | 3069 | no |
| GRAND TOTAL | | | | 30923 | |

**Fig. 1.** Analysis of Date Forms in the Corpus

result of semi-automatic processing of data that originated from a database.In
the database, dates were entered as free-form strings (13,470 of them). In all,
information has been provided about 30,923 dates; these were entered by many
different people over a significant time period. Figure 1 shows an analysis of
the patterns found in this data. The "general class" is a rough indication of the
precision of the information and the patterns indicate the rough syntactic forms
used, with "y", "m" and "d" indicating numbers for years, months and dates,
"M" indicating a month expressed as a string, "-" indicating a separator, such as
"-", "/", "." or a space and some patterns just having vaguely mnemonic names.
Interestingly, only the first general pattern (total frequency 13,954, about 45% of
the data) represents exact dates: all the rest are inexact to some extent. Amongst
the inexact dates, we have distinguished those which are exact within a specified
range from those that are *vague*, in that there is scope for argument about the
intended boundaries. Since all of the relevant properties are intended to indicate
specific dates, not periods of time, this inexactness must be due to uncertainty
on the part of the person entering the information. The inexact dates in our
collection represent a substantial amount of information - a significant semantic
resource that would be lost if we left the information as free-form strings.

Our example is likely to be typical of other projects seeking to exploit cultural heritage data. What is needed is a general mechanism capable of representing (most of) the types of inexactness encountered here in a way that supports some semantic reasoning. This mechanism needs to be *lightweight*, in that:

1. Expressing information about inexact dates should involve as few changes as possible to the ontological decisions already made in the original data;
2. It should be possible easily to query and update information about inexact dates using standard languages such as SPARQL 1.1. In particular, given that the underlying data is uncertain, it should be possible to represent queries that are: **high precision** - in that results returned definitely satisfy the search criteria; and **high recall** - in that all results that might possibly satisfy the criteria are returned.

where criteria for the information seached for can be expressed either as an inexact date or as a specific date.

## 2    Previous Work

Naturally the first question to ask about inexact dates is whether they can be represented by standard XML Scheme datatypes; for this, the two possibilities are `xsd:date` and `xsd:dateTime`. Unfortunately, SPARQL 1.1 has no support for `xsd:date`. Although `xsd:dateTime` is supported by SPARQL 1.1, "dateTime uses the date/time SevenPropertyModel, with no properties except -timezoneOffset- permitted to be absent" [2]. Therefore this property cannot be used on its own to represent inexact dates.

There has been a significant amount of work on ontologies for representing time (and hence dates). In the cultural heritage domain, CIDOC CRM [3] defines a class of entities `E2 Temporal Entity` which describes "objects characterised by a certain condition over a time-span". `E2 Temporal Entitys` include `E4 Periods`, which include examples such as "Jurassic". Although the philosophical position of the OWL-Time ontology [4] is somewhat different to that of CIDOC CRM (e.g., time instants are believed to exist), the treatment of dates is rather similar to the above. In this case, the analogue of `E2 Temporal Entity` is `owltime:TemporalEntity`, which has subclasses `owltime:Instant` and `owltime:Interval`. As above, the actual `TemporalEntity` has to be distinguished from the information stated about it – the latter is included within the former. The latter can be directly associated with the `owltime:Instants` which are its beginning and end. These can in turn be associated with `xsd:dateTimes`.

When it comes to the representation of relatively simple inexact dates these proposals probably satisfy our criteria about access via standard query mechanisms. However, all of them involve significant ontological commitments that may not be made by the rest of the ontology. Whereas for many applications "date of birth" is thought of as a simple property of a person, akin to "name" or "gender", CIDOC CRM requires the existence of 2 extra individuals and OWL-Time 4 extra individuals to express the information, apart from the basic `xsd:dateTime` values involved. Although both are highly principled and in the end very flexible ways of incorporating inexact dates, CIDOC CRM or OWL-Time would significantly break this way of viewing the world, and we suspect they would be similarly disruptive in terms of other ontologies. The question is whether there is a lighter weight alternative available.

## 3   The Proposal

Our proposal is a very simple one that caters for many kinds of inexact dates and complicates the ontology only minimally. The proposal involves the introduction of one new class, `hc:dateRange`[1], on which the `xsd:dateTime`-valued datatype properties `hc:dateFrom` and `hc:dateTo` are defined. An `hc:dateRange` represents a specific date/time which is within the range between the `hc:dateFrom` and the `hc:dateTo`, *not* the period of time between those values. Figure 2 shows

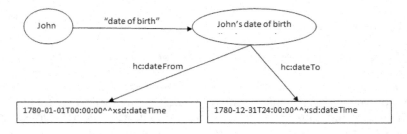

**Fig. 2.** "John was born in 1780" in Hebridean Connections

how this looks for John's birthday. Expressing the information this way commits the ontology author to John being born at a specific time; however, it does not require them to specify that time precisely. Notice the inclusion of time information in the representation – this is forced by the requirements for the completeness of `xsd:dateTime` instances mentioned above. In terms of the dates in our corpus, this proposal provides a way of directly encoding all "Exact to ..." patterns (13,954 exact dates and 13,210 inexact dates, in total about 88% of the corpus). With an appropriate user interface to apply or elicit the relevant cultural norms (see Section 3.2 below), we believe that it can also provide some coverage of the vague patterns in our corpus as well.

### 3.1   Queries Involving Inexact Dates

An `hc:dateRange` represents the start and end of an interval of time. A query to the dataset (e.g. "find all people born in 1780") also involves an interval of time (even a specific day involves two different times for its start and end). In general,

---

[1] We abbreviate the namespace used in the Hebridean Connections ontology, http://www.hebrideanconnections.com/hebridean.owl#, by `hc:`.

then, in a querying situation there is an interval $i$ provided by the query and an interval $j$ provided by data that has been retrieved and is being considered as a possible solution. Two possible types of query then involve particular required relationships between $i$ and $j$ – we name these using the relations defined by [5]:

**high recall query** - this is a query guaranteed to return any results which might satisfy the query. The required relationship is $Overlaps(i, j)$, i.e. the lower bound of $i$ is less than or equal to the upper bound of $j$, and the upper bound of $i$ is greater than or equal to the lower bound of $j$.

**high precision query** - this is a query guaranteed only to return results which definitely satisfy the query. The required relationship is $Contains(i, j)$, i.e. the lower bound of $j$ is greater than or equal to that of $i$, and the upper bound of $j$ is less than or equal to that of $i$.

```
SELECT ?id ?name ?from ?to WHERE
{   ?id hc:born ?dr .
    ?dr hc:dateFrom ?from .
    ?dr hc:dateTo ?to .
    FILTER (?to >= "1780-01-01T00:00:00"^^xsd:dateTime &&
        ?from <= "1780-12-31T24:00:00"^^xsd:dateTime)   .
     ?id hc:englishName ?name}
SELECT ?id ?name ?from ?to WHERE
{   ?id hc:born ?dr .
    ?dr hc:dateFrom ?from .
    ?dr hc:dateTo ?to .
    FILTER (?from >= "1780-01-01T00:00:00"^^xsd:dateTime &&
        ?to <= "1780-12-31T24:00:00"^^xsd:dateTime)   .
     ?id hc:englishName ?name}
```

**Fig. 3.** High Recall and High Precision SPARQL queries for "Find people born 1780"

Figure 3 shows example SPARQL queries of these two types, both giving answers to "find people born in 1780". In these examples, `hc:born` is the property that relates a person to their date of birth and `hc:englishName` is the property that provides the English name of a person.

## 3.2  User Interface Issues

Our simple proposal is still unable to handle directly the *vague* dates found in our corpus (those without clear boundaries). In fact, the actual usage of vague date expressions in English can be very specific to particular communities. For instance, for one of our partners, plus or minus 4 years is a good interpretation for circa dates from 1850 on, but prior to 1850 they were more vague and plus or minus 10 years would be more realistic. However in the linked data world we have to represent information in such a way that anybody can access and understand it. We therefore believe that it is the role of user interfaces to intervene between specific cultural communities and the unbiased data that they create and access.

In particular, the mapping between a phrase such as *c.1987* and a particular
`hc:DateRange` has to be managed by an interface aware of the characteristics of
the particular user community it is serving.

We are building a new website based on the Hebridean Connections data (with
the intention of replacing the earlier database-based website), using generic software for the semi-automatic construction of a Drupal website which supports the
management and presentation of information from a semantic web ontology [6].
A part of this is a user interface for this community, to be used for the acquisition of inexact date information. The user can decide to enter an "exact" date (in
which case they are directed to a standard calendar widget and end up creating
an `hc:dateRange` where the two endpoints correspond to the start and end of the
specified day) or a "circa" date (which here means "inexact"). On selecting the
latter, they are then presented with five possible patterns (Figure 4). "Covered"

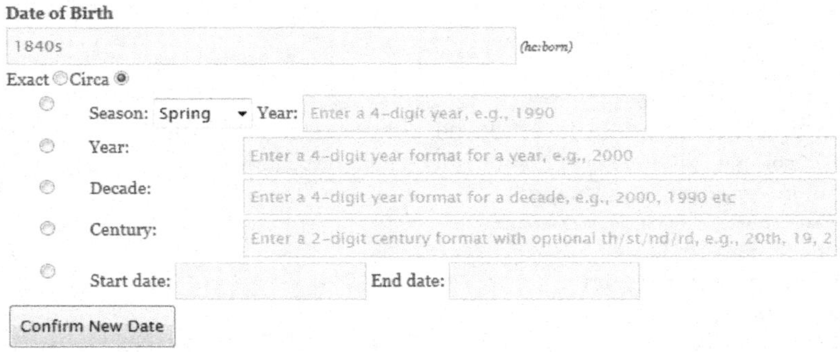

**Fig. 4.** Subsequent interface for entering an inexact date

column in Figure 1 shows the classes of dates in the corpus can be covered by
this interface. Although the first pattern cannot be found anywhere in our corpus (though it is somewhat similar to the pattern **yend**), it was suggested by our
project partners as being particularly appropriate when one is attempting to date
a photograph. In this case, the partners proposed a particular interpretation of
the seasons in terms of dates. The next three patterns correspond to three "exact
to ..." classes that we noted above. The remaining pattern allows for the entry
of an arbitrary date range where we aim to cover dates in the "vague ..." classes.
We imagine that this is a last resort as mostly one of the previous patterns will
apply.

## 4   Future Work

Although our lightweight representation of inexact dates covers very well the
inexact date data we have collected so far, further work is needed to see how
well it copes with the entry and use of new data in the future. In particular, we
may find it useful to add further input patterns to the user interface where a
clear convention emerges.

# References

1. Lukasiewiwc, T., Straccia, U.: Managing uncertainty and vagueness in description logics for the semantic web. Web Semantics **6**(4), 291–308 (2008)
2. Peterson, D., Gao, S., Malhotra, A., Sperberg-McQueen, C.M., Thompson, H.S.: W3C XML Schema Definition Language (XSD) 1.1 Part 2: Datatypes, April 2012. http://www.w3.org/TR/xmlschema11-2/
3. Boeuf, P.L., Doerr, M., Ore, C.E., Stead, S.: Definition of the CIDOC Conceptual Reference Model, May 2013. http://www.cidoc-crm.org/docs/cidoc_crm_version_5.1-draft-2013May.pdf
4. Hobbs, J., Pan, F.: Time Ontology in OWL, September 2006. http://www.w3.org/TR/owl-time/
5. Allen, J.F., Ferguson, G.: Actions and events in interval temporal logic. J Logic Computation **4**(5), 531–579 (1994)
6. Taylor, S., Jekjantuk, N., Mellish, C., Pan, J.Z.: Reasoning Driven Configuration of Linked Data Content Management Systems. In: Kim, W., Ding, Y., Kim, H.-G. (eds.) JIST 2013. LNCS, vol. 8388, pp. 429–444. Springer, Heidelberg (2014)

# Learning and Discovery

# A Multi-strategy Learning Approach to Competitor Identification

Tong Ruan[1]([✉]), Yeli Lin[1], Haofen Wang[1], and Jeff Z. Pan[2]

[1] East China University of Science and Technology, Shanghai 200237, China
{ruantong,whfcarter}@ecust.edu.cn, lin_yeli@163.com
[2] University of Aberdeen, Aberdeen, Scotland
jeff.z.pan@abdn.ac.uk

**Abstract.** Competitor identification tries to find competitors of some entity in a given field, which is the key to the success of market intelligence. Manually collecting competitors is labor-intensive and time consuming. So automatic approaches are proposed for this purpose. However, these approaches suffer from the following two main challenges. Competitor information might not only be contained in semi-structured sources like lists or tables, but also be mentioned in free texts. The diversity of its sources make competitor identification quite difficult. Also, these competitors might not always occur in form of their full names. The occurrences of name variants further increase the diversity, and make the task more challenging. In this paper, we propose a novel unsupervised approach to identify competitors from prospectuses based on a multi-strategy learning algorithm. More precisely, we first extract competitors from lists using some predefined heuristic rules. By leveraging redundancies among competitor information in lists, tables, and texts, these competitors are fed as seeds to distantly supervise the learning process to find table columns and text patterns containing competitors. The whole process is iteratively performed. In each iteration, the newly discovered competitors of high confidence from various sources are treated as new seeds for bootstrapping. The experimental results show the effectiveness of our approach without human intentions and external knowledge bases. Moreover, the approach significantly outperforms traditional named entity recognition approaches.

**Keywords:** Competitor mining · Unsupervised learning · Distant supervision · Wrapper induction

## 1 Introduction

Competitor mining tries to identify competitors of certain companies in a given domain. It is quite important to the success of market intelligence. The information of competitors is not only useful for individual companies, but also vital for market analyzers and investors. In the new economic environment of China, stockbroking companies and stock exchanges begin to have rights to recommend

© Springer International Publishing Switzerland 2015
T. Supnithi et al.(Eds.): JIST 2014, LNCS 8943, pp. 197–212, 2015.
DOI: 10.1007/978-3-319-15615-6_15

or approve companies to IPO (Initial Public Offering) in a brand new board. These agencies are very active in looking for companies with great potentials. Given that a competitor of a well-known company in a specific field is a good candidate, the above mentioned company finding problem can be treated as a competitor identification task.

Due to the large number of companies on markets and the emergence of new companies, it is labor-intensive and time consuming to manually collect competitors. Therefore, automatic approaches have been proposed for this purpose [1, 2]. However, these approaches suffer from the following two main challenges. Competitor information might not only be contained in semi-structured sources like lists or tables, but also be mentioned in free texts. We call it the *source diversity* challenge. Traditional information extraction methods only focus on a particular type of data sources. For example, Ciravegna et al. [3], Milne et al. [4], and Limaye et al. [5] studied how to extract knowledge from lists, tables, or Semantic Web. On the other hand, Web-based IE methods and systems like Snowball [6], OpenIE/TextRunner [7], and KnowItAll [8] mainly extract data from texts. While systems such as LODIE [9] extract information from both free texts and structured data, how to fully utilize different kinds of data sources is not fully investigated especially for competitor identification. In addition, competitors might not always be mentioned in form of their full names. The occurrences of name variants further increase the diversity, and make the task more challenging. We call it the *expression diversity* challenge.

In this paper, we focus on competitor identification in prospectuses. Each prospectus contains competitors of a particular company occurring in tables, lists or free texts with different name descriptions. In order to tackle both challenges, we propose a novel unsupervised approach based on a multi-strategy learning algorithm. More precisely, we first extract competitors from lists using some predefined heuristic rules. By leveraging redundancies among competitor information in lists, tables, and texts, these competitors are fed as seeds to distantly supervise the learning process to find table columns and text patterns containing competitors. The whole process is iteratively performed. We carried out experiments on prospectuses of Chinese listed companies obtained from Shanghai Stock Exchange. The experimental results show the effectiveness of our approach without human interventions and external knowledge bases. Moreover, the approach significantly outperforms traditional named entity recognition approaches.

The rest of the paper is organized as follows. Section 2 lists several aspects of related work. Section 3 analyzes the structure as well as the competitor occurrences of prospectus. It then gives a overview of multi-strategy learning for the competitor identification task. Section 4 presents our approach in details. Section 5 shows experiment results of our work. Finally, Section 6 concludes the paper and points out the future direction of our work.

## 2   Related Work

While our work is the first to identify competitors with combined strategies on prospectuses, there exist several aspects of related work.

## 2.1   Competitor and Competitive Mining

Lappas et al. [2] defined a formal definition of competitiveness between products sold on B2C Web sites. They developed an algorithm called CMiner to find top-$k$ competitive items for a given product. They run their algorithm on different datasets ranging from Amazon.com, Booking.com to TripAdvisor.com. The results show that the algorithm is effective and can be applied to different domains. Cominer [1] extracted competitors of an object (a company, a sports team etc) from Web. Given the name of an object, it queried the search engine with predefined linguistic patterns to gather its competitor name and rank these competitors accordingly. Cominer also mined competitive domain and competitive evidence. Since competitors are expressed in different ways on the Web, the linguistic patterns cannot cover all situations, and thus Cominer can only mine competitive relationship between well-known companies whose information is very redundant on the Web.

## 2.2   Information Extraction

Competitor identification is an application of information extraction which combines relation extraction with named entity recognition. Information extraction has been studied intensively over the past few years. *Wrapper induction* is a sort of information extraction, which extracts knowledge from semi-structured data. Multi-view learner [10] and Vertex! [11] use supervised learning algorithms to learn data extraction rules from manually labeled training examples. Other systems like SKES [12] and LODIE [9] use unsupervised methods. Moreover, Dalvi et al. [13] presented a generic framework to learn wrappers across Web sites. Another kind of information extraction is to extract structured information from texts, which is called *text mining*. Snowball [6] and TextRunner [7] are two typical examples. The input to Snowball is a corpus of text documents and a small set of seeds. Extracted patterns can be learned by summarizing the occurrence patterns of seeds in the corpus. TextRunner learns all relations in a corpus without any predefined rules or hand-tagged seeds, thus the method is called "Open Information Extraction".

One trend of information extraction is to utilize various data published on the Web including Web pages, Linked Open Data as well as lists and tables on dynamic Web sites. Gentile et al. [14] proposed a methodology called *multi-strategy learning* which combines text mining with wrapper induction to extract knowledge from tables, lists and Web pages. While the method seems promising, there are no clear evaluation results in their paper. On the other hand, *distant supervision* is an effective mean to leverage redundancies among different sources, which has been used in [15,16]. Mintz et al. [15] leveraged entity pairs of a certain relation from Freebase as seeds and collected sentences containing these pairs as *weakly labeled data* from a text corpus. Then they extracted syntactic and semantic features of context words around entity pairs to train a multiclass logistic classifier for relation extraction. As a result, 10,000 instances of 102 relations were extracted at a precision of 67.6%. Roth et al. [16] used distant

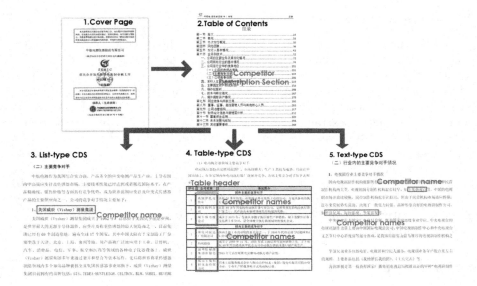

**Fig. 1.** Observations of Competitor Information in Prospectuses

supervision in pattern learning and built a system called RelationFactory [17]. RelationFactory RelationFactory achieved top ranked F1-Score at 37.3% in TAC KBP 2013 English Slot Filling evaluation. To the best of our knowledge, distant supervision has not been used for competitor identification.

## 3   Approach Overview

### 3.1   Problem Analysis

After we analyze more than 800 prospectuses about companies in different fields from the Chinese stock market, we have the following observations.

– **Observation 1.** Almost every prospectus has a specific section to describe the competitors of a company. The section is called *Competitor Description Section (CDS)*. Each CDS contains one or more paragraphs, and nearly 80% titles of CDS contain the word "竞争对手(Competitors)". The right upper part (i.e. "Table of Contents") of Figure 1 shows a CDS called " 主要竞争 对手 (Major Competitors)".
– **Observation 2.** Competitors mentioned in a CDS might appear in different forms (e.g., a list, a table or free text). Moreover, a CDS may contain competitor information of more than one form. As shown in the lower parts of figure 1, competitor names are listed as titles of subsections in a list-type CDS, are contents of the same column of a table in a certain table-type CDS, and are mentioned in a text-type CDS respectively.

**Table 1.** Distribution of Different CDS Types in One Prospectus

|        | List   | Text   | Table  | List+Text | List+Table | Table+Text | All    |
|--------|--------|--------|--------|-----------|------------|------------|--------|
| Amount | 246    | 147    | 142    | 148       | 44         | 94         | 18     |
| Ratio  | 29.32% | 17.52% | 16.92% | 17.64%    | 5.24%      | 11.20%     | 2.15%  |

- **Observation 3.** Competitor information is redundant in prospectuseses. Redundancies not only exist in different forms of the same prospectus, but also can be found different prospectuses of the same domain. The former is called *intra redundancy*, and the later is called *inter redundancy*.

Table 1 shows the distribution of different CDS types in one prospectus. From the table, we can see a large proportion of prospectuses contain at least two forms of CDS, which indicates big intra redundancies. We then check the inter redundancies between different prospectuses. 99 appear in list-type CDSs of at least two prospectuses, 85 come from table-type CDSs of different prospectuses, and another 85 are from text-type CDSs. Furthermore, 209 competitors occur in a list-type CDS of one prospectus but in a table-type CDS or a text-type CDS of another prospectus with the same names. 203 is the number of inter redundancies for the situation when one is from a table-type CDS and we find matches from another type of CDS in a different prospectus. Similarly, 195 is the answer for the third situation. Observation 2 and 3 are the basis of our multi-strategy learning algorithm. Redundant competitors from CDSs of one type can be used to annotate their occurrences in CDSs of other two types.

### 3.2   Overall Architecture of Our Approach

The objective of our approach is to find a way to extract competitor information in a language-independent way without the use of any named entity recognition (NER) tools or any prior knowledge about company information. According to observations introduced in Section 3.1, competitors are mentioned in different kinds of CDSs (list-type, table-type, and text-type). We also find rich intra- and inter-redundancies between different types. In such circumstances, competitors are first extracted from structured sources using specific wrappers. As far, the main concern is whether we can use a limited set of heuristic rules or some automatic mechanism to get these wrappers which can cover most cases. Then the extracted competitors are further used as seeds to help competitor identification from free texts, which can be modeled as a distant supervision process. The overall architecture of our approach is shown in Figure 2. We have two main steps namely *Competitor Description Section Detection* and *Multi-Strategy Learning*.

- **Competitor Description Section Detection.** As mentioned in Observation 1, there exists a specific section describing competitors and competitive information in each prospectus called CDS. Compared with the whole prospectus, a CDS is more focused and is thus more appropriate for competitor identification. So the first step of our work is to find the CDS for each

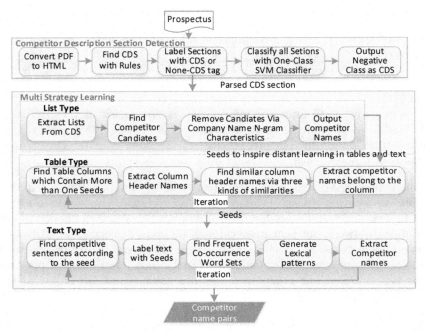

**Fig. 2.** Overall Architecture of Our Approach

prospectus. Since titles of a large number of CDSs contain the word "竞争对手 (Competitors)", heuristic rules can be used to find these CDSs easily. In order to find CDSs for the remaining prospectuses, we use a classification-based method. The details will be described in Section 4.1.

– **Multi-Strategy Learning.** According to Observation 2, competitor names may occur in lists, tables or free texts in CDSs. Since list-type CDSs are the easiest to deal with, we first identify competitive lists and extract competitor names from these lists, as shown in Figure 1. Then these competitor names are served as seeds to extract more competitors from table-type CDSs and text-type CDSs. Furthermore, the extracted competitors of high confidence from table-type CDSs and text-type CDSs can also be fed as seeds to each other. For table-type CDSs, we detect competitive columns in tables based on the input seeds, and then header names of detected columns are used to find similar column headers. In the iterative process, cell contents within the detected columns are extracted as competitor names. The iteration will not terminate until the newly discovered column contents do not conform to a n-gram model of company names. For text-type CDSs, competitive sentences containing seeds are collected. Competitive lexical patterns are then learned from contexts of these seeds in sentences, and competitor names are extracted by applying the above patterns to new sentences in other text-type CDSs. The process is also iterative. More details of multi-strategy learning are introduced in Section 4.2.

# 4  Approach Details

## 4.1  Competitor Description Section Detection

If the title of a section has the word "竞争对手(Competitors)", the section is a CDS. Since each prospectus has a well-organized table of contents, the CDS section is easy to locate. However, still 20 percent prospectuses cannot be covered. Since the heuristic rule can find CDSs with high precision, we use CDSs found by the rule as positive examples while other sections in these prospectuses are treated as the background corpus. Then the CDS detection task is modeled as a one-class classification problem, and we use one-class SVM (Support Vector Machine) as the classification model. Each section is represented as a word feature selection. When applying the model to each section of a prospectus, CDSs will be detected. For feature selection, *Information Gain* is used to select words which can distinguish CDSs with the background corpus for performance improvements. Since prospectuses are in PDF formats, certain preprocessing steps are required. Firstly, prospectuses are converted into HTML formats, then HTML tags are removed and texts are extracted. At last, texts are segmented into words using natural language processing tools, and stop words are removed. The overall accuracy of CDS detection is higher than 95%.

## 4.2  Multi-Strategy Learning

**Seeds Extraction from List-type CDSs.** Lists containing competitor names in list-type CDSs have the following characteristics. These lists are parallel structures in CDSs and each item in such a structure starts with sequence numbers. There are two types of such structure. One is subsections in a CDS, as shown in Figure 1. The other is in a text paragraph whose precedent words are "竞争对手是 (competitors are)" or "竞争对手有 (competitors include)". Since some lists might contain false positives, the extracted texts from these lists have no relationship with competitor names. In order to filter out unrelated strings, we use three rules to check whether a string is an organization name. If any of the rules does not hold, the corresponding string does not refer to an organization. The rules are learned by calculating statistics collected from the organization list provided by Shanghai Bureau of Public Security [18].

- *Lengths of organization names.* Organization names have a limited number of characters. The length distributions of organization names, including full names and abbreviations, are shown in Figure 3 and Figure 4 respectively. In most cases, the lengths of full organization names are between 4 and 24. If the length of a string is out of the range, we can filter it out.
- *Character occurrences in organization names.* The distribution of commonly used 3500 Chinese characters is calculated from the organization list, which shows the user preferences to name organizations. In particular, we collect positive characters that occur frequently with organizations, and negative characters which never appear in organization names. For instance, negative

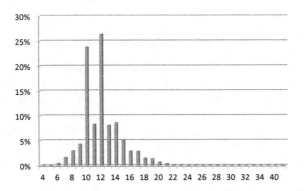

**Fig. 3.** Length Distribution of Full Names

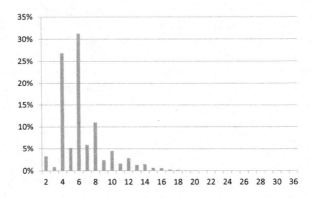

**Fig. 4.** Length Distribution of Abbreviated Names

words such as "篡 (tamper)" and "瘟 (pestilence)" have no occurrences in the organization list. A language model is built using these characters to predict the probability whether the string is an organization. If the probability is very low, it can be treated as a non-organization of high confidence.

– *Length Difference of items in a list.* If items in a same list are organization names, their lengths should not differ too much. In contrast, the word lengths might vary a lot. When the length difference exceeds 6 characters, we can safely remove it.

We extract competitors from a list-type CDS as follows. First, we find all lists in the CDS of a prospectus. Then we remove those parallel structures in text paragraphs whose precedent words do not contain "竞争对手(competitor)".

After that, the first string after each sequence number is extracted as a competitor name candidate. We further remove names from those candidates which do not follow any of the above three rules. Finally, the remaining ones are returned as competitors, which are used as seeds for further processing.

**Competitive Table Column Detection.** The following observations are used to detect competitors in table-type CDSs.

- While the formats of tables are diverse in real world, tables in prospectuses are much simpler and table headers always refer to columns. Even for complex tables which have nested headers across multiple columns, we can still find headers strictly aligned to one column as our targets.
- Contents in a table column have the same sort of data. For instance, if more than two cells of a column in a table contain competitor names, then all cells of that column correspond to competitors.
- Table columns contain similar contents if they have similar headers. For example, the column "主要竞争对手 (major competitors)" is similar to the column titled " 竞争企业 (competitive companies)", and both column contents refer to competitors.
- Column headers are similar if they have similar contexts. The context of a column header is defined as names of all other headers in the table. For example, If a table has four columns in form of $(h1, h2, h3, h4)$, then $(h2, h3, h4)$ is the context of a header $h1$.

The process of detecting competitors from table-type CDSs is as follows.

1. Extracting tables. All tables in CDSs are extracted.
2. Finding competitive table columns. For a table column, if it has more than two cells and the contents of these cells are recognized as competitor names by results extracted from list-type CDSs as seeds, the column is a possible competitive table column.
3. Finding similar table columns. We then find all columns whose header names are similar to the names in $C_{names}$. There are three ways to calculate the similarity between two table columns. They are **Cosine Similarity**, **Edit Distance**, and **Context Similarity**. The former two captures the header name similarity while the last one is the distributional similarity between the contexts of two headers. We use the combined results of the three types of similarities to find additional competitive table column candidates. This step is inspired by the work done by Limaye et al. [5].
4. Extracting competitor names. For a column returned in the previous two steps, we check its contents to see whether they conform to the three rules of organization names. If yes, we take contents of the column as competitor names, and add these names to the seed set, and finally add the column header name to the name set of competitive columns denoted as $C_{names}$.
5. Iterating the whole process until no new column headers can be found.

The details of the three types of similarity calculations are as follows.

- **Cosine Similarity.** It is a measure of similarity between two vectors. The words occurring in the two table header names are used as features, and the times of occurrences are used as weights of features. For example, to compute the similarity between "竞争对手" and "竞争企业", after word segmentation, three features "竞争", "对手", and "企业" are selected. Then the two header names are converted into vectors as $< 1, 1, 0 >$ and $< 1, 0, 1 >$, and thus the similarity score is 0.5.

- **Edit Distance.** It is a way to quantify how two strings are similar to each other by counting the minimum number of operations (insert, delete, and substitution) required to transform one string into the other. The smaller the `Edit Distance` is, the more similar the two headers are. For example, "厂商" can be transformed to "厂家" by one substitution operation, namely "商" replaced by "家". Therefore, the `Edit Distance` is 1.

- **Context Similarity.** The context of each header is also represented as a vector, and `Cosine Similarity` is used to calculate the similarity between two header contexts. If the name of a header name appears in more than one table, the two "local" contexts are merged and the occurrences of some headers are accumulated. For example, one table column header list is $(h1, h2, h4, h5)$ and the other is $(h1, h2, h3, h4)$. The context vector of $h1$ is $< h2, h3, h4, h5 >$ by merging two local context vectors $< h2, h4, h5 >$ and $< h2, h3, h4 >$. The weight of the context is $< 2, 1, 1, 1 >$. If there are altogether eight header names for all tables, the context vector of $h1$ is $< 0, 2, 1, 1, 1, 0, 0, 0 >$.

**Distant Supervision for Competitor Patterns in Free Texts.** Competitor identification on text-type CDSs requires competitor name annotations in sentences to learn patterns. These patterns are further used in other sentences to extract competitors. The quality of the extracted competitors heavily depends on the number of annotated sentences while manual annotation costs too many human efforts. Here, we leverage competitors extracted from previous steps to label free texts automatically. Such kind of distant supervision can save manual efforts of labeling sentences significantly. We first collect sentences that contain seed competitor names and label these sentences. Then we generate frequent co-occurrence word sets from the labeled corpus. Extraction patterns are further generated from the word sets and the annotated sentences. Finally, we use the generated patterns to extract new competitors from the other text-type CDSs. The whole process is iterative until there are no new patterns found. We describe the details of each step as follows:

1. **Labeling Text with Seeds**
   We use a triple <L, seed, R>to express the occurrences of a seed. The "L" is the left context of a seed with a few words before the seed, and the "R" is the right context having several words after the seed. We do not allow triples spanning across multiple sentences separated by punctuations like full stops, commas, and semi-commas.

   Sometimes, there are more than one competitor names in a sentence and these names occur continuously. For example, "竞争对手有微软、索尼与苹果等 (competitors are Microsoft, Sony, and Apple etc..)". Such sentences are called *multi-slot sentences*, and sentences that contain only one competitor name are called *single-slot sentences*. For the above example, if the seed is "SONY", the triple looks like < "competitors are Microsoft", seed, "and

Apple etc..">. Since the left side or the right side of a seed may also be an organization name, it is slightly different to generate patterns for multi-slot sentences, which will be discussed later.

2. **Generating Frequent Co-occurrence Word Sets**

   In multi-slot sentences, competitor names are delimited by "slight-pause marks". We neglect seeds which have "slight-pause mark" directly before or after them at their left contexts or right contexts. That is to say, we only select sentences labeled with at least two seeds where the left side of the most left seed should not be a "slight-pause mark", and the right side of the most right seed is not a "slight-pause mark" either. Back to the above example, if "微软 (Microsoft)" and "苹果(Apple)" become seeds, the whole sentence can be labeled, and the triple is <"竞争对手有 (competitors are)", "微软、索尼与苹果 (Microsoft, Sony, and Apple)", "等 (etc..) ">.

   In this way, multi-slot sentences and one-slot sentences can be processed in the same way as follows. First, we perform word segmentation for $S$ where $S$ is the set of strings of all "L"s and "R"s in <L, seed, R>triples. Each string in $S$ is segmented into words using segmentation tools. Then we select words occurring more than 5 times as words of high frequencies. Finally, we construct the high frequency co-occurrence word sets by selecting those co-occur in <L, seed, R>triples for more than twice.

3. **Generating Extraction Patterns**

   Frequent co-occurrence word sets are used to generate extraction patterns. Take $WS$ as the word sets, and an element $ws$ in $WS$ would be a word set $\{w_1, w_2...w_n\}$. All sentences which contain $\{w_1, w_2...w_n\}$ are returned, and each sentence forms a distinguished pattern. Patterns are aligned and consolidated based on the order of occurrences of elements in $ws$. Some words are replaced by wildcards, others are unioned together. We also retain the boundary words of the occurrences in the merged pattern. The pattern generalization is similar to that of Snowball [6]. The novelty lies on the previous steps especially the first step to label sentences with seeds extracted from other types of CDS corpus automatically. The distant supervision part as well as the multi-strategy learning part have not been covered in traditional seed-based pattern learning methods.

   For example, if "企业 (Company)" and "竞争对手 (Competitor)" are within one frequent word set. Given two sentences "企业的主要竞争对手有微软与甲骨文。(The major competitors of the company are Microsoft and Oracle.)", and "主要竞争对手企业是索尼与苹果。 (Major competitive company are SONY and Apple.)", the distinguished pattern for each sentence is "企业*竞争对手*有 (Company*Competitor*include)", and "竞争对手*企业*是 (Competitor *Company* are)" respectively. The merged pattern is "(企业*竞争对手—竞争对手*企业)*(有—是)". It is more general to cover both situations. Here * denotes any number of any characters, | means the union of several characters, and boundary strings such as "有 (include)" and "是 (are)" are included in the pattern.

**Table 2.** Classification of Identified Competitors

|  | Is-competitor | Not-competitor |
|---|---|---|
| Extracted | A | B |
| Not-Extracted | C | D |

## 4. Extracting Competitor Names Using Patterns

Based on patterns generated in the previous step, strings are extracted as competitor candidates. Note that not all candidates refer to organizations. Thus, we reuse the above mentioned three rules to check whether a string is an organization name, and thus filter out irrelevant ones.

## 5 Experiments

### 5.1 Experiment Setup

All prospectuses used in the experiment were crawled from the Web site of Shanghai Stock Exchange (http://www.sse.com.cn). Although our approach is unsupervised, labeled data is required to assess the quality and the coverage of extracted results. 836 prospectuses are manually labeled, which result in 3000 competitor pairs in total. Precision and Recall are used as the evaluation metrics. As shown in Table 2, we use A to represent the number of correctly extracted competitor names, B is the number of incorrectly extracted competitor names, C indicates the number of competitor names that are not extracted. In this way, precision can be defined as $A/(A+B)$, and Recall is $A/(A+C)$.

Precision and recall are defined at two levels: the micro level and the macro level. The micro level evaluates on prospectuses while the macro level evaluates on competitor names. If the corpus has $n$ prospectuses, and for each prospectus $d_i$, we can get the corresponding $A_i$, $B_i$, and $C_i$ as defined in Table 2. Precision and Recall at two levels are defined as follows.

$$Micro\ Precision = \frac{\sum_1^n precison(d_i)}{n} \tag{1}$$

$$Micro\ Recall = \frac{\sum_1^n Recall(d_i)}{n} \tag{2}$$

$$Macro\ Precision = \frac{\sum_1^n A_i}{\sum_1^n A_i + \sum_1^n B_i} \tag{3}$$

$$Macro\ Recall = \frac{\sum_1^n A_i}{\sum_1^n A_i + \sum_1^n C_i} \tag{4}$$

**Table 3.** Results of Each Type

|  | List | Table | Text |
|---|---|---|---|
| Micro Recall | 0.9157 | 0.9691 | 0.7440 |
| Micro Precision | 0.9734 | 0.9143 | 0.9029 |
| Macro Recall | 0.9285 | 0.9493 | 0.7497 |
| Macro Precision | 0.9814 | 0.9557 | 0.9279 |

**Table 4.** Iterations for Table-type CDSs

|  | 1 | 2 | 3 |
|---|---|---|---|
| #extracted candiates | 874 | 984 | 1015 |
| #new competitor names | 869 | 98 | 3 |
| Micro Recall | 0.9030 | 0.9686 | 0.9691 |
| Micro Precision | 0.9770 | 0.9663 | 0.9143 |
| Macro Recall | 0.8868 | 0.9433 | 0.9493 |
| Macro Precision | 0.9943 | 0.9827 | 0.9557 |

## 5.2  Results Evaluation

**Results for Each Type of CDSs.** Table 3 shows the evaluation results for competitor identification in the corpus of each CDS type. Here, precision and recall are calculated against the corpus of each CDS type instead of the whole corpus. For instance, regarding the recall $A/(A + C)$ for the list type, $A$ is the number of correctly extracted competitor names in list-type CDSs, and $C$ is number of competitor names that are not extracted in the list-type corpus. The other metrics can be calculated in the similar way. From the table, we can find that our approach achieves very high precision for each type of CDSs. The recall for text-type CDSs is lower than that on other type of corpus, but is still about 0.74. This is because we only capture frequent patterns but some sentences about competitors are described by ad hoc lexical patterns that seldomly occur. We also find the gap between micro-level and macro-level is small, which means our approach is stable without very poor performance on some prospectus.

Since competitor identification on table-type CDSs and text-type CDSs are both iterative, we show the results after each iteration for the two types of corpus in Table 4 and Table 5 respectively. For both types, after a small number of iterations (3 for table-type, and 2 for text-type), the whole process terminates. As shown in Table 4, for each iteration of table-type corpus extraction, more similar table headers are found, but fewer competitor names are extracted. For example, after the second Iteration, 984 candidates are found, but only 98 of them are recognized as organization names. This is due to the fact that most candidates are from headers similar to the headers which do not represent actual competitive table columns. Recalls increase after each iteration, but the precisions might drop a bit especially for the late iterations. This is a signal to tell that we should set more strict threshold values to ensure the quality of extracted competitors. We can have similar findings in Table 5.

**Table 5.** Iterations for Text-type CDSs

|  | 1 | 2 |
|---|---|---|
| #extracted candidates | 724 | 735 |
| #new competitor names | 675 | 7 |
| Micro Recall | 0.7345 | 0.7440 |
| Micro Precision | 0.9098 | 0.9029 |
| Macro Recall | 0.7272 | 0.7497 |

**Table 6.** Multi-strategy Learning Results

|  | List | Table | Text | List+Table | List+Text | Table+Text | All |
|---|---|---|---|---|---|---|---|
| Micro Recall | 0.4715 | 0.3791 | 0.3406 | 0.7838 | 0.6544 | 0.5611 | **0.8423** |
| Micro Precision | 0.9688 | 0.9147 | 0.8944 | 0.9433 | 0.9109 | 0.8807 | **0.9090** |
| Macro Recall | 0.4114 | 0.4470 | 0.3142 | 0.7892 | 0.5814 | 0.6058 | **0.8487** |
| Macro Precision | 0.9814 | 0.9557 | 0.9279 | 0.9698 | 0.9540 | 0.9444 | **0.9437** |

**Results of Multi-strategy Learning.** We also carry out experiments to compare the performance using multi-strategy learning with that based on single-strategy learning. The detailed experimental results are shown in Table 6. The column "All" represents our final results, which combines the results from different types of CDSs. The column in form of "A+B" refer to the combined results from the A-type corpus and the B-type corpus. Unlike the computation of recall values in Table 3, recalls are computed against the whole corpus instead of one specific type of CDS corpus. This is why for the first three columns (i.e. List, Table, and Text), their recalls are much lower than those reported in Table 3.

From the table, we can see that recall values are greatly improved through multi-strategy learning. The micro recall value for list-type, table-type and text-type CDSs is 0.4715, 0.3791 and 0.3406 respectively. They are all below 0.5. However, the combined result is about 0.8423, almost 200% increases. In addition, even we use multi-strategy learning on two types of CDS corpus, the recall value improvements are obvious. Meanwhile, the precision values are still very high. All these findings show the effectiveness of multi-strategy learning.

## 5.3   Comparison with Traditional NER-Based Methods

In order to identify competitors, we can also use named entity recognition (NER) methods to find organization mentions in the CDSs. Here, we select some popular NLP tools namely NLPIR[1], FudanNLP[2], and Stanford NER[3] to extract competitors from the corpus. For the Stanford NER, we further distinguish whether it uses distributional similarity features, denoted as Stanford NER with dist-Sim, and Stanford NER without distSim. We use these tools as the baselines to

---

[1] http://ictclas.nlpir.org/

[2] http://code.google.com/p/fudannlp/

[3] http://nlp.stanford.edu/software/CRF-NER.shtml

**Table 7.** The Results of Different Methods

|  | NLPIR | FudanNLP | Stanford NER with distSim | Stanford NER without distSim | Our Approach |
|---|---|---|---|---|---|
| Precision | 34.21% | 28.97% | 68.02% | 63.85% | **94.37%** |
| Recall | 1.02% | 2.43% | 47.30% | 40.80% | **84.87%** |

compare with our approach. Macro precisions and macro recalls of all methods are shown in Table 7.

From the table, NLPIR and FudanNLP perform worst with low precisions (around 30%) and pretty low recalls (between 1% and 3%). When using distributional similarity features, Stanford NER can achieve more than 5% increases in terms of precision and recall. Compared with these baselines, our approach has much more promising results. The precision almost reaches 94.37% and the recall is also higher than 80%.

# 6    Conclusions and Future Work

In this paper, we provide a multi-strategy learning approach to extract competitors from Chinese prospectuses. Different kinds of competitive description sections (list-type, table-type, and text-type) require different extraction methods and have different levels of difficulties. We extract competitors from list-type CDSs first, and the extraction results are fed as seeds to boost the extraction process from other two CDS types. Distant supervised learning is used in these processes to avoid manual labeling efforts. One benefit of our approach is that the named entity recognition (NER) step is not required to identify competitors. Experimental results show our approach achieves higher precision and recall than those of the traditional NER methods. As for the future work, we plan to try our approach on English prospectuses and then extend to other corpus like company Web sites for mining competitors.

**Acknowledgments.** This work is funded by the National Key Technology R&D Program through project No. 2013BAH11F03.

# References

1. Bao, S., Li, R., Yu, Y., Cao, Y.: Competitor mining with the web. IEEE Transactions on Knowledge and Data Engineering **20**(10), 1297–1310 (2008)
2. Lappas, T., Valkanas, G., Gunopulos, D.: Efficient and domain-invariant competitor mining. In: Proceedings of the 18th ACM SIGKDD International Conference on Knowledge Discovery and Data Mining, pp. 408–416. ACM (2012)
3. Ciravegna, F., Chapman, S., Dingli, A., Wilks, Y.: Learning to harvest information for the semantic web. In: Bussler, C.J., Davies, J., Fensel, D., Studer, R. (eds.) ESWS 2004. LNCS, vol. 3053, pp. 312–326. Springer, Heidelberg (2004)

4. Milne, D., Witten, I.H.: Learning to link with wikipedia. In: Proceedings of the 17th ACM Conference on Information and Knowledge Management, pp. 509–518. ACM (2008)
5. Limaye, G., Sarawagi, S., Chakrabarti, S.: Annotating and searching web tables using entities, types and relationships. Proceedings of the VLDB Endowment **3**(1–2), 1338–1347 (2010)
6. Agichtein, E., Gravano, L.: Snowball: extracting relations from large plain-text collections. In: Proceedings of the Fifth ACM Conference on Digital Libraries, pp. 85–94. ACM (2000)
7. Banko, M., Cafarella, M.J., Soderland, S., Broadhead, M., Etzioni, O.: Open information extraction for the web. IJCAI **7**, 2670–2676 (2007)
8. Etzioni, O., Cafarella, M., Downey, D., Kok, S., Popescu, A.M., Shaked, T., Soderland, S., Weld, D.S., Yates, A.: Web-scale information extraction in knowitall: (preliminary results). In: Proceedings of the 13th International Conference on World Wide Web, pp. 100–110. ACM (2004)
9. Ciravegna, F., Gentile, A.L., Zhang, Z.: Lodie: Linked open data for web-scale information extraction. SWAIE **925**, 11–22 (2012)
10. Hao, Q., Cai, R., Pang, Y., Zhang, L.: From one tree to a forest: a unified solution for structured web data extraction. In: Proceedings of the 34th International ACM SIGIR Conference on Research and Development in Information Retrieval, pp. 775–784. ACM (2011)
11. Gulhane, P., Madaan, A., Mehta, R., Ramamirtham, J., Rastogi, R., Satpal, S., Sengamedu, S.H., Tengli, A., Tiwari, C.: Web-scale information extraction with vertex. In: IEEE 27th International Conference on Data Engineering (ICDE 2011), pp. 1209–1220. IEEE (2011)
12. He, J., Gu, Y., Liu, H., Yan, J., Chen, H.: Scalable and noise tolerant web knowledge extraction for search task simplification. Decision Support Systems **56**, 156–167 (2013)
13. Dalvi, N., Kumar, R., Soliman, M.: Automatic wrappers for large scale web extraction. Proceedings of the VLDB Endowment **4**(4), 219–230 (2011)
14. Gentile, A.L., Zhang, Z., Ciravegna, F.: Web scale information extraction with lodie. In: 2013 AAAI Fall Symposium Series (2013)
15. Mintz, M., Bills, S., Snow, R., Jurafsky, D.: Distant supervision for relation extraction without labeled data. In: Proceedings of the Joint Conference of the 47th Annual Meeting of the ACL and the 4th International Joint Conference on Natural Language Processing of the AFNLP, vol. 2, pp. 1003–1011. Association for Computational Linguistics (2009)
16. Roth, B., Barth, T., Wiegand, M., Singh, M., Klakow, D.: Effective slot filling based on shallow distant supervision methods. arXiv preprint arXiv:1401.1158 (2014)
17. Roth, B., Barth, T., Chrupała, G., Gropp, M., Klakow, D.: Relationfactory: a fast, modular and effective system for knowledge base population. In: EACL 2014, p. 89 (2014)
18. Xue, C., Wang, H., Jin, B., Wang, M., Gao, D.: Effective chinese organization name linking to a list-like knowledge base. In: Zhao, D., Du, J., Wang, H., Wang, P., Ji, D., Pan, J.Z. (eds.) CSWS 2014. CCIS, vol. 480, pp. 97–110. Springer, Heidelberg (2014)

# Mining Type Information from Chinese Online Encyclopedias

Tianxing Wu[1]([✉]), Shaowei Ling[1], Guilin Qi[1], and Haofen Wang[2]

[1] Southeast University, Nanjing, China
{wutianxing,lingshaowei,gqi}@seu.edu.cn
[2] East China University of Science and Technology, Shanghai, China
whfcarter@ecust.edu.cn

**Abstract.** Recently, there is an increasing interest in extracting or mining type information from Web sources. Type information stating that an instance is of a certain type is an important component of knowledge bases. Although there has been some work on obtaining type information, most of current techniques are either language-dependent or to generate one or more general types for a given instance because of type sparseness. In this paper, we present a novel approach for mining type information from Chinese online encyclopedias. More precisely, we mine type information from abstracts, infoboxes and categories of article pages in Chinese encyclopedia Web sites. In particular, most of the generated Chinese type information is inferred from categories of article pages through an attribute propagation algorithm and a graph-based random walk method. We conduct experiments over Chinese encyclopedia Web sites: Baidu Baike, Hudong Baike and Chinese Wikipedia. Experimental results show that our approach can generate large scale and high-quality Chinese type information with types of appropriate granularity.

**Keywords:** Mining type information · Category attributes generation · Data of online encyclopedias · Knowledge base

## 1 Introduction

Linking Open Data (LOD)[1] is the largest community effort for semantic data publishing which converts the Web from a Web of document to a Web of interlinked knowledge. There have been over 200 datasets within the LOD project. Among these datasets, DBpedia [1] and Yago [18] serve as hubs to connect others. Similar to DBpedia, Zhishi.me [13] has been developed as the first effort of Chinese LOD. It extracted RDF triples from three largest Chinese encyclopedia Web sites namely Baidu Baike[2], Hudong Baike[3] and Chinese Wikipedia[4].

---

[1] http://linkeddata.org/
[2] http://baike.baidu.com/
[3] http://www.hudong.com/
[4] http://zh.wikipedia.org/

© Springer International Publishing Switzerland 2015
T. Supnithi et al.(Eds.): JIST 2014, LNCS 8943, pp. 213–229, 2015.
DOI: 10.1007/978-3-319-15615-6_16

Recently, there is an increasing interest in extracting or mining type information from Web sources. Type information stating that an instance is of a certain type (e.g. *"China"* is an instance of *"country"*) is an important component of knowledge bases. Type information plays an important role in many applications, such as query understanding [5,19], question answering [10,21] and product recommendation [9,11].

Among the datasets in current LOD, the number of Chinese type information is limited. Very little or no effort has been devoted to obtaining type information in Chinese LOD. For example, Zhishi.me only uses the SKOS vocabulary[5] to represent the category system and does not strictly define the *IsA* relation between instances and categories, which makes it less comprehensive than DBpedia and Yago. Besides, most of current techniques on obtaining type information are either language-dependent [3,4,7,8,18,22] or to generate one or more general types for a given instance because of type sparseness [1,14,15]. Therefore, with the purpose of generating large scale and high-quality Chinese type information with types of appropriate granularity, we make the first effort to mine type information from Chinese online encyclopedias.

In Chinese online encyclopedias, we discover that lots of fine-grained types exist in categories of article pages, e.g., given the article page of *"China"*, its categories are *"Asia country"*, *"country with an ancient civilization"*, *"socialist country"*, *"East Asia"*, etc. Obviously, some categories can be regarded as correct types, but the noise does exist (i.e. *"East Asia"*). Thus, we take the categories of one given instance as its candidate types and try to filter out the noise. We argue that attributes are critical in filtering out the noise in categories of article pages. Intuitively, when given attributes *"actors, release date, director"* of a certain instance, people may infer that it is an instance of *"movie"*, but when given *"name, foreign name"*, people cannot infer the instance type because too many categories have the attribute *"name"* or *"foreign name"*. Here, we assume if an instance contains the representative attributes (e.g. *"actors"*, *"release date"* and *"director"* of the type *"movie"*) of one candidate type, the instance probably belongs to this type. However, category attributes are not abundantly available. Therefore, we need to generate attributes for as many categories as possible.

In this paper, we propose a three-step approach for mining type information from Chinese online encyclopedias. Our approach first identifies explicit *InstanceOf* relations (i.e. type information) and *SubclassOf* relations with several heuristics. It then applies an attribute propagation algorithm leveraging existing category attributes, instance attributes, identified *InstanceOf* and *SubclassOf* relations to generate new category attributes. Finally, it constructs a weighted directed graph for each instance which has been enriched with attributes and categories, and applies a graph-based random walk method to discover more type information. We conduct experiments over Chinese encyclopedia Web sites: Baidu Baike, Hudong Baike and Chinese Wikipedia. Experimental results show that our approach can generate large scale and high-quality Chinese type information with types of appropriate granularity.

---

[5] http://www.w3.org/2004/02/skos/

The contributions in this paper are summarized as follows:

- We present an approach for mining type information from Chinese online encyclopedias, which harvests large scale and high-quality Chinese type information as an important complementary part of Chinese LOD.
- We generate high-quality attributes for 42,934 categories using an attribute propagation algorithm.
- We present an evaluation for the quality of category attributes and type information generated by our approach. The category attributes and type information respectively achieve the precision of more than 88% and 91%.
- We compare the generated Chinese type information with that of DBpedia, Yago and BabelNet [12]. It shows that our generated type information has not only the largest number of Chinese types, typed instances and type statements (i.e. *InstanceOf* relations), but also instance types with more appropriate granularity.

The rest of this paper is organized as follows. Section 2 gives an overview of previous work that is related to mining type information. Section 3 introduces the proposed type mining approach in detail. In Section 4, we evaluate our approach on different Chinese encyclopedia Web sites and make a comparison between our obtained Chinese type information and that of other knowledge bases. At last, we conclude the paper and describe the future work in Section 5.

## 2    Related Work

Since type information has been identified as an important component of knowledge bases, researchers have used different techniques to mine type information on Wikipedia or the Web from scratch. Furthermore, there has been some work on completing the missing type information due to the low coverage of typed instances in many knowledge bases.

**Wikipedia-Based Type Mining:** Yago [18] applies a language-dependent rule to infer instance types with categories of article pages in Wikipedia. Type information in DBpedia [1] is obtained by an infobox-based method. However, only several hundred infobox-based types are used for typing millions of instances. Tipalo [7] extracts natural language definitions of instances in Wikipedia. These definitions are parsed with FRED [16] and mapped to WordNet [6] and DULplus[6] for finding appropriate types. BabelNet [12] harvests type information by the means of mapping Wikipedia instances and concepts to WordNet.

**Web-Based Type Mining:** Lexico-syntactic patterns are widely used for extracting type information from the Web. This is first proposed in [8] for acquiring hyponyms from large text corpora, and later followed by many successful systems such as PANKOW [3] and KnowItAll [4]. More recently, Probase [22]

---

[6] http://www.ontologydesignpatterns.org/ont/wn/dulplus.owl

obtains large scale type information from Web pages leveraging a fixed set of syntactic patterns as well as the existing knowledge.

**Type Information Completion:** In order to mine types for DBpedia untyped instances, machine learning and rule-based techniques are proposed in [14] to complete the missing type information with existing DBpedia instance types. SDType [15] is another approach for completing instance types based on statistic distribution. This approach is capable of dealing with noisy data as well as faulty schemas or unforeseen usage of schemas and can be applied to any RDF knowledge base.

In this paper, we focus on mining type information from encyclopedia Web sites rather than the whole Web, because encyclopedia Web sites contain many high-quality and comprehensive types of structures (e.g. infobox, categories of article pages, category system, etc.) fitting for mining instance types. Thus, mining type information in Yago and DBpedia are the closest work to ours. However, the heuristic rule Yago applied can not be extended to other languages and the method used in DBpedia generates only one general type for a given instance because of type sparseness. Here, we aim to generate large scale and high-quality type information with types of appropriate granularity and do not pay attention to the problem of the low coverage of typed instances.

## 3    Approach

In this section, we introduce our approach for mining Chinese type information from Chinese online encyclopedias. The workflow is implemented as outlined in Figure 1. At first, *Explicit IsA Relation Detector* identifies explicit *InstanceOf* relations from infoboxes and abstracts of article pages, and *SubclassOf* relations from the category system. Then, existing category attributes and instance attributes are respectively derived from infobox templates and infoboxes of article pages. For example, the attributes in infobox template of *"people"* are *"name"*, *"birthdate"*, *"job"*, etc. and the infobox of *"Steven Spielberg"* contains attributes *"birthdate"*, *"occupation"*, *"education"*, etc. The input of *Category Attributes Generator* consists of the identified *InstanceOf* relations, existing category attributes, instance attribues and a *Category Graph* composed of all categories with identified *SubclassOf* relations. *Category Attributes Generator* tries to generate attributes for as many categories as possible leveraging an attribute propagation algorithm. Afterwards, we organize each given instance, its attributes and categories (i.e. candidate types) of the corresponding article page into an *Instance Graph* as the input of *Instance Type Ranker*. *Instance Type Ranker* infers high-quality types for each given instance through a graph-based random walk method. Finally, we acquire Chinese type statements by converting all instance types into RDF triples. In the following, we introduce each component of the proposed approach in more detail.

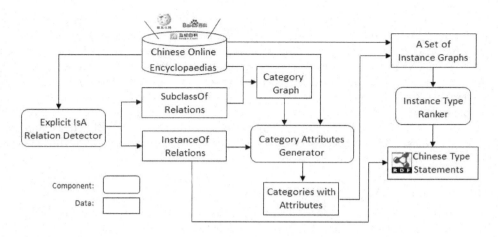

**Fig. 1.** The Workflow of Our Approach

### 3.1  Explicit IsA Relation Detector

This section describes the heuristics for detecting explicit *IsA* relations consisting of explicit *InstanceOf* relations and *SubclassOf* relations, which can be used to help generate category attributes.

**Explicit InstanceOf Relation Detection:** We detect explicit *InstanceOf* relations from infoboxes and abstracts appearing in Chinese online encyclopedias. (1) In infoboxes, we find that some of (attribute, value) pairs are likely to be (concept, instance) pairs, e.g., (director, Steven Spielberg). In encyclopedia Web sites, articles and categories are used to construct an instance set and a concept set (or a type set) respectively. We assume if an attribute exists in the concept set and the instance set contains its value, then there is an *InstanceOf* relation between the value and attribute. (2) Generally, each abstract of article pages gives the specific definition of its describing instance in the first sentence. Natural Language Processing (NLP) technologies are adopted to extract *InstanceOf* relations. We perform dependency parsing with FudanNLP [17] on the first sentence of each abstract. If the subject is the instance described by the given article page, the predicate is word "是 (a 'be' verb)", and the object belongs to the concept set, then an *InstanceOf* relation exists between the subject and the object.

**Explicit SubclassOf Relation Detection:** Explict *SubclassOf* relations are identified from categories in Chinese online encyclopedias. Though categories are organized in a thematical manner as a thesaurus, there are many *SubclassOf* relations among them. We first generate candidate *SubclassOf* category pairs in the form of (sub-category, category) based on the category system. Then, we detect *SubclassOf* relations in these category pairs with two heuristics. The first heuristic method labels *SubclassOf* relations for category pairs sharing the same lexical head, e.g., 江苏学校 (school in Jiangsu) *SubclassOf* 中国学校 (school in China). According to the language characteristics of Chinese, we take the last noun as the lexical head after performing POS tagging on each category with FudanNLP, e.g., we parse the category "中国足球运动员 (Chinese football player)" to get the result "中国 (Chinese)/LOC 足球 (football)/NN 运动员 (player)/NN", then, the word "*player*" is treated as the lexical head. After implementing the first heuristic method, for each (sub-category, category) pair remained, if the category is a parent concept of the sub-category in Zhishi.schema [20] (a Chinese concept taxonomy with large scale *SubclassOf* relations), the category pair has a *SubclassOf* relation.

## 3.2    Category Attributes Generator

Attributes in infobox templates and infoboxes of article pages are respectively extracted as existing category attributes and instance attributes, but the infobox templates in encyclopedia Web sites are not abundantly available, e.g., the number of infobox templates in Baidu Baike, Hudong Baike and Chinese Wikipedia is 0, 214, 812, respectively. The lack of category attributes has an adverse effect on inferring type information with attributes from categories of article pages, because lots of categories without attributes which can be taken as the correct instance types have to be discarded. In this section, a *Category Graph* composed of all categories with *SubclassOf* relations is constructed at first. Subsequently, we try to propagate attributes over the *Category Graph* leveraging existing category attributes, instance attributes, identified *InstanceOf* and *SubclassOf* relations.

**Category Graph:** We define a *Category Graph* as a directed acyclic graph $G = (N, E)$, where $N$ is the set of nodes representing all categories and a directed edge $< c_1, c_2 > \in E$ represents a *SubclassOf* relation between category $c_1$ and $c_2$, where $c_1, c_2 \in N$.

**Attribute Propagation Algorithm:** Attribute propagation over the *Category Graph* follows the rules below:

- **Rule 1**: If a category $c \in N$ has attributes from infobox templates, these attributes should remain unchanged.
- **Rule 2**: If a category $c \in N$ has some instances with attributes, the attributes should be propagated to c when they are shared by more than half of these instances.
- **Rule 3**: If a category $c \in N$ has some child categories with attributes, the attributes should be propagated to c when they are shared by more than half of these child categories.
- **Rule 4**: If parent categories of a category $c \in N$ have attributes, all the attributes should be inherited by c.

Here, **Rule 2** and **Rule 3** are based on the idea of majority voting. Thus, we choose 0.5 as the threshold. Before introducing the attribute propagation algorithm, several related definitions are given at first, we define:

- **CA** $= \{CA_1, ..., CA_k\}$, where $CA_i$ is a set of category attributes, $i \in [1, k]$, $k \leq |N|$ and $|N|$ is the number of all categories containing attributes; $\mu_{CA}$ : $N \rightarrow$ **CA** is a mapping function.
- **TCA** $= \{TCA_1, ..., TCA_l\}$, where $TCA_i$ is a set of category attributes from infobox templates, $i \in [1, l]$, $l \leq |N|$ and $|N|$ is the number of the categories, which contains attributes deriving from infobox templates. $\mu_{TCA} : N \rightarrow$ **TCA** is a mapping function.
- **CI** $= \{CI_1, ..., CI_m\}$, where $CI_i$ is a set of category instances, $i \in [1, m]$, $m \leq |N|$ and $|N|$ is the number of the categories containing instances.
- **IA** $= \{IA_1, ..., IA_o\}$, where $IA_i$ is a set of instance attributes, $i \in [1, o]$, and $o$ is the number of the instances with attributes.
- $\nu_{IA}$ is a function returning the instance attributes based on **Rule 2**.
- $\nu_{CA}$ is a function returning the category attributes based on **Rule 3**.

---

**Algorithm 1.** Attribute Propagation

---

**Input:** $G, \mathbf{TCA}, \mathbf{CI}, \mathbf{IA}$
**Output:** **CA**

1  $\mathbf{CA} \leftarrow \emptyset, S_{root} \leftarrow \emptyset, Queue_{\mathbf{CA}} \leftarrow \emptyset, Queue_{top\_down} \leftarrow \emptyset, Queue_{bottom\_up} \leftarrow \emptyset$;
2  $EnQueue(Queue_{\mathbf{CA}}, \mathbf{CA})$;
3  **for** *each* $c \in N$ **do**
4      **if** $getParents(c) = \emptyset$ **then**
5          $EnQueue(Queue_{top\_down}, c)$;
6          $S_{root} \leftarrow S_{root} \bigcup \{c\}$;
7  Sort $c \in N$ by the maximum depth in $G$ by desc;
8  Put each sorted $c \in N$ into $Queue_{bottom\_up}$;
9  **while** *true* **do**
10      $\mathbf{CA} \leftarrow TopDownPropagation(G, \mathbf{CA}, \mathbf{TCA}, \mathbf{CI}, \mathbf{IA}, S_{root}, Queue_{top\_down})$;
11      $\mathbf{CA}' \leftarrow DeQueue(Queue_{\mathbf{CA}})$;
12      **if** $CA = CA'$ **then**
13          **return CA**;
14      **else**
15          $EnQueue(Queue_{\mathbf{CA}}, \mathbf{CA})$;
16      $\mathbf{CA} \leftarrow BottomUpPropagation(G, \mathbf{CA}, \mathbf{TCA}, Queue_{bottom\_up})$;
17      $\mathbf{CA}' \leftarrow DeQueue(Queue_{\mathbf{CA}})$;
18      **if** $CA = CA'$ **then**
19          **return CA**;
20      **else**
21          $EnQueue(Queue_{\mathbf{CA}}, \mathbf{CA})$;

---

Algorithm 1 gives a high level overview of our attribute propagation algorithm. It repeats the top-down (*line* 10) and bottom-up (*line* 16) process of attribute propagation over the constructed *Category Graph*, until convergence, i.e., **CA** no longer changes (*line* 12-13, 18-19). During the process of top-down attribute propagation (i.e. Algorithm 2), it checks whether a category $c \in N$ has attributes from infobox templates, if so, these attributes belonging to $c$ should remain unchanged (i.e. **Rule 1**) (*line* 3-6). Otherwise, instance attributes and parent category attributes can be propagated to c with **Rule 2** (*line* 8-11) and **Rule 4** (*line* 13-21). The attributes of child categories are propagated to $c$ with **Rule 3** unless **Rule 1** (*line* 3-9) is satisfied in the process of bottom-up attribute propagation (i.e. Algorithm 3).

---

**Algorithm 2.** TopDownPropagation

---

**Input:** $G, \mathbf{CA}, \mathbf{TCA}, \mathbf{CI}, \mathbf{IA}, S_{root}, Queue$
**Output:** **CA**

1  **while** $Queue \neq \emptyset$ **do**
2     $c \leftarrow DeQueue(Queue)$;
3     **if** $c$ *satisfies* **Rule 1** **then**
4        **if** $\mu_{CA}(c) = \emptyset$ **then**
5           $\mathbf{CA} \leftarrow \mathbf{CA} \bigcup \{\mu_{TCA}(c)\}$;
6           $\mu_{CA} : c \rightarrow \mu_{TCA}(c)$;

7     **else**
8        **if** $c$ *satisfies* **Rule 2** *with* **CI** *and* **IA** **then**
9           **if** $\mu_{CA}(c) = \emptyset$ **then**
10             $\mathbf{CA} \leftarrow \mathbf{CA} \bigcup \{\nu_{IA}(c)\}$;
11             $\mu_{CA} : c \rightarrow \nu_{IA}(c)$;

12       //*check whether c satisfies* **Rule4** *or not*;
13       **if** $c \notin S_{root}$ **then**
14          **for** *each* $c^{'} \in getParents(c)$ **do**
15             **if** $\mu_{CA}(c^{'}) \neq \emptyset$ **then**
16                **if** $\mu_{CA}(c) = \emptyset$ **then**
17                   $CA* \leftarrow \emptyset, CA* \leftarrow \mu_{CA}(c^{'})$;
18                   $\mathbf{CA} \leftarrow \mathbf{CA} \bigcup \{CA*\}$;
19                   $\mu_{CA} : c \rightarrow CA*$;

20                **else**
21                   $\mu_{CA}(c) \leftarrow \mu_{CA}(c) \bigcup \mu_{CA}(c^{'})$;

22    **for** *each* $c^{'} \in getChildren(c)$ **do**
23       $EnQueue(Queue, c^{'})$;

24 **return CA**

---

**Algorithm 3.** BottomUpPropagation

---
**Input**: $G, \mathbf{CA}, \mathbf{TCA}, Queue$
**Output**: $\mathbf{CA}$

1 **while** $Queue \neq \emptyset$ **do**
2 $\quad$ $c \leftarrow DeQueue(Queue)$;
3 $\quad$ **if** $c$ *does not satisfy* **Rule 1** *and satisfies* **Rule 3** **then**
4 $\quad\quad$ **if** $\mu_{CA}(c) = \emptyset$ **then**
5 $\quad\quad\quad$ $CA^* \leftarrow \emptyset, CA^* \leftarrow \nu_{CA}(c)$;
6 $\quad\quad\quad$ $\mathbf{CA} \leftarrow \mathbf{CA} \bigcup \{CA^*\}$;
7 $\quad\quad\quad$ $\mu_{CA} : c \rightarrow CA*$;
8 $\quad\quad$ **else**
9 $\quad\quad\quad$ $\mu_{CA}(c) \leftarrow \mu_{CA}(c) \bigcup \nu_{CA}(c)$;

10 **return CA**

---

### 3.3 Instance Type Ranker

After generating attributes for as many categories as possible, we organize each given instance, its attributes and categories (i.e. candidate types) of the corresponding article page into an *Instance Graph*. Then, a graph-based random walk method is applied to rank the candidate types for filtering out the noise.

**Instance Graph:** Each *Instance Graph* is defined as a weighted directed graph $G = (N, E, P; \varphi)$, where

- $N = N_I \uplus N_A \uplus N_C$ is the set of nodes, consisting of $N_I = \{i_j\}$, the singleton set of the given instance $i_j$, $N_A$, the nodes representing attributes of $i_j$, and $N_C$ of $i_j$, the category nodes;
- $E = E_{IA} \uplus E_{AC}$ is the set of edges, where $E_{IA}$ represents directed edges from the given instance to its attributes s.t. $< i_j, a_k > \in E_{IA}$ iff $i_j \in N_I$ and $a_k \in N_A$, $E_{AC}$ stands for directed edges from attributes to categories of the corresponding article page s.t. $< a_k, c_l > \in E_{AC}$ iff $a_k \in N_A$ and $c_l \in N_C$;
- $P = P_{IA} \bigcup P_{AC}$ is the set of probabilities, where $P_{IA}$ represents the set of probabilities of walking from the given instance to its attributes, $P_{AC}$ stands for the set of probabilities of walking from the attributes to categories of the corresponding article page;
- $\varphi : E \rightarrow P$ is a mapping function.

An example of *Instance Graph* is constructed in Figure 2. Since all the attributes are derived from encyclopedia Web sites, people are free to use different labels to represent an attribute with the same meaning (e.g., 生日 (birthday) and 出生日期 (birth date)). This may cause the absence of many edges between attributes and categories when constructing each *Instance Graph* without the synonym sets of attributes. For a given instance $i_j \in N_I$, if an instance attribute $a_k \in N_A$ does not belong to an category $c_l \in N_C$ of the corresponding article page but $c_l$ has a synonymous attribute of $a_k$, then a directed edge

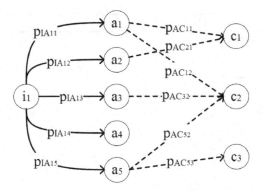

**Fig. 2.** An Example of *Instance Graph*

should be connected $a_k$ to $c_l$. Therefore, we group synonymous attributes with BabelNet before constructing all *Instance Graphs*.

**Graph-Based Random Walk:** We assume that the fewer categories an attribute belongs to, the more representative the attribute is. If an instance contains such representative attributes belonging to a category, the category may probably be the type of the instance. Considering all the categories, the weight for each attribute is fixed and defined as:

$$Weight(a_i) = \frac{1}{Count(a_i)} \tag{1}$$

where $a_i$ is an attribute and $Count(a_i)$ is the amount of categories that have $a_i$ or its synonymous attributes. Hence, the attributes in the same synonym set have identical weight.

The random walk process starts from the given instance $i_j \in N_I$, then, executes a random step to one of its attributes $a_k \in N_A$ with the probability $pIA_{jk} \in P_{IA}$, finally, executes a random step from an attribute to one of categories $c_l \in N_C$ in the article page of $i_j$, using the probability $pAC_{kl} \in P_{AC}$. $pIA_{jk}$ and $pAC_{kl}$ are defined as follows:

$$pIA_{jk} = \frac{Weight(a_k)}{\sum_{t=1}^{N} Weight(a_t)} \tag{2}$$

$$pAC_{kl} = \frac{1}{Count^*(a_k)} \tag{3}$$

where $N$ is the total number of attributes and $Count^*(a_k)$ means the amount of the directed edges from $a_k$ to categories. During the process of the graph-based random walk, when executing a random step from the given instance to one of its attributes, the walk tends to choose the most representative attribute (i.e. the attribute belonging to the fewest categories) in order to walk to the

correct categories (i.e. types). When executing a random step from an attribute
to the one of the categories in the article page, the categories containing this
attribute have equal opportunity, e.g., in Figure 2, there are two directed edges
from attribute $a_1$ to categories $c_1$ and $c_2$, hence, both $p_{AC_{11}}$ and $p_{AC_{12}}$ are 0.5.
We define the probability of one category being the instance type as follows:

$$P(c_l|i_j) = \frac{Count'(c_l)}{t} \qquad (4)$$

where $i_j \in N_I$ is the given instance, $c_l \in N_C$ is one of the categories in the article
page of $i_j$, $Count'(c_l)$ is the total number of times for $c_l$ as the destination and
$P(c_l|i_j)$ converges after $t$ times graph walk. Note that when executing a random
step from an attribute to one of the categories, if there is no directed edge from
this attribute(e.g., $a_3$ in Figure 2) to any category in the article page, then this
graph walk will terminate but also count.

## 4    Experiments

We apply the approach proposed in Section 3 to mine type information from
three largest Chinese encyclopedia Web sites: Baidu Baike, Hudong Baike and
Chinese Wikipedia. In this Section, we first determine the parameters of *Instance
Type Ranker*. Then, we evaluate the accuracy of the generated category attribu-
tes and type information. At last, we make a comparison between our obtained
Chinese type information and that of other knowledge bases.

### 4.1    Parameters Determination

As mentioned in Section 3.3, a category $c_l \in N_C$ in the article page of the given
instance $i_j \in N_I$ has a probability $P(c_l|i_j)$ of being the instance type. The prob-
ability depends on the number of times that the graph-based random walk is
executed until convergence. If $P(c_l|i_j)$ is greater than $\theta$, we take $c_l$ as the type
of $i_j$. In our experiments, we used a conservative strategy to ensure the precision.
First, we randomly selected 500 instances in the results of the graph-based random walk from each encyclopedia Web sites and labelled candidate types of each
instance. Then, we took the highest probability of the types labelled "*Incorrect*"
as the value of $\theta$. For Baidu Baike, Hudong Baike and Chinese Wikipedia, $\theta$ is
set to 0.0513, 0.0399, 0.009, respectively.

### 4.2    Accuracy Evaluation

**Accuracy of Category Attributes:** After implementing the attributes propagation algorithm, high-quality attributes are generated for 42,934 distinct categories from Baidu Baike, Hudong Baike and Chinese Wikipedia. In order to
evaluate the accuracy of the generated category attributes, we apply a similar
labeling process as that used in Yago due to the lack of ground truths. We randomly selected 500 (category, attribute) pairs from each encyclopedia Web site

to form three subsets and invited six postgraduate students who are familiar with linked data to participant in the labeling process. Each annotator chose *"Correct"*, *"Incorrect"*, or *"Unknown"* to label (category, attribute) pairs in three subsets mentioned above. To generalize findings on each subset to the whole (category, attribute) pairs of each encyclopedia Web site, we computed the Wilson intervals [2] for $\alpha = 5\%$. The evaluation results of category attributes is shown in Table 1. All (category, attribute) pairs from different encyclopedia Web sites have achieved the average precision of more than 88%.

**Table 1.** Evaluation Results of Category Attributes

| Web Site | Category Number | Precision |
|---|---|---|
| Baidu Baike | 5,636 | 90.68% ± 2.51% |
| Hudong Baike | 12,531 | 91.67% ± 2.39% |
| Zh-Wikipedia | 34,956 | 94.06% ± 2.03% |

**Accuracy of Type Information:** We obtained 1,326,625 distinct Chinese type statements by converting all instance types into RDF triples, including explicit *InstanceOf* relations and instance types inferred from categories of article pages. According to the statistics, totally 661,680 distinct Chinese instances have been typed with 31,491 types and each of them has two types on average. The labeling process used to evaluate the accuracy of category attributes is also applied to evaluating the accuracy of the obtained type information. Each annotator labelled the same subsets, each of which contains 500 randomly selected type statements from different sources in encyclopedia Web sites. Table 2 shows the evaluation results. According to the results, all type statements from different sources in encyclopedia Web sites have achieved the average precision of more than 91%. In particular, 70.97% of the type statements are inferred from categories of article pages, achieving the high precision of more than 96%. This shows the effectiveness of our attribute propagation algorithm and graph-based random walk method.

**Table 2.** Evaluation Results of Type Statements

| Web Site | Source | Number | Precision |
|---|---|---|---|
| Baidu Baike | infobox | 96979 | 94.66% ± 1.93% |
|  | abstract | 61951 | 93.66% ± 2.10% |
|  | category | 97258 | 98.03% ± 1.16% |
| Hudong Baike | infobox | 93064 | 96.24% ± 1.62% |
|  | abstract | 73424 | 96.64% ± 1.53% |
|  | category | 734545 | 98.03% ± 1.16% |
| Zh-Wikipedia | infobox | 24567 | 97.63% ± 1.28% |
|  | abstract | 50854 | 94.46% ± 1.67% |
|  | category | 148084 | 97.63% ± 1.28% |

### 4.3   Comparison with Other Knowledge Bases

**Overlap of Type Information:** We compared all obtained Chinese type information from different encyclopedia Web sites (i.e. 1,326,625 distinct Chinese type statements) with that of other well-known knowledge bases namely DBpedia, Yago and BabelNet. Table 3 not only gives the number of types, typed instances and type statements in each dataset, but also shows the overlap of types, typed instances and type statements between our obtained Chinese type information and that of other knowledge bases. BabelNet contains Chinese type information but DBpedia and Yago do not. Since DBpedia and Yago have multilingual versions, we mapped the English type statements to Chinese ones (both instance and type in one type statement can be mapped to the Chinese labels).

**Table 3.** Overlap between Our Type Information and that of Other Knowledge Bases

|                        | Our Data  | DBpedia | Yago   | BabelNet |
|------------------------|-----------|---------|--------|----------|
| Type Number            | 31,491    | 155     | 1,719  | 318      |
| Type Overlap           | /         | 74      | 133    | 94       |
| Typed Instance Number  | 661,680   | 150,827 | 42,818 | 142      |
| Typed Instance Overlap | /         | 39,093  | 13,629 | 20       |
| Type Statements Number | 1,326,625 | 263,765 | 45,947 | 563      |
| Type Statements Overlap| /         | 2,689   | 167    | 0        |

According to the comparison results, the number of types, typed instances and type statements generated by our approach is significantly more than those in other knowledge bases. As for the overlap of types, typed instances and type statements, our dataset does not cover much of them in other knowledge bases. It indicates that current knowledge bases or the datasets in linked data are really short of Chinese type information, and our obtained Chinese type information is an effective complementary part of Chinese LOD.

**Comparison for the Granularity of Instance Types:** High-quality type information is not merely of high precision, but also has instance types with appropriate granularity. As shown in Table 3, all of DBpedia, Yago and Babel-Net have a small number of Chinese instance types. A large amount of instances are typed with general types because of type sparseness, e.g., *"Tang Wei"* is typed with *"person"* in DBpedia, but the fine-grained type *"actress"* or *"Chinese actress"* may be more appropriate and useful. Here, we made a manual comparison between the granularity of instance types in our obtained type information and other knowledge bases. 500 instances were randomly selected from each overlap of typed instances, but the number of typed instances in the overlap between our obtained type information and BabelNet is only 20, thus, all of them were taken as the labelling data. We presented all types from the corresponding datasets for each selected instance to six postgraduate students, who compared the granularity of instance types in our type information with that in other knowledge bases. Three choices (i.e. *"Better"*, *"Poorer"* and *"Similar"*) are provided to label each sample. For example, *"Tang Wei"* is typed with *"Chinese actress"* and *"people from Hangzhou"* in our type information, but in DBpedia, *"Tang Wei"* is only typed with *"person"*. In this circumstance, all of the annotators chose *"Better"*. After labelling, we computed the average proportions of three choices for each pairwise comparison.

Figure 3 gives the results of each pairwise comparison. Compared with DBpedia and Yago, the types of more than half of the instances in our type information have better or more appropriate granularity. In contrast, only the types of less than 18% instances are of poor granularity. Though the number of typed instance overlap between our type information and BabelNet is only 20, the average proportion of *"Better"* is still larger than that of *"Poorer"*, which is less than 30%. Hence, based on such comparison results, we hold that the instance types generated by our proposed approach have relatively appropriate granularity.

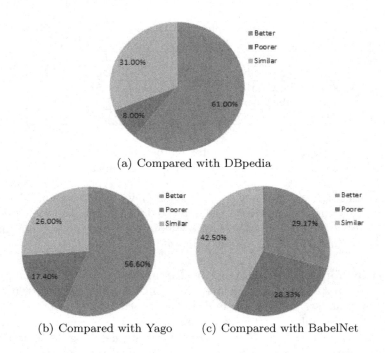

(a) Compared with DBpedia

(b) Compared with Yago    (c) Compared with BabelNet

**Fig. 3.** Average Proportions of Three Choices for Each Pairwise Comparison

## 5    Conclusions and Future Work

In this paper, we proposed a novel approach to mine type information from Chinese online encyclopedias. Our approach mines type information from categories as well as infoboxes and abstracts. In particular, more than 70% of the generated Chinese type statements are inferred from categories of article pages leveraging attributes. In order to mine type information from categories, we presented an attribute propagation algorithm to generate attributes for as many categories as possible and a graph-based random walk method to infer instance types from categories of article pages. Unlike traditional approaches focusing on mining English type information in LOD, our proposed approach has effectively mined large scale and high-quality Chinese type information with types of appropriate granularity as an important complementary part of Chinese LOD.

The experimental results have shown that the generated category attributes and type information have respectively achieved the precision of more than 88% and 91%. Totally, our approach has generated attributes for 42,934 categories as well as 661,680 typed instances and 1,326,625 type statements. Compared with well known knowledge bases including DBpedia, Yago and BabelNet, our obtained type information has not only the largest number of Chinese types, typed instances and type statements, but also instance types with more appropriate granularity.

As for the future work, we consider generalizing our approach to make it language independent and quantifying the granularity of instance types to automatically determine which type is more appropriate for a given instance. Since our approach depends largely on the categories of article pages, instance attributes and category attributes, it may lead to the low coverage of typed instances. Therefore, we also plan to solve this problem from two aspects. First, the attributes for more instances and categories can be further extracted with technologies of information extraction. Second, we will attempt to find out the latent article categories (i.e. candidate types) in encyclopedia Web sites for the instances, which contain few categories in the corresponding article pages.

# References

1. Bizer, C., Lehmann, J., Kobilarov, G., Auer, S., Becker, C., Cyganiak, R., Hellmann, S.: Dbpedia-a crystallization point for the web of data. Web Semantics: Science, Services and Agents on the World Wide Web **7**(3), 154–165 (2009)
2. Brown, L.D., Cai, T.T., DasGupta, A.: Interval estimation for a binomial proportion. Statistical Science, 101–117 (2001)
3. Cimiano, P., Handschuh, S., Staab, S.: Towards the self-annotating web. In: Proceedings of the 13th International Conference on World Wide Web (WWW 2004), pp. 462–471 (2004)
4. Etzioni, O., Cafarella, M., Downey, D., Kok, S., Popescu, A.M., Shaked, T., Soderland, S., Weld, D.S., Yates, A.: Web-scale information extraction in knowitall: (preliminary results). In: Proceedings of the 13th International Conference on World Wide Web (WWW 2004), pp. 100–110 (2004)
5. Fang, Y., Si, L., Somasundaram, N., Al-Ansari, S., Yu, Z., Xian, Y.: Purdue at trec 2010 entity track: a probabilistic framework for matching types between candidate and target entities. In: Proceedings of the 18th Text REtrieval Conference (TREC 2010) (2010)
6. Fellbaum, C. (ed.): WordNet: An electronic lexical database. MIT Press, Cambridge (1998)

7. Gangemi, A., Nuzzolese, A.G., Presutti, V., Draicchio, F., Musetti, A., Ciancarini, P.: Automatic typing of DBpedia entities. In: Cudré-Mauroux, P., Heflin, J., Sirin, E., Tudorache, T., Euzenat, J., Hauswirth, M., Parreira, J.X., Hendler, J., Schreiber, G., Bernstein, A., Blomqvist, E. (eds.) ISWC 2012, Part I. LNCS, vol. 7649, pp. 65–81. Springer, Heidelberg (2012)

8. Hearst, M.A.: Automatic acquisition of hyponyms from large text corpora. In: Proceedings of the 14th Conference on Computational Linguistics (COLING 1992), pp. 539–545 (1992)

9. Hepp, M.: GoodRelations: An ontology for describing products and services offers on the web. In: Gangemi, A., Euzenat, J. (eds.) EKAW 2008. LNCS (LNAI), vol. 5268, pp. 329–346. Springer, Heidelberg (2008)

10. Kalyanpur, A., Murdock, J.W., Fan, J., Welty, C.: Leveraging community-built knowledge for type coercion in question answering. In: Aroyo, L., Welty, C., Alani, H., Taylor, J., Bernstein, A., Kagal, L., Noy, N., Blomqvist, E. (eds.) ISWC 2011, Part II. LNCS, vol. 7032, pp. 144–156. Springer, Heidelberg (2011)

11. Lee, T., Chun, J., Shim, J., Lee, S.G.: An ontology-based product recommender system for b2b marketplaces. International Journal of Electronic Commerce 11(2), 125–155 (2006)

12. Navigli, R., Ponzetto, S.P.: Babelnet: Building a very large multilingual semantic network. In: Proceedings of the 48th Annual Meeting of the Association for Computational Linguistics (ACL 2010), pp. 216–225 (2010)

13. Niu, X., Sun, X., Wang, H., Rong, S., Qi, G., Yu, Y.: Zhishi.me - weaving Chinese linking open data. In: Aroyo, L., Welty, C., Alani, H., Taylor, J., Bernstein, A., Kagal, L., Noy, N., Blomqvist, E. (eds.) ISWC 2011, Part II. LNCS, vol. 7032, pp. 205–220. Springer, Heidelberg (2011)

14. Nuzzolese, A.G., Gangemi, A., Presutti, V., Ciancarini, P.: Type inference through the analysis of wikipedia links. In: Proceedings of WWW 2012 Workshop on Linked Data on the Web (LDOW 2012) (2012)

15. Paulheim, H., Bizer, C.: Type inference on noisy RDF data. In: Alani, H., Kagal, L., Fokoue, A., Groth, P., Biemann, C., Parreira, J.X., Aroyo, L., Noy, N., Welty, C., Janowicz, K. (eds.) ISWC 2013, Part I. LNCS, vol. 8218, pp. 510–525. Springer, Heidelberg (2013)

16. Presutti, V., Draicchio, F., Gangemi, A.: Knowledge extraction based on discourse representation theory and linguistic frames. In: ten Teije, A., Völker, J., Handschuh, S., Stuckenschmidt, H., d'Acquin, M., Nikolov, A., Aussenac-Gilles, N., Hernandez, N. (eds.) EKAW 2012. LNCS, vol. 7603, pp. 114–129. Springer, Heidelberg (2012)

17. Qiu, X., Zhang, Q., Huang, X.: Fudannlp: A toolkit for Chinese natural language processing. In: Proceedings of the 51th Annual Meeting of the Association for Computational Linguistics (ACL 2013) (2013)

18. Suchanek, F.M., Kasneci, G., Weikum, G.: Yago: a core of semantic knowledge. In: Proceedings of the 16th International Conference on World Wide Web (WWW 2007), pp. 697–706 (2007)

19. Tonon, A., Catasta, M., Demartini, G., Cudré-Mauroux, P., Aberer, K.: *TRank*: Ranking entity types using the web of data. In: Alani, H., Kagal, L., Fokoue, A., Groth, P., Biemann, C., Parreira, J.X., Aroyo, L., Noy, N., Welty, C., Janowicz, K. (eds.) ISWC 2013, Part I. LNCS, vol. 8218, pp. 640–656. Springer, Heidelberg (2013)

20. Wang, H., Wu, T., Qi, G., Ruan, T.: On publishing chinese linked open schema. In: Mika, P., Tudorache, T., Bernstein, A., Welty, C., Knoblock, C., Vrandečić, D., Groth, P., Noy, N., Janowicz, K., Goble, C. (eds.) ISWC 2014, Part I. LNCS, vol. 8796, pp. 293–308. Springer, Heidelberg (2014)
21. Welty, C., Murdock, J.W., Kalyanpur, A., Fan, J.: A comparison of hard filters and soft evidence for answer typing in Watson. In: Cudré-Mauroux, P., Heflin, J., Sirin, E., Tudorache, T., Euzenat, J., Hauswirth, M., Parreira, J.X., Hendler, J., Schreiber, G., Bernstein, A., Blomqvist, E. (eds.) ISWC 2012, Part II. LNCS, vol. 7650, pp. 243–256. Springer, Heidelberg (2012)
22. Wu, W., Li, H., Wang, H., Zhu, K.Q.: Probase: a probabilistic taxonomy for text understanding. In: Proceedings of the 2012 ACM SIGMOD International Conference on Management of Data (SIGMOD 2012), pp. 481–492 (2012)

# G-Diff: A Grouping Algorithm for RDF Change Detection on MapReduce

Jinhyun Ahn[1,2], Dong-Hyuk Im[3]($\boxtimes$), Jae-Hong Eom[1,2]($\boxtimes$), Nansu Zong[1], and Hong-Gee Kim[1,2]($\boxtimes$)

[1] Biomedical Knowledge Engineering Laboratory,
Seoul National University, Seoul, Republic of Korea
{jhahncs,zpage,zongnansu1982,hgkim}@snu.ac.kr
[2] Dental Research Institute, Seoul National University, Seoul, Republic of Korea
[3] Department of Computer and Information Engineering, Hoseo University,
Cheonan, Republic of Korea
dhim@hoseo.edu

**Abstract.** Linked Data is a collection of RDF data that can grow exponentially and change over time. Detecting changes in RDF data is important to support Linked Data consuming applications with version management. Traditional approaches for change detection are not scalable. This has led researchers to devise algorithms on the MapReduce framework. Most works simply take a URI as a Map key. We observed that it is not efficient to handle RDF data with a large number of distinct URIs since many Reduce tasks have to be created. Even though the Reduce tasks are scheduled to run simultaneously, too many small Reduce tasks would increase the overall running time. In this paper, we propose G-Diff, an efficient MapReduce algorithm for RDF change detection. G-Diff groups triples by URIs during Map phase and sends the triples to a particular Reduce task rather than multiple Reduce tasks. Experiments on real datasets showed that the proposed approach takes less running time than previous works.

## 1 Introduction

RDF (Resource Description Framework) is a standard for the Semantic Web to represent facts by triples (subject, predicate and object). Linked Data is a collection of RDF data that is updated over time [4]. Triples in a dataset in Linked Data might overlap with previously published ones. Instead of storing every version of RDF data, keeping only changes would benefit RDF Triple Stores in terms of version management [9,10]. Detected changes can also tell us some interesting facts. For example, in DBPedia 3.7, we have a triple (*San Marcos Baptist Academy, affiliation, Southern Baptist Convention*). However, in DBPedia 3.8, the same triple doesn't exist but another triple does (*San Marcos Baptist Academy, affiliation, Baptist General Convention of Texas*). We can say that *San Marcos Baptist Academy* has changed its *affiliation* from *Southern Baptist Convention* to *Baptist General Convention of Texas*. RDF change detection

© Springer International Publishing Switzerland 2015
T. Supnithi et al.(Eds.): JIST 2014, LNCS 8943, pp. 230–235, 2015.
DOI: 10.1007/978-3-319-15615-6_17

can be defined as follows: given a source triple set and a target triple set, return added triples which are in the target but not in the source and also output deleted triples which are in the source but not in the target. The changed triple is then automatically identified by grouping two triples in added triples and deleted triples by a combination of subject, predicate and object.

**Table 1.** A comparison of selected tools that can be used for RDF changed detection. Only G-Diff and Link-Diff are based on distributed environment. Balanced distribution indicates the mechanism proposed by this paper.

| | RDF Triple support | Inference support | Distributed processing | Balanced distribution |
|---|---|---|---|---|
| G-Diff | Yes | No | MapReduce | Yes |
| Link-Diff[2] | Yes | No | MapReduce | No |
| SemVersion[5] | Yes | Yes | No | No |
| GNU diff[11] | No (Text) | No | No | No |
| X-Diff[3] | No (XML) | No | No | No |

Table 1 lists existing tools for RDF change detection and G-Diff. GNU diff utility is one of the most popular tools to detect changes of text [11]. But it cannot directly be used for RDF triples. X-Diff [3] loads DOM (Document Object Model) trees of two input XML documents into main memory and tries to find edit scripts, a sequence of operations that convert one XML document into another one. SemVersion [5] and Delta Function [6] proposed algorithms to find RDF deltas (changes). They tried to minimize the size of RDF deltas by not including added or deleted triples that can be inferred by reasoning rules (e.g. transitive relation). These works run on a single machine for which it is hard to handle large data. In this paper, we compared with Link-Diff, SemVersion and GNU diff whose implementation is available. Recently, the MapReduce (MR) framework is widely used for distributed processing. To the best of our knowledge, few researches proposed RDF change detection algorithms based on MR, except Im et al. [2]. They simply take the subject URI of input triples as a Map key. Each Reduce task[1] then receives a list of triples, all of which have the same subject. The drawback of the approach is that too many Reduce tasks must be created to process a RDF dataset with many distinct subject URIs. The cumulative time for creating many Reduce tasks could increase the overall running time in the MR framework [8]. The similar issues has been tackled by researchers from the database field. Okcan and Riedewald [1] proposed a MR algorithm for theta-join that determines Map keys in a balanced fashion. The approach motivates us. In this paper, we propose a Map key grouping algorithm on MR for RDF change detection.

---

[1] Reduce task in this paper refers to a dynamically created task that follows Map tasks in running time. Multiple Reduce tasks run on a "Reducer" on a slave machine.

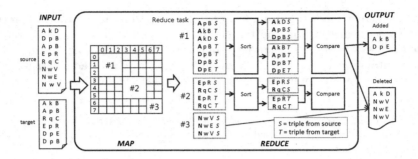

**Fig. 1.** An example work-flow of G-Diff. The fourth letter followed by triple stands for its origin, $S$ for source and $T$ for target.

## 2   A Subject URI Grouping Algorithm

We implemented G-Diff on the MR framework, which takes two RDF files (source and target) as input and outputs two files containing added triples and deleted triples. The input RDF files are assumed to be in a N-triple format that represents one triple in a single line. Figure 1 depicts an example of work-flow of G-Diff. In the preprocessing step, the number of triples in both source and target are counted by simply examining the number of lines in two input files. Of the two values, the bigger one becomes $Z$ in Equation 1. The value of $M$ is determined by the memory size available in a single machine, which is assigned the maximum number of triples that can be processed in a single Reduce task. The *hash* function takes a string as input and returns an integer.

$$MapKey(\text{triple}) = \frac{hash(\text{getSubjectURI(triple)}) \bmod Z}{M} \tag{1}$$

Each triple encountered during Map phase is passed to the Equation 1. The result value would be an integer ranged from 0 to $Z$-1 divided by $M$. To make it easier to understand, a $Z$ by $Z$ table is drawn in Figure 1. The row corresponds to source and the column to target. For example, suppose that the source triples having subject "A" or "D" is mapped into a row indexed from 0 to 2 and the target triples having subject "A" or "D" is mapped into a column indexed from 0 to 2. The intersection region means that the Map key called #1 is assigned to these triples. The triples will then be sent to the Reduce task called #1. In the Reduce task, the triples are sorted alphabetically and compared to emit deleted and added triples. Note that the procedure does not guarantee that the same number of triples is sent to different Reduce tasks. We argue that we don't have to force Reduce tasks to process exactly the same amount of data. To do so, the histogram of subject URIs must be available which is not the typical case. Moreover, it will not help decrease much running time on a MR environment where different size of Reduce tasks are automatically scheduled to run concurrently. Rather, decreasing the number of Reduce tasks to some degree helps reduce the overall running time. We discuss this issue in the next section.

**Table 2.** Datasets come from DBPedia 3.7 and 3.8. The 3rd & 4th column represent the number of triples in the version of 3.7 and 3.8, respectively. The 5th & 6th column represent the number of added and deleted triples, respectively, detected by G-Diff taking 3.7 as source and 3.8 as target.

|    | DBpedia datasets | 3.7 | 3.8 | added | deleted |
|----|---|---|---|---|---|
| D1 | specific_mappingbased_properties | 522,142 | 635,830 | 248,599 | 135,015 |
| D2 | geo_coordinates | 1,771,100 | 1,900,004 | 397,621 | 268,717 |
| D3 | short_abstracts | 3,550,567 | 3,769,926 | 1,926,581 | 1,707,220 |
| D4 | persondata | 4,504,182 | 5,959,455 | 2,123,038 | 668,375 |
| D5 | article_categories | 13,610,094 | 15,115,484 | 4,732,120 | 3,225,874 |
| D6 | mappingbased_properties | 17,520,158 | 20,516,859 | 7,316,827 | 4,370,869 |

## 3   Experiments

Experiments were carried out on 10 machines, each of which has a 3.1GHz Quad Core CPU and 4GB RAM. We configured `Apache Hadoop 1.2.1` with default settings [7]. Datesets used are listed in Table 2. It might be interesting to see that the number of deleted triples in `persondata` is relatively smaller than the other cases. We can say that data about person, in particular, is not likely to be removed but accumulated over time. The running time shown in Figure 2, 3 in this paper is averaged over 5 trials. Standard deviation is also depicted by error bars.

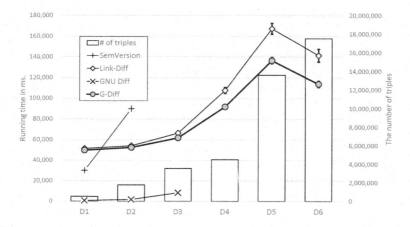

**Fig. 2.** The running time of each approach. The rectangle corresponding to the Y-axis in the right represents the number of triples in the datasets. X-axis represents datasets whose label is in Table 2. Due to the memory error, SemVersion failed from D3 to D6 and `GNU diff` on from D4 to D6.

Figure 2 shows the running time of G-Diff compared with previous works. Link-Diff is from [2] that takes a subject URI as a Map key. We used

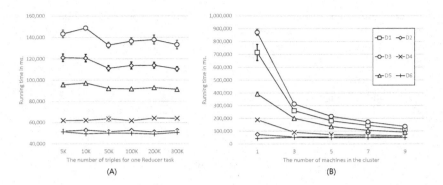

**Fig. 3.** (A) The running time for various values of parameter $M$. For example, 100K means that maximum 100,000 triples are sent to one Reduce task. (B) The running time of G-Diff on clusters with different number of slaves. For example, 3 means that the cluster has one master machine and three slave machines.

SemVersion 1.0.3 [2]. GNU `diff` refers to a Linux script that firstly issues `sort` command to sort two input files and then pass the two sorted files to `diff` command to output added and deleted lines (triples). It should be noted that GNU `diff` cannot correctly detect triple (line) changes if duplicate triples (lines) exist. We include GNU `diff` to see the scalability of existing diff tools. For G-Diff, we set the memory capacity parameter $M$ in Equation 1 as 100,000 (the value is chosen from the best ones in Figure 3). Both Link-Diff and G-Diff run on 10 machines and SemVersion and GNU `diff` on a single machine. It is shown that SemVersion and GNU `diff` failed to process large size of data as these are designed to run on a single machine. In the case of Link-Diff and G-Diff, the running time does not increase dramatically as the data size increases, which means that the algorithm is scalable. Further, we notice that G-Diff is more efficient than Link-Diff. Note that it took less time to process D6 than D5. The reason is that in D6 there exists only one predicate ⟨http://purl.org/dc/terms/subject⟩.

Figure 3 depicts the running time of G-Diff for various numbers of triples per reduce task and the number of slaves. We have configured Hadoop to make a slave machine to have maximum 1GB memory. The good performances are in the cases of 50K, and 100K. We can observe an improvement from 5K to 50K but not much from 50K to 300K. It is because that according to Equation 1, the bigger the $M$ value, the more chances to have a particular Reduce task to receive more triples. The existence of such a *big* Reduce task could increase the overall running time. The lesson we learned from this experiment is that the parameter $M$ needs to be assigned according to the subject URI histogram and the size of available memory, which is beyond the scope of this paper. In addition, one can see that more machines lead to less running time. The effect is more dramatic for larger size of datasets.

---

[2] http://mavenrepo.fzi.de/semweb4j.org/repo/org/semanticdesktop/semversion/

# 4   Conclusion

We proposed G-Diff, a grouping algorithm for RDF change detection on MapReduce frameworks. Triples having distinct subject URI are grouped and sent to a single Reducer to avoid creating many small Reduce tasks. We showed that G-Diff is scalable compared to previous works and more efficient than Link-Diff with no grouping strategy introduced in this paper. As a future direction we plan to devise a way of determining the memory capacity parameter suitable for given number of machines. It might also be interesting to make it to only extract changed triples in schema (e.g. added or deleted properties).

**Acknowledgement.** This research was supported by Basic Science Research Program (BSRP) through the National Research Foundation of Korea (NRF) funded by the Ministry of Education (NRF-2013R1A1A2065656) and BSRP through the NRF funded by the Ministry of Science, ICT & Future Planning (MSIP, Korea) (NRF-2014R1A1A1002236). This work was also supported by the Industrial Strategic Technology Development Program (10044494, WiseKB: Big data based self-evolving knowledge base and reasoning platform) funded by MSIP, Korea.

# References

1. Okcan, A., Riedewald, M.: Processing theta-joins using MapReduce. In: Proceedings of the 2011 ACM SIGMOD International Conference on Management of Data, New York, USA (2011)
2. Im, D.-H., Ahn, J., Zong, N., Jung, J., Kim, H.-G.: Link-Diff: change detection tool for linked data using MapReduce framework. In: The Workshop on Big data for Knowledge Engineering in JIST 2012, Nara, Japan, December 2–4, 2012
3. Wang, Y., DeWitt, D.J., Cai, J.-Y.: X-Diff: an effective change detection algorithm for XML documents. In: The Proceeding of Data Engineering (2003)
4. Bizer, C., Heath, T., Berners-Lee, T.: Linked Data The story so far. International Journal on Semantic Web and Information Systems **5**(3) (2009)
5. Volkel, M., Groza, T.: SemVersion: an RDF-based ontology versioning system. In: Proceedings of the IADIS International Conference WWW/Internet (2006)
6. Zeginis, D., Tzitzikas, Y., Christophides, V.: On the foundations of computing deltas between RDF models. In: Aberer, K., et al. (eds.) ASWC 2007 and ISWC 2007. LNCS, vol. 4825, pp. 637–651. Springer, Heidelberg (2007)
7. Apache Hadoop. http://hadoop.apache.org
8. Husain, M., McGlothlin, J., Masud, M.M., Khan, L., Thuraisingham, B.M.: Heuristics-Based Query Processing for Large RDF Graphs Using Cloud Computing. TKDE **23**(9), 1312–1327 (2011)
9. Cassidy, S., Ballantine, J.: Version control for RDF triple store. In: ICSOFT (ISDM/EHST/DC), vol. 512 (2007)
10. Vander Sande, M., Colpaert, P., Verborgh, R., Coppens, S., Mannens, E., Van de Walle, R.: R&Wbase: git for triples. In: Proceedings of the 6th Workshop on Linked Data on the Web (2013)
11. GNU diff. http://www.gnu.org/software/diffutils

# RDF and SPARQL

# Graph Pattern Based RDF Data Compression

Jeff Z. Pan[1]([⊠]), José Manuel Gómez Pérez[2], Yuan Ren[1], Honghan Wu[1,3],
Haofen Wang[4], and Man Zhu[5]

[1] Department of Computing Science, University of Aberdeen, Aberdeen, UK
jeff.z.pan@abdn.ac.uk
[2] iSOCO, Barcelona, Spain
[3] Nanjing University of Information and Technology, Nanjing, China
[4] East China University of Science and Technology, Shanghai, China
[5] School of Computer Science, Southeast University, Nanjing, China

**Abstract.** The growing volume of RDF documents and their inter-linking raise a challenge on the storage and transferring of such documents. One solution to this problem is to reduce the size of RDF documents via compression. Existing approaches either apply well-known generic compression technologies but seldom exploit the graph structure of RDF documents. Or, they focus on minimized compact serialisations leaving the graph nature inexplicit, which leads obstacles for further applying higher level compression techniques. In this paper we propose graph pattern based technologies, which on the one hand can reduce the numbers of triples in RDF documents and on the other hand can serialise RDF graph in a data pattern based way, which can deal with syntactic redundancies which are not eliminable to existing techniques. Evaluation on real world datasets shows that our approach can substantially reduce the size of RDF documents by complementing the abilities of existing approaches. Furthermore, the evaluation results on rule mining operations show the potentials of the proposed serialisation format in supporting efficient data access.

## 1 Introduction

The digital universe is booming, especially in terms of the amount of metadata and user-generated data available. Studies like IDC's Digital Universe[1] estimate that the size of the digital universe turned 1Zb (1 trillion Gb) for the first time in 2010, reached 1.8Zb just one year later in 2011 and will go beyond 35Zb in 2020. Some interesting figures include that 70% of such data is user-generated through several channels like social networks, mobile devices, wikis and other content publication approaches. Even more interestingly, 75% of such data results from data transformation, copying, and merging while metadata is the fastest growing data category. This is also the trend in semantic data, where datasets are increasingly being dynamically and automatically published, e.g. by semantic

---

[1] http://www.emc.com/leadership/digital-universe

© Springer International Publishing Switzerland 2015
T. Supnithi et al.(Eds.): JIST 2014, LNCS 8943, pp. 239–256, 2015.
DOI: 10.1007/978-3-319-15615-6_18

sensor networks [3], and consumed, e.g. by silico experiments in the form of scientific workflows [2]. In these domains a large number of distributed data sources are considered as opposed to classic data integration scenarios. This means that the amount of data available is growing at an exponential rate but also that data is not statically stored in their datasets. Combined with the growing size of the overall Linked Open Data cloud, with more than 30 billion triples, and of its individual datasets, with some of its hubs e.g. DBPedia exceeding 1,2 billion triples, the need of effective RDF data compression techniques is clear.

This raises serious data management challenges. Semantic data need to be compact and comprehensible, saving storage and communication bandwidth, while preserving the data integrity. Several approaches can be applied to achieve lossless RDF document compression. They can be categorised into either application-dependent or application-independent approaches: Application-dependent approachs include Michael Meier's rule-based RDF graph minimisation [10] and Reinhard et. al.'s approach [12]. They are usually semi-automatic, requiring human input. Application-independent approaches are more generic. First of all, universal file compression techniques [4], such as bzip2 [2] and LZMA [3], can be applied on RDF document. Such approaches alter the file structure of RDF documents and can significantly reduce file size. Alternative RDF serialisations, such as HDT serialisation [5], lean graphs [8] and K2-triples [1] can be used to reduce file size. Such techniques preserve the structured nature of RDF documents. Another approach is based on logical compression, such as the rule-based RDF compression [9], which can be used to substantially reduce the number of triples in an RDF document. Ontology redundancy elimination [6] can also be regarded as logical RDF compression in which the RDF documents are intepreted with OWL (Web Ontology Language [4]) semantics.

Despite the compression results achieved by existing works, they make little or no use of the graph structure of RDF datasets. For example, universal compression techniques usually exploit the statistical redundancy in a document and the document is treated as a series of ordered characters. However an RDF document is essentially a graph in which the ordering in which nodes and edges are presented is irrelevant to the semantics of the data. Even the few approaches that leverage this kind of information are constrained to simple and fixed graph structures. This makes them less effective when reducing the size of compressed file. For example, logical compression [9] compresses re-occuring star-shaped graph structures of varying center nodes in an RDF document with single triples:

*Example 1.* In an RDF document, if it contains the following triples, where both $m$ and $n$ are large numbers:

$$< s_1, p_1, o_1 >, \ldots, < s_1, p_n, o_n >,$$

$$\ldots$$

$$< s_m, p_1, o_1 >, \ldots, < s_m, p_n, o_n >$$

---

[2] http://www.bzip.org/

[3] http://www.7-zip.org/

[4] http://www.w3.org/TR/owl2-overview/

then it can be compressed with the following triples:

$$< s_1, p_1, o_1 >, \ldots, < s_m, p_1, o_1 > \tag{1}$$

And a rule $<?s, p_1, o_1 > \rightarrow <?s, p_2, o_2 >, \ldots, <?s, p_n, o_n >$, where $?s$ is a variable, can be applied to recover all the removed triples.

This approach works well when the document contains many different nodes sharing many same "neighbours". But it is not applicable when such graph structures are not observed, e.g. when $n = 1$. In fact, its results, the triples in (1) is an example of such a scenario. This is because the logical compression presented by Joshi et al. [9] is constrainted to graph patterns with only 1 variable. By extending to more generic graph structures, improvement of compression rate can be easily achieved. In fact, triples in (1) can be compressed by exploiting a graph pattern with 2 variables:

*Example 2.* Without lose of generality, we assume $m$ is an even number. We can further compress triples in (1) with the following ones

$$< s_1, p_x, s_2 >, \ldots, < s_{m-1}, p_x, s_m >$$

where $p_x$ is a fresh predicate introduced for this graph pattern. And we can use a rule $<?s, p_x, ?o > \rightarrow <?s, p_1, o_1 >, <?o, p_1, o_1 >$ to decompress the triples.

Apparently, the number of triples is halved and such a further compression is guaranteed applicable and better on any results obtained by Joshi et al.'s approach. This shows that logical compression with lower compression rate can be achieved when more syntactic and semantic information of RDF datasets are better exploited.

In addition to semantic redundancies, the ways in which RDF graphs are serialised as sequences of bytes can also introduce another type of redundancies, i.e. the syntactic one. Existing approaches including textual serialisation syntaxes, e.g. RDF/XML, and binary ones, e.g. HDT [5] only deal with syntactic redundancies in concrete graph structures (defined as intra-structural redundancy in section 3.2). Without the knowledge of graph patterns in RDF graphs, they are not able to make use of the common graph structure (defined as inter-structural redundancy in section 3.2) shared by many instances of one graph pattern.

In this work, we aim at exploiting the graph structure of RDF datasets as a valuable source of information in order to increase the data compression gain in both the semantic level and syntactic level. Main contributions of the paper include:

1. we develop application-independent graph pattern-based logical compression and serialisation technologies for RDF documents;
2. we implement a framework that combines different compression techologies, including logical compression and serialisation;
3. we show that implementations of our approach can complement the compression abilities of existing solutions in semantic ans syntactic levels. The potentials of efficient data access are also revealed in the evaluation.

The rest of the paper is organised as follows. Section 2 will introduce basic notions, such as RDF graphs and graph patterns. Section 3 presents techniques of graph pattern based approaches for removing both semantic and syntactic redundancies. Sections 4 and 5 present our implementation and evaluations respectively.

## 2    Preliminaries

Resource Description Framework (RDF) [7] is the most widely used data interchange format on the Semantic Web. It makes web content machine readable by introducing annotations. Given a set of URI reference $\mathcal{R}$, a set of literals $\mathcal{L}$ and a set of blank nodes $\mathcal{B}$, an RDF statement is a triple ¡$s, p, o$¿ on $(\mathcal{R} \cup \mathcal{B}) \times \mathcal{R} \times (\mathcal{R} \cup \mathcal{L} \cup \mathcal{B})$, where $s, p, o$ are the subject, predicate and object of the triple, respectively. An RDF document is a set of triples.

With these notions, to facilitate RDF data compression, we define a graph as follows:

**Definition 1.** *(Graph) A <u>labeled, directed multiple graph</u> (graph for short) $G = \langle N, E, M, L \rangle$ is a four-tuple, where $N$ is a set of URI references, blank nodes, variables and literals, $E$ is the set of edges, $M : E \to N \times N$ maps an edge to an ordered pair of nodes, $L$ is the labelling function that for each edge $e \in E$, its label $L(e)$ is a URI reference.*

It is apparent that every RDF document can be converted to a graph whose nodes are not variables, and vice versa. In the following we use the notions RDF document, RDF triple set and RDF graph interchangably. Given a set $T$ of RDF triples, we use $G(T)$ to denote the graph of the triple set. Given a RDF graph $G$, we use $T(G)$ to denote the set of triples that $G$ represents.

**Definition 2.** *(Graph Operations) A graph $\langle N_1, E_1, M_1, L_1 \rangle$ is a <u>sub-graph</u> of another graph $\langle N_2, E_2, M_2, L_2 \rangle$ IFF $N_1 \subseteq N_2$, $E_1 \subseteq E_2$, $\forall e \in E_1$, $M_1(e) = M_2(e)$ and $L_1(e) = L_2(e)$.*

*Two graphs $\langle N_1, E_1, M_1, L_1 \rangle$ and $\langle N_2, E_2, M_2, L_2 \rangle$ have a <u>union</u> IFF $\forall e \in E_1 \cap E_2$, $M_1(e) = M_2(e)$, $L_1(e) = L_2(e)$. Their <u>union</u> is a graph $\langle N, E, M, L \rangle$ such that $N = N_1 \cup N_2$, $E = E_1 \cup E_2$, $\forall e \in E_1 \cup E_2$, $M(e) = M_1(e)$ or $M(e) = M_2(e)$ and $L(e) = L_1(e)$ or $L(e) = L_2(e)$.*

**Definition 3.** *(Graph Pattern) A <u>graph pattern</u> is a graph in which some nodes represent variables.*

*The none-variable nodes in a graph pattern are called <u>constants</u> of the graph pattern.*

In this paper we are not concerned with the direction of triples in a graph pattern. For conciseness, we use ¡$?x, p, o$¿ to represent a triple with $?x$ as either the subject, or the object.

**Definition 4. (Instance)** *A <u>substitution</u> $\S = (v_1 \to v_2)$ replaces a vector of variables $v_1$ with a vector of URI references/literals/blank nodes $v_2$.*

*A graph $G$ is an <u>instance</u> of a graph pattern $G'$, denoted by $G : G'$, IFF there exists a sustitution $\S$ such that $G'_\S = G$. Given an RDF graph $\mathcal{D}$, we use $I_\mathcal{D}(G')$ to denote the set of all sub-graphs of $\mathcal{D}$ that are instances of $G'$. Obviously, $G \in I_\mathcal{D}(G)$ since an empty substitution $(\emptyset \to \emptyset)$ exists. And $I_\mathcal{D}(G) = \{G\}$ when $G$ contains no variable.*

*When the $\mathcal{D}$ is clear from context, we omit it in the notations.*

An RDF graph can be considered as the union of instances of several graph patterns.

**Definition 5. (Rule)** *Let $GP$ and $GP'$ be two graph patterns, $GP \to GP'$ is a rule.*

*Let $\mathcal{D}$ be an RDF document, results of applying $GP \to GP'$ on $\mathcal{D}$, denoted by $GP \to_\mathcal{D} GP'$, is another RDF document $\mathcal{D}' = \bigcup_{GP_\S \in I_\mathcal{D}(G)} GP'_\S$.*

In other words, $\mathcal{D}'$ is the union of $GP'$ instances with substitutions that are used by instances in $I_\mathcal{D}(GP)$.

With these notations, we invesitgated different graph pattern-based in for RDF and their relation to the compression problem.

# 3   Graph Pattern-Based Approaches

In this section, we propose graph pattern-based approaches to deal with both semantic redundancies and syntactic redundancies in RDF data. They complement existing approaches by either generalising existing semantic compression techniques, i.e. rule based approaches, or extending serialisation approaches, e.g. RDF/XML or HDT, to deal with new type of syntactic redundancies.

## 3.1   Semantic Compression: Graph Pattern-Based Logical Compression

As we mentioned in the previous section, an RDF graph can be expanded from a smaller graph with the help of rules, whose body and head are both RDF graph patterns. This essentially means that the instances of the bigger graph pattern can be replaced by smaller instances of the smaller graph pattern. Below is an example:

*Example 3.* In the DBpedia dataset, the following graph pattern has a large number of instances:

$$GP_1 :<?x, a, foaf : Person>, <?x, a, dbp : Person>$$

Such a graph pattern can be replaced by a smaller graph pattern. For example, we can use one type $T$ to represent the two types in the above graph pattern, and replace the two triples with a single triple

$$GP_2 :<?x, a, T>.$$

In this example, $GP_1$ is compressed by $GP_2$. As a consequence, $I(GP_1)$ is compressed by $I(GP_2)$. This will reduce the number of triples in the original RDF document by 50%. Such a compression can be achieved by applying rule $GP_1 \rightarrow GP_2$ on the RDF document. Decompression is achieved by applying rule $GP_2 \rightarrow GP_1$ on the compressed data set.

During logical compression, the fact that $GP_2$ contains less triples than $GP_1$ is exploited to ensure that $I(GP_2)$ contains less triples than $I(GP_1)$. Such a reduction of triple number is the main focus in logical compression.

**A Unified Model for Graph Pattern-based Logical Compression.** We generalise the above compression mechanism to support more variables with the following unified model:

**Definition 6. (Graph Pattern-based Logical Compression of RDF)** *Let $\mathcal{D}$ be a RDF document, its graph pattern-based logical compression consists of an RDF document $\mathcal{D}'$ and a rule set $S$, such that the following holds:*

$$\mathcal{D} = \bigcup_{GP_2 \rightarrow GP_1 \in S} (\mathcal{D}' \setminus I_{\mathcal{D}'}(GP_2)) \cup (GP_2 \rightarrow_{\mathcal{D}'} GP_1)$$

*We call $\mathcal{D}$ the original RDF document, $\mathcal{D}'$ the compressed RDF document, $S$ the decompression rule set. And for each $GP_2 \rightarrow GP_1 \in S$, we say that $GP_1$ is compressed by $GP_2$.*

In this procedure, triples in $I_{\mathcal{D}}(GP_1)$ are replaced by triples in $I_{\mathcal{D}'}(GP_2)$. Note that all triples in $\mathcal{D}'$ that involve new resources introduced in $GP_2$ but not $GP_1$ are removed during decompression.

It's worth mentioning that variables in the decompression rules bind only to explicitly named entities in the RDF document. Hence the compression results $\mathcal{D}'$ can directly be used in RDF reasoning and SPARQL query answering, e.g. by reasoners supporting DL-safe rules [11].

In the above definition, for each $GP$ compressed by $GP'$, the original $|GP|$ triples in $\mathcal{D}$ are replaced by $|GP'|$ new triples in $\mathcal{D}'$. The extra cost of compression is the maintenance of the rule, which consists of $|GP| + |GP'|$ triples. To characterise the effect of compression we define the following notions:

**Definition 7. (Compression Quantification)** *For $GP$ compressed with $GP'$ using the graph pattern-based compression defined in Def. 6, let $T_O$, $T_C$ and $T_R$ be the total number of different triples in $I(GP)$, in $I(GP')$ and in $R$, respectively, then the $\underline{compression\ gain}$ is $T_O - T_C - T_R$, the $\underline{compression\ ratio}$ is $\frac{T_C + T_R}{T_O}$.*

Note that when several instances of $GP$ share a triple, this triple will be compressed multiple times but the actual size of the document $\mathcal{D}$ will only be reduced by at most 1. Similarly, when several instances of $GP'$ share a triple, this triple only needs to be included in the compression result $\mathcal{D}'$ once. With these considerations, the compression quantification of a single graph pattern can be characterised as follows:

**Lemma 1. (Compression Ratio)** *For a GP having $n$ variables and $N$ instances, assuming triples are shared redundantly by instances of GP for $S$ times, then the compression quantification of compressing GP with $GP'$ as defined in Def. 6 is as follows, where $I$ is the redudant number of triples shared by instances of $GP'$:*

$$T_O = N * |E| - S$$
$$T_C = N * m - I$$
$$T_R = m + |E|$$

The lemma is quite straightfoward, $N*|E|$ is the total number of triples in all the $GP$ instances, $S$ is the redundant number of shared triples, which should be removed when calculating $T_O$. Similarly $I$ should be removed when calculating $T_C$.

With the above quantification, graph pattern-based RDF logical compression is a problem of finding appropriate graph patterns that yield best (highest) compression gain. To achieve that, we should look for graph patterns with the following criterias:

- larger $N$, i.e. more instances;
- larger $|E|$, i.e. more triples in the original pattern;
- larger $I$, i.e. more triples shared by instances of the compressed graph pattern;
- smaller $m$, i.e. less triples in the compressed pattern;
- smaller $S$, i.e. less shared triples among instances of the original graph pattern;

These criterias are not easy to satisfy at the same time as the values of these parameters are related to one another. Yet they can already help us to identify "good" or to eliminate "bad" patterns. For example, a compressed graph pattern should not contain circle (removing an edge to eliminate the circle will only reduce the value of $m$ but not the others). More interestingly, we have the following observations:

**Lemma 2.** *Let GP be a graph pattern containing a constant triple $t$, whose subject, predicate and object are all constants, then compression gain of GP being compressed by any $GP'$ is no higher than the compression gain of $GP \setminus \{t\}$ being compressed by $GP' \setminus \{t\}$.*

This lemma is quite obvious. Assuming compressing $GP$ with $GP'$ yields compression gain $T_O - T_C - T_R$, because $t$ is shared by all instances of $GP$, it is maintained only once in $T_O$. If $t$ is also in $GP'$, it is similarly maintained in $T_C$ only once. In this case, compressing $GP \setminus \{t\}$ with $GP' \setminus \{t\}$ will yield compression gain $(T_O - 1) - (T_C - 1) - (T_R - 2)$; If $t$ is not in $GP'$, it is not included in $T_C$. Then the compression gain of compressing $GP \setminus \{t\}$ with $GP'$ will be $(T_O - 1) - T_C - (T_R - 1)$. In both cases, compressing $GP$ with $GP'$ is not more benificial than compressing $GP \setminus \{t\}$ with $GP' \setminus \{t\}$. Although there is only minor, it implies that in graph-pattern based logical compression, we only need to focus on graph patterns with no such "constant triples".

Another observation from the quantification is as follows:

**Lemma 3.** *Let $GP_1$, $GP_2$ be two disconnected graph patterns, then the compression gain of compressing them into a single connected graph pattern is lower than compressing $GP_1$ and $GP_2$ separately.*

This lemma is also quite straight-forward. Compressing two disconnected graph patterns into a connected one will only require addition triples to connect previously disconnected instances. This implies that we should always compress each connected graph pattern separately.

A further observation is that the decompression rules themselves, particularly the head part, may also contain redundancies that can be exploited:

**Lemma 4.** *For $N$ ($N \geq 2$) graph patterns whose decompression rules share the same head triples $t_1, \ldots, t_M$ ($M \geq 2$), the compression gain will be higher if we replace $t_1, \ldots, t_M$ with a single triple $t'$ and further compress with a decompression rule $t' \rightarrow t_1, \ldots, t_M$.*

This lemma actually indicates that we can first compress all $t_1, \ldots, t_M$ with $t'$, and then further compress the compressed graph pattern with the replaced rules. The reason is also straight-forward: by changing the rules, we do not change the original triples, nor the triples in the final compression results, but only replace the $M * N$ triples in the original compression rules with $N + 1 + M$ triples. This replacement will be beneficial when either $M$ or $N$ is large. In worst case, we only reduce the gain by 2, which is negligible. This implies a practical simplification of compression: when a graph pattern contains multiple triples of form $<?x, p_i, o_i>$ associated to the same $?x$, we can first compress them into a single triple $<?x, p, o>$ before compressing the rest of the graph pattern.

It is worth mentioning that the logical compression approach can be applied on different syntactic forms of the RDF documents, as long as graph patterns and triples can be accessed. In fact, in our compression solution we also apply it on the compressed serialisation we will introduce in the next section because it has much smaller physical size than the origianl RDF document.

## 3.2   Syntactic Compression: Graph Pattern-Based Serialisation

Logical compression reduces the number of triples in an RDF graph by eliminating semantic redundancies. Nevertheless, with the same set of triples, redundancies can arise depending on how triples are serialised in an RDF file. For example, one RDF graph can be represented as two different RDF/XML files of $F1$ and $F2$ in Figure 1. In $F1$ (cf. the bold and red texts in the upper part), URIs of *jeff-z-pan* and *iSWC09_423* appear twice; in $F2$ both of them appear only once (cf. the bold and blue texts in the lower part). While the two files convey the same meaning, $F2$ is more concise by using RDF/XML's abbreviation and striping syntaxes[5].

---

[5] http://www.w3.org/TR/2002/WD-rdf-syntax-grammar-20020325/

```
<rdf:Description rdf:about="http://data.semanticweb.org/person/jeff-z-pan">
    <foaf:name rdf:datatype="http://www.w3.org/2001/XMLSchema#string">Jeff Z. Pan</foaf:name>
</rdf:Description>
<rdf:Description rdf:about="http://data.semanticweb.org/person/jeff-z-pan">
    <foaf:creator rdf:resource="http://data.semanticweb.org/conference/iswc/2009/paper/research/423"/>
</rdf:Description>
<rdf:Description rdf:about="http://data.semanticweb.org/conference/iswc/2009/paper/research/423">
    <rdfs:label>Concept and Role Forgetting in ALC Ontologies</rdfs:label>
</rdf:Description>
```
*RDF/XML F1*

*RDF/XML F2*
```
<rdf:Description rdf:about="http://data.semanticweb.org/person/jeff-z-pan">
    <foaf:name rdf:datatype="http://www.w3.org/2001/XMLSchema#string">Jeff Z. Pan</foaf:name>
    <foaf:creator>
        <rdf:Description rdf:about="http://data.semanticweb.org/conference/iswc/2009/paper/research/423">
            <rdfs:label>Concept and Role Forgetting in ALC Ontologies</rdfs:label>
        </rdf:Description>
    </foaf:creator>
</rdf:Description>
```

**Fig. 1.** Syntactic Redundancy

The above phenomenon emerging from RDF file *serialisation* can be characterised with the following equation, in which $F$ is a file, $|F|$ is the file size in terms of bits, $r_b$ is the average number of bits needed to encod a resource and $N_c$ is the total number of resource occurrences.

$$|F| = N_c \times r_b \qquad (2)$$

A serialisation $F$ of an RDF graph $g$ with resource occurrences of $N_c$ contains syntactic redundancy if there is another serialisation $\bar{F}$ of $g$ with resources occurrences $\bar{N_c}$, s.t. $\bar{N_c} < N_c$. Let $n$ be the number of triples in an RDF graph. The worst case is $N_c = 3 \times n$, which means that the serialisation is to store triples one by one[6].

Most RDF serialisation approaches provide syntaxes to avoid the worst case e.g., the RDF abbreviation and striping syntax. Similar ideas are also adopted in other RDF serialisation syntaxes like Turtle[7] and Notation 3[8]. Beside the textual serialisations, Fernández and et. al. [5] introduced a binary serialisation approach which applies similar ideas by using bitmaps to record the resource occurrences.

Generally speaking, in the RDF graph, there are two types of syntactic redundancies. The first type is the intra-structure redundancies, which denotes the multiple occurrences of the same resources within the same structures (subgraphs) of an RDF graph. For example, in Figure 1 *jeff-z-pan* has 2 occurrences in $F_1$. One of them is redundant and can be committed in $F_2$. The second type of syntactic redundancies is the inter-structure redundancies, which denotes the multiple occurrences of the same resources across different structures. Suppose there is another author *Jose* in ISWC09 dataset. He might be described using the same graph pattern of *jeff-z-pan*: $GP_{author} = \{<?x, foaf:name, ?n>, <?x, foaf:made, ?p>, <?x, rdf:type, foaf:Person>\}$. In such a graph pattern, resources

---

[6] For the sake of simplicity, collection constructs like rdf:collection, rdf:list or rdf:bags are viewed as single resources in this calculation.

[7] http://www.w3.org/TeamSubmission/turtle/

[8] http://www.w3.org/DesignIssues/Notation3

such as *foaf:name*, *foaf:made*, *foaf:Person*, etc. do not have to be repeated for both *jeff-z-pan* and *Jose* in a serialisation. According to the above categorisation, existing work only deals with intra-structure redundancies leaving inter-structure ones untouched. We will show (cf. Sec. 3.2) how these inter-structure ones can be dealt with in our approach.

**Grouping Triples by Graph Patterns.** We introduce a graph pattern-based serialisation method which can remove both intra-structural redundancies and inter-structural redundancies.

$$I_{GP} \xrightarrow{serialised} GP + ((r_{I_1,1}, ..., r_{I_1,k})...(r_{I_N,1}, ..., r_{I_N,k})) \tag{3}$$

As shown in formula 3, given the instances $I_{GP}$ of a graph pattern $GP$, the serialisation method generates a sequence of bits which is composed of two components. The first component is the graph pattern its self. Such graph pattern is essentially a structure shared by its instances. The second component is a sequence of instances of $GP$ and each instance is a list of resource IDs. By using this graph pattern-based serialisation, we can serialise an RDF graph $G$ as a file $F$ which takes the form as follows.

$$F \rightarrow I_{GP_1}, I_{GP_2}, ..., I_{GP_i}, ... \tag{4}$$

Given an graph pattern $GP$, the serialisation size of its instances can be calculated as follows:

$$|I_{GP}| = (|GP| + N \times |V|) \times b$$

In the above formula, $|GP|$ is the number of resources (edges and constant nodes) in $GP$, $N$ is the number of instances of $GP$ and $V$ is the number of variable nodes in $GP$. The size of serialisation file $F$ is calculated as follows:

$$|F| = ( \sum_{GP_i \in \{GP\}} |I_{GP_i}|) + |Dictionary| \tag{5}$$

Compared to triple based serialisation (formula 2), the storage space saved by graph pattern based serialisation can be calculated as:

$$\sum_{GP_i \in \{GP\}} |GP_i| \times (N_{GP_i} - 1) \times b$$

## 4   Implementation

In this section, we discuss the implementation details. Firstly we overview our methods by a framework. Following which are the technical details of the graph pattern-based serialisation and the logical compression method respectively.

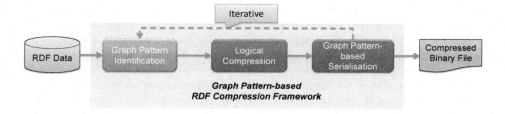

**Fig. 2.** RDF Data Compression Framework

### 4.1 Framework

The framework (cf. Figure 2) is composed of 3 different steps, including graph pattern identification, logical compression and data serialisation, to exploit both semantic and syntactic redundancies discussed in previous sections.

We describe the main components in this framework as follows:

1. **Graph Pattern Identification** component implements an efficient and incremental graph pattern identification method. The identified graph patterns are not only utilised in semantic compression by the <u>logical compression</u> component, but also used to remove syntactic redundancies by the <u>graph pattern-based serialisation</u> components.
2. **Logical Compression** component implements the graph pattern-based logical compression techniques we presented in the Sec. 3.1. Given the identified graph patterns, it tries to reduce the size of the graph patterns so that the total entities encoded in the instances of graph patterns can be reduced.
3. **Graph Pattern Serialisation** component implements the graph pattern-based serialisation techniques we presented in Sec. 3.2. It produces a compact serialisation of the documents by grouping triple blocks with similar structures, called Entity Description Blocks (EDBs), into graph patterns, and then applying the serialisation techniques;

As shown as dashed line in Figure 2, an iterative compression method is implemented in our framework. After each iteration, the serialisation will result with graph pattern headed files (cf. formula 2 in section 3.2), which can be utilised to identify larger graph patterns efficiently. Larger graph patterns can be used to further reduce redundancies so that repetitive appearances of entities in different graph patterns can potentially be reduced. Below we explain the implementation details of the first two components. The <u>graph pattern-based serialisation</u> has been introduced in section 3.2.

### 4.2 Graph Pattern Identification

In this paper, we introduce an approach to find and generate graph patterns efficiently. The basic idea is based on an observation that most RDF dump files were generated by following some specific patterns. Such patterns lead to data patterns in the dumped file e.g. triples about an entity are often put together or putting the same relation triples together. Obviously, such data patterns are useful for the serialisation task, or at least good sources for finding more useful graph patterns.

Following this idea, we propose an incremental serialisation approach, which is composed of two stages. In the first stage, it utilises the graph patterns in the RDF file directly. Such existing graph patterns are called direct graph patterns. The second stage moves on from the results of the first stage by manipulating the direct graph patterns to get better patterns for compression. The second stage can be iterative by applying different graph pattern processing techniques.

**Stage 1 Finding Direct Patterns.** This stage is composed of two steps of generating entity description block (EDB shortly) and grouping EDBs. When iterating the triples of an RDF file, we group the triples as an EDB, if these triples share the same subject and also form a continuous sequence in the RDF file.

$$(..., < s_1, p_1, o_1 >, < s_1, p_2, o_2 >, ..., < s_1, p_n, o_n >, ...)$$
$$\xrightarrow{grouped} (..., EDB(s_1), ...)$$

With this method, a sequence of triples is converted into a sequence of EDBs. In the second step, we group EDBs by their schema information which is called entity description pattern as follows.

$$EDP(EDB(s)) = (C, P),$$

where C is the types of s in this EDB and P is the properties of s in the EDB. Hence, basically the grouping operation is to put all EDBs with the same structure together so that we can apply techniques discussed in section 3.2 to store them. The EDP based serialisation approach is called Level 0 method, LV0 shortly.

**Stage 2 Merge Graph Patterns.** In this stage, the main problem to be dealt with is how to *merge* existing graph patterns from previous stage(s) to get a better pattern which can remove more redundancies. Hence, the key is how to define the *merge* operator. Generally speaking, a bigger graph pattern will always be better because it can avoid storing the same resources multiple times in smaller graph patterns. One possible *merge* operation can be merge the EDBs of the same entities together so that we do not have to store the same entity IDs in multiple places in a file. Another possible way is to utilise the linking nature of RDF graph i.e. merge EDBs by there relations. In this paper, we apply the second strategy in the evaluations, where we call it Level 1 method., LV1 shortly.

**Later Iterations.** The graph patterns identified stage 2 or later stages can be further enlarged by applying merge operation on the results of current stage. Obviously, more iterations require more processing time. Finding a trade-off between the compression gain and costs of the processing timing is critical in this iterative process. In this paper, we focus only on LV1, i.e. stage 2.

### 4.3   Logical Compression

We implement the logical compression approach with one variable, which is the basis for logical compression with multiple variables. As discussed in Sec. 4.2, the new graph pattern proposed in this paper eliminates 50% of triples by combining a pair of triples needed to be saved into one triple. This subsection discusses how this kind of compression rules are found.

The logical compression algorithm is shown in Algorithm 1. After direct patterns are constructed, subjects with same predicates will be grouped together into same direct patterns, making finding the instances of each graph pattern very easily. In order to facilitate the logical compression, we support $candidates(GP, threshold)$ as an atomic operation, which is calculated through fast indexing of the graph patterns, and will list the candidate graph patterns whose number of solutions is greater than threshold.

We find the compression graph patterns with the following procedure: We list all candidate graph patterns with enough instances using $candidates(GP, threshold)$. Each candidate graph pattern $GP'$ is renamed as an object property $p_{GP'}$ in the compression. Suppose values of the variable in $GP'$ are $a_1, a_2, \ldots, a_n$, we save the following triples in the compressed dataset:

$$\{< a_1, p_{GP'}, a_2 >, < a_3, p_{GP'}, a_4 >, \ldots, < a_{n-1}, p_{GP'}, a_n >\}$$

---

**Algorithm 1** Logical Compression Algorithm

---
1:   **procedure** LOGICAL_COMPRESSION($GPs, threshold$)           ▷ $GPs$ are direct patterns
2:      **for** each direct pattern $GP$ **do**
3:         $NEWGPs \leftarrow candidates(GP, threshold)$
4:         **for** each graph pattern $GP'$ in $NEWGPs$ **do**
5:            $IGP' \leftarrow$ instantiations of variables in $GP'$
6:            rename $GP'$ as $p_{GP'}$
7:            $compressedSet \leftarrow < a_i, p_{GP'}, a_{i+1} >$      ▷ $a_i \in IGP', a_{i+1} \in IGP'$
8:      **return** $compressedSet$

---

## 5   Evaluation

As introduced in the previous section, we implemented both RDF serialisation and logical compression based graph patterns. And we can support different incremental solutions. In this section, we evaluate their performance and compare against existing technologies.

**Datasets.** The main strategy of our dataset selection is to use real world datasets with various size and from different domains. The idea is that LV0 of our incremental approach tries to use the direct graph patterns in the dumped RDF files. A heterogeneous datasets might reveal how our approach can work in different situations.

As aforementioned, LV0 utilises the data patterns in the dumped RDF file directly. One might be interested to such direct graph patterns. The first concern would be how many numbers of graph patterns one dataset could have. If there were too many graph patterns, the compression method might not work as expected. For example, in the worst case, each triple is a distinct pattern. In such case, the first level compression would degrade to be each triple based serialisation. The second concern might be the data distributions among such patterns. Our approach prefers leanly distributed data patterns. One of the main reason is that our method would be much more efficient when most of the data only reside in a small number of graph patterns.

**Table 1.** Dataset Statistics and Direct Graph Patterns

| Dataset | Archive Hub | Jamendo | linkedMDB | DBLP2013 |
|---------|-------------|---------|-----------|----------|
| #Triples | 431,088 | 1,047,950 | 6,148,121 | 94,252,254 |
| Plain File Size | 71.8M | 143.9M | 850.3M | 14G |
| Compressed Size | 2.5M | 6M | 22M | 604M |
| #*Direct* GP | 623 | 34 | 119 | 77 |
| Top 5 GP | 35% | 78% ˙ | 54% | 72% |

Table 2 gives the statistics of the four datasets used in our experiments. The datasets are sorted ascendantly by size from left to right. The last two rows show the statistics of direct graph patterns. The numbers of direct graph patterns in each dataset are displayed in the fifth row. In most datasets, the pattern numbers are quite small. In addition, the number does NOT increase with the dataset size. This is understandable because graph patterns are more related to the complexity of data schemas instead of the individual numbers. The last row list the ratio of entity numbers in top 5 largest graph patterns to the number of all entities in the dataset. As we can see, in the three large datasets, most data reside in the top 5 graph patterns. The exception is the Archive Hub dataset which also has the largest number of graph patterns. One reason is that the dataset is a gateway of collections in UK. This means that it might cover a large number of concepts.

A quick conclusion from the pattern analysis is that in most datasets the direct graph patterns might be good resources which can be utilised to remove redundancies in them. As aforementioned, LV0 utilises the data patterns in the dumped RDF file directly. One might be interested to such direct graph patterns. The first concern would be how many numbers of graph patterns one dataset could have. If there were too many graph patterns, the compression method might not work as expected. For example, in the worst case, each triple is a distinct pattern. In such case, the first level compression would degrade to be each triple based serialisation. The second concern might be the data distributions among such patterns. Our approach prefers leanly distributed data patterns. One of the main reason is that our method would be much more efficient when most of the data only reside in a small number of graph patterns.

Table 2 gives the statistics of the four datasets used in our experiments. The datasets are sorted ascendantly by size from left to right. The last two rows show the statistics of direct graph patterns. The numbers of direct graph patterns in each dataset are displayed in the fifth row. In most datasets, the

**Table 2.** Dataset Statistics and Direct Graph Patterns

| Dataset | Archive Hub | Jamendo | linkedMDB | DBLP2013 |
|---|---|---|---|---|
| #Triples | 431,088 | 1,047,950 | 6,148,121 | 94,252,254 |
| Plain File Size | 71.8M | 143.9M | 850.3M | 14G |
| Compressed Size | 2.5M | 6M | 22M | 604M |
| #*Direct* GP | 623 | 34 | 119 | 77 |
| Top 5 GP | 35% | 78% | 54% | 72% |

pattern numbers are quite small. In addition, the number does NOT increase with the dataset size. This is understandable because graph patterns are more related to the complexity of data schemas instead of the individual numbers. The last row list the ratio of entity numbers in top 5 largest graph patterns to the number of all entities in the dataset. As we can see, in the three large datasets, most data reside in the top 5 graph patterns. The exception is the Archive Hub dataset which also has the largest number of graph patterns. One reason is that the dataset is a gateway of collections in UK. This means that it might cover a large number of concepts.

A quick conclusion from the pattern analysis is that in most datasets the direct graph patterns might be good resources which can be utilised to remove redundancies in them.

**Logical Compression Evaluation.** In section 3.1, we propose a general model of RDF logical compression. In this subsection, we focus on the evaluation of one particular type of graph patterns i.e. $<?x, p, o>$. This type of patterns, or in other word rules covers the intra-property and inter-property rules of [9]. As we pointed out in section 3.1, the compression techniques proposed in [9] can be further optimized by grouping two instances together and compress with 1 triple.

In table 4, we compare such optimised results with the results reported in [9]. As shown there, the optimized results outperforms existing work quite well. In addition to the instance grouping optimization, the results also benefits from frequent data values. As we mentioned, our logical compression, i.e. LV2 method is based on GP-LV0 result. In the first level serialisation, the data values are also assigned with an ID value based on their MD5 hashes. Hence, if some values are frequent, they will be treated similarly as frequent instances.

In section 3.1, we propose a general model of RDF logical compression. In this subsection, we focus on the evaluation of one particular type of graph patterns i.e. $<?x, p, o>$. This type of patterns, or in other word rules covers the intra-property and inter-property rules of [9]. As we pointed out in section 3.1, the

**Table 3.** The optimized results of one variable patterns

| Data Set | # Total triples | Optimized Compression | | RB Comp. ratio |
|---|---|---|---|---|
| | | #Removed triples | Comp. ratio | |
| Archive Hub | 431,088 | 187,887 | **1.77** | 1.41 |
| Jamendo | 1,047,950 | 436,101 | **1.72** | 1.22 |
| LinkedMDB | 6,148,121 | 2,679,593 | **1.77** | 1.33 |
| DBLP | 94,252,254 | 61,383,224 | **2.86** | 1.16 |

compression techniques proposed in [9] can be further optimized by grouping two instances together and compress with 1 triple.

In table 4, we compare such optimised results with the results reported in [9]. As shown there, the optimized results outperforms existing work quite well. In addition to the instance grouping optimization, the results also benefits from frequent data values. As we mentioned, our logical compression, i.e. LV2 method is based on GP-LV0 result. In the first level serialisation, the data values are also assigned with an ID value based on their MD5 hashes. Hence, if some values are frequent, they will be treated similarly as frequent instances.

**Table 4.** The optimized results of one variable patterns

| Data Set | # Total triples | Optimized Compression | | RB Comp. ratio |
|---|---|---|---|---|
| | | #Removed triples | Comp. ratio | |
| Archive Hub | 431,088 | 187,887 | **1.77** | 1.41 |
| Jamendo | 1,047,950 | 436,101 | **1.72** | 1.22 |
| LinkedMDB | 6,148,121 | 2,679,593 | **1.77** | 1.33 |
| DBLP | 94,252,254 | 61,383,224 | **2.86** | 1.16 |

**Graph Pattern-Based Serialisation Evaluation.** The proposed serialisation approach can deal with inter-structure syntactic redundancies which are not touched by most existing approaches. Table 5 gives an idea about the volume of such redundancies (removed by basic EDP patterns) in test datasets. The syntactic redundancies [9] removed by our approach can be quantified as $SRR_{syntac} = \sum_{EDP_i} |EDP_i| \times (f_{EDP_i} - 1)$, where $f_{EDP_i}$ is the frequency or number of instances of $EDP_i$.

The second row of Table 5 lists the $SRR_{syntac}$ of four datasets. The third row shows the ratios of the redundancies over the whole datasets by $SRR_{syntac}/(3 \times \#Triples_G)$. The fourth row shows the syntactic redundancies our approach can further remove from the results of approaches only dealing with intra-structure redundancies like HDT, which are the inter-structure redundancies. It is interesting to see that in the first three datasets most syntactic redundancies are not dealt with by existing serialisation approaches. The situation in DBLP2013 is different which means that there are more intra-structure redundancies in it. The last row shows the improvements in compression ratio by removing inter-structural redundancies.

**Table 5.** Inter-structure redundancies removable

| Dataset | Archive Hub | Jamendo | linkedMDB | DBLP2013 |
|---|---|---|---|---|
| Total syntactic redundancies | 370,389 | 999,353 | 5,939,980 | 79,399,947 |
| Ratio over the original data | 28.6% | 31.8% | 32.2% | 28.1% |
| Inter-structure redundancies | 355,917 | 855,893 | 5,583,975 | 46,208,641 |
| Compression Ratio Improvement | 38.49% | 39.93% | 44.65% | 22.74% |

---

[9] The syntactic redundancies are calculated by # (unnecessary) resource occurrences.

In general, these statistics show that compared to intra-structure redundancies the inter-structure ones constitute the major part of syntactic redundancies in all test datasets. This indicates that our graph pattern-based serialisation can improve compression ratio significantly.

**Rule Mining Operation Evaluation.** Mining rules from RDF graph directly might be expensive, when the dataset is large. One of the biggest advantage of our incremental approach is that the first serialisation results can provide efficient graph pattern manipulation operations. Firstly, its a compacted representation of the original RDF graph. This makes it more efficient in disk IO and RAM processing. Secondly, and more importantly, the GP-LV0 results are EDP based. From the EDP definition $EDP = (C, P)$, one can figure out that instance types are already treated as constant nodes. Such patterns can be used directly in mining compression rules. Finally, the graph pattern based serialisation makes it very convenient to get pattern based index which can be very useful for mining rules. Given these advantages, we propose a rule mining method based on GP-LV0 results. Figure 3 illustrates the mining time of our approach by comparing to RB [9] compression time. It can be figured out that we can get the compression rules in less than 10 seconds in 3 datasets. For DBLP dataset, we can get the results in about 3 minutes.

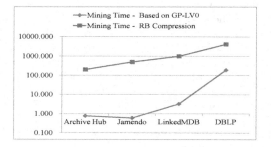

**Fig. 3.** Rule mining cost

# 6   Conclusion and Future Work

In this paper, we investigated the problem of application-independent, lossless RDF compression based on graph patterns. By considering the graph nature of RDF and its semantics, we focused on two types of redundancies, namely semantic redundancy and syntactic redundancy. We developed graph pattern-based logical compression and novel serialisation technologies for RDF data. Evaluation results showed that our approach can complement existing technologies such as HDT and rule-based compression significantly in both semantic and syntactic levels on benchmark datasets. In addition, the evaluation of the rule mining task shows the potentials of the graph pattern-based serialisation in supporting efficient data accesses.

In the future, we will put special focuses on efficient data access over the proposed serialisation formats, e.g. extending the results with dedicated index structures to support SPARQL query answering. In addition, we will also further look into the redundancies of RDF data with special interests in the linked data environment, where the redundancies might be different when different data sources are linked together or different vocabularies are reused.

# References

1. Álvarez-García, S., Brisaboa, N.R., Fernández, J.D., Martínez-Prieto, M.A.: Compressed k2-triples for full-in-memory rdf engines. arXiv preprint arXiv:1105.4004 (2011)
2. Chen, H., Yu, T., Chen, J.Y.: Semantic web meets integrative biology: a survey. Briefings in Bioinformatics **14**(1), 109–125 (2013)
3. Compton, M., Barnaghi, P., Bermudez, L., García-Castro, R., Corcho, O., Cox, S., Graybeal, J., Hauswirth, M., Henson, C., Herzog, A., et al.: The ssn ontology of the w3c semantic sensor network incubator group. Web Semantics: Science, Services and Agents on the World Wide Web (2012)
4. Fernández, J.D., Gutierrez, C., Martínez-Prieto, M.A.: Rdf compression: basic approaches. In: Proceedings of the 19th International Conference on World Wide Web, pp. 1091–1092. ACM (2010)
5. Fernández, J.D., Martínez-Prieto, M.A., Gutierrez, C.: Compact representation of large rdf data sets for publishing and exchange. In: Patel-Schneider, P.F., Pan, Y., Hitzler, P., Mika, P., Zhang, L., Pan, J.Z., Horrocks, I., Glimm, B. (eds.) ISWC 2010, Part I. LNCS, vol. 6496, pp. 193–208. Springer, Heidelberg (2010)
6. Grimm, S., Wissmann, J.: Elimination of redundancy in ontologies. In: Antoniou, G., Grobelnik, M., Simperl, E., Parsia, B., Plexousakis, D., De Leenheer, P., Pan, J. (eds.) ESWC 2011, Part I. LNCS, vol. 6643, pp. 260–274. Springer, Heidelberg (2011)
7. Hayes, P.: RDF Semantics. Technical report, W3C. W3C recommendation (February 2004). http://www.w3.org/TR/rdf-mt/
8. Iannone, L., Palmisano, I., Redavid, D.: Optimizing rdf storage removing redundancies: an algorithm. In: Ali, M., Esposito, F. (eds.) IEA/AIE 2005. LNCS (LNAI), vol. 3533, pp. 732–742. Springer, Heidelberg (2005)
9. Joshi, A.K., Hitzler, P., Dong, G.: Logical Linked Data Compression. In: Cimiano, P., Corcho, O., Presutti, V., Hollink, L., Rudolph, S. (eds.) ESWC 2013. LNCS, vol. 7882, pp. 170–184. Springer, Heidelberg (2013)
10. Meier, M.: Towards rule-based minimization of rdf graphs under constraints. In: Calvanese, D., Lausen, G. (eds.) RR 2008. LNCS, vol. 5341, pp. 89–103. Springer, Heidelberg (2008)
11. Motik, B., Sattler, U., Studer, R.: Query answering for owl-dl with rules. Web Semantics: Science, Services and Agents on the World Wide Web **3**(1), 41–60 (2005)
12. Pichler, R., Polleres, A., Skritek, S., Woltran, S.: Redundancy elimination on rdf graphs in the presence of rules, constraints, and queries. In: Hitzler, P., Lukasiewicz, T. (eds.) RR 2010. LNCS, vol. 6333, pp. 133–148. Springer, Heidelberg (2010)

# Optimizing SPARQL Query Processing on Dynamic and Static Data Based on Query Time/Freshness Requirements Using Materialization

Soheila Dehghanzadeh[1]([✉]), Josiane Xavier Parreira[2], Marcel Karnstedt[3], Juergen Umbrich[4], Manfred Hauswirth[5], and Stefan Decker[1]

[1] Insight Centre for Data Analytics, National University of Ireland, Galway, Republic of Ireland
{soheila.dehghanzadeh,stefan.decker}@deri.ie
http://www.insight-centre.org
[2] Siemens AG Sterreich, Wien, Austria
josiane.parreira@siemens.com
[3] Bell Labs, Dublin, Ireland
marcel.karnstedt@alcatel-lucent.com
[4] Vienna University of Economics and Business, Vienna, Austria
juergen.umbrich@wu.ac.at
[5] TU Berlin and Fraunhofer FOKUS, Berlin, Germany
manfred.hauswirth@tu-berlin.de

**Abstract.** To integrate various Linked Datasets, data warehousing and live query processing provide two extremes for optimized response time and quality respectively. The first approach provides very fast responses but with low-quality because changes of original data are not immediately reflected on materialized data. The second approach provides accurate responses but it is notorious for long response times. A hybrid SPARQL query processor provides a middle ground between two specified extremes by splitting the triple patterns of SPARQL query between live and local processors based on a predetermined coherence threshold specified by the administrator. Considering quality requirements while splitting the SPARQL query, enables the processor to eliminate the unnecessary live execution and releases resources for other queries. This requires estimating the quality of response provided with current materialized data, compare it with user requirements and determine the most selective sub-queries which can boost the response quality up to the specified level with least computational complexity. In this work, we propose solutions for estimating the freshness of materialized data, as one dimension of the quality, by extending cardinality estimation techniques. Experimental results show that we can estimate the freshness of materialized data with a low error rate.

**Keywords:** Freshness estimation · Hybrid SPARQL Querying · RDF Data Warehouse · View Materialization

© Springer International Publishing Switzerland 2015
T. Supnithi et al.(Eds.): JIST 2014, LNCS 8943, pp. 257–270, 2015.
DOI: 10.1007/978-3-319-15615-6_19

# 1    Introduction

There is a huge amount of RDF data published in the Linked Data and many companies want to publish their data to maximize their participation in the extracted knowledge. However, processing queries over all published data is challenging. One approach to integrate Linked Datasets for query processing is to materialize all available data sources in one endpoint and respond queries from the materialized cache [1],[16]. This approach mainly suffers from providing low quality responses because changes of the original data are not immediately reflected on the materialized local cache. On the other hand, the main advantage of this approach is providing very fast responses as the data has been stored in the local cache and can be accessed without latency.

Another approach for integrating Linked Datasets for query processing is to fetch the relevant data on demand which is called live querying [8]. Live querying approaches process queries by dereferencing URIs and following relevant links on-demand and naturally incur very slow response times but with better quality (fresh and complete). Thus, the more data fetched from sources the higher the response quality and vice versa. This represents an inherent trade-off between the response quality and time which can be considered as a spectrum from a response with high quality and long retrieval time to a response with low quality but short retrieval time.

To mitigate the time consuming live querying, a hybrid approach for the Linked Data information integration has been proposed in a previous work [18]. The idea is to combine the data warehousing with the live query processing techniques so that part of the query is retrieved from the materialized data and rest of the query is executed on fresh, live data. To do so the query engine needs to make the critical decision of when to redirect a triple pattern to the live or the local query processor. For that the notion of coherence for each predicate was introduced which represents the dynamicity of that predicate and is defined as the ratio between the cardinality of live results that exist in the materialized data and the total cardinality of live results. The query processor splits triple patterns of a query based on their coherence for live or local processing using a pre-specified coherence threshold. Changing the coherence threshold in query splitting leads to different points of the response time/quality trade-off spectrum.

The main purpose of hybrid query processing is to achieve faster response than live approach and fresher response than the data warehousing approach. However some queries require high level of quality and do not care about response time and some other queries require fast response as far as a specific quality level is guaranteed. To address this issue the query processor should be able to adaptively split the query between live and local processors based on response quality requirements. This will also prevent unnecessary live execution, provide fast response and release resources for other queries. This problem boils down to two sub problems: First, estimating the quality of the fastest response which is provided with current materialized data and doesn't require any live execution. Second, if the local store couldn't satisfy user quality requirements, then the query processor needs to decide among various query splitting strategies to find the one which satisfies user quality requirements with the lowest live execution time. To the best of our knowledge these problems have not been addressed so far in the context of Linked Data query processing and is the main focus of this paper.

The remainder of the paper is structured as follows: In Section 2 we discuss related work. We define the problem and make it concrete in the context of Linked Data in Section 3. In Section 4 we propose our solution for estimating join freshness. We explain the experimental set up in Section 5. Finally, the result are presented in Section 6. we conclude the paper in Section 7 and provide insights for future work.

## 2    Related Work

The problem of efficiently processing queries by exploiting the materialized data requires a comprehensive *view management* procedure. This includes the view selection, the view maintenance, the view exploitation and the cost modelling [7]. The view selection phase is choosing a set of views for materialization to either respond queries without accessing the original data, or partially respond queries and fetch the query residual from original sources. The view maintenance phase mainly deals with the update processing or the live execution when the update stream doesn't exist. The view exploitation depends on two former phases. That is, if selected views fully cover queries and they are fresh then there is only a unique way for the exploitation; fully execute query on fresh materialized data. But if any of the above lacks, i.e, queries are partially covered or updates are postponed or no update stream exists at all, then the cost modelling phase is responsible to identify the best exploitation and maintenance strategy based on the required response time and quality.

Table 1 summarizes four possible situations of the view management in terms of the view maintenance and the view selection. It shows if query requirements can be used to fine-tune the view management for efficient query processing, phases that contributes in response time/quality trade-off and related works that target each particular settings. Various strategies for view selection (i.e. full or partial materialization) and view maintenance (i.e. immediate or deferred maintenance) can change the response time/quality trade-off. DBToaster [12] fully materializes data and immediately applies updates. Thus, there is no option to adjust the time/quality trade-off. It minimizes the cost of processing updates by converting the maintenance task to an efficient code for execution in a relational data model. [12] processes the update stream one-by-one and maintains materialized data as incrementally as possible. As alluded before, in this case only one execution plan exists which directly executes the query on the high-quality materialized data.

To evaluate SPARQL queries without accessing the original RDF data, [6] addresses view selection based on a fixed query set. They proposed a cost model to choose a particular view set that minimizes the overall cost of space, re-writing and maintenance. They assumed an update stream to quantify the maintenance cost. Since queries can be fully re-written from materialized views, execution

**Table 1.** Various states in a view management scenario : view maintenance can be done immediately which can be generalized to cases where update was not considered at all, or deferred which can be generalized to cases where no update stream exists and live execution is required to get the up-to-date version. View selection can fully materialize all queries or partially materialize common sub-queries.

| View selection | Full | Full | Partial | Partial |
|---|---|---|---|---|
| View maintenance | Immediately | Postpone | Immediately | Postpone |
| fine-tune view management based on query requirements | NOT applicable | applicable | applicable | applicable |
| Influential phases in trade-off | - | view maintenance (Coherence) | view Selection | view selection& view maintenance |
| Related work | DBToaster [12] for a fixed query set | Hybrid[18] for ad-hoc querying& Data warehouse[6] for a fixed query set | RDFMatView [4] for a fixed queryset | online view selection[11] & deferred view maintenance[9] both for ad-hoc querying |

time/freshness of the response can only be influenced by postponing the maintenance which is not discussed in [6].

The hybrid approach introduced by [18] considers the existing materialized data in a data warehouse as a predefined set of materialized data to be exploited for responding queries. Thus, the hybrid approach actually relaxes the view selection. They use a coherence value to split the query for local or live execution(i.e, maintenance). As alluded before, to maintain the views according to response requirements, we need to adaptively refine the coherence threshold which is not addressed by [18] and is the main focus of this paper. On the other hand, [4] recommended RDF indices to materialize based on a given workload aiming to improve performance of the query evaluation. However, in contrast to [6], queries still need to access the original data set because indices are partially covering queries. This approach assumes the original data are not changing and materialized data never gets out-of-date.

In order to optimally adjust the trade-off among response quality dimensions (i.e. maximize the other quality dimensions under the restriction on some of them), we need to explicitly define them to be able to estimate and measure them for each query execution plan and choose the best execution strategy. Freshness as one of the quality dimensions has been the topic of many research papers. [3] has categorized all freshness metrics in the literature into time-based freshness and cardinality based freshness. The problem of scheduling update processing based on response freshness requirements has been addressed previously under the time based definition of freshness using quality contracts [10] and OVIS algorithm for online view selection[11]. However in a Linked Data query processing environment, individual endpoints are not equipped to report every single update. Hence, query processor has to collect dynamics of materialized views and re-execute them lively to refresh the content of materialized data. The type of statistics to collect for estimating the quality of the response provided with materialized data and the decision of when to refresh a materialized view has not been addressed thoroughly in a Linked Data query processing environment and is the main focus of this paper. Less research has been done to adjust the time/freshness trade-off using the cardinality based definition of freshness. There has been research to estimate the quality of query response provided by materialized data in relational data model for cardinality based freshness which requires accurate cardinality estimation and accuracy of involved attributes [5,14]. This provides solutions to the first sub-problem (i.e. estimating the quality of the response provided with the current materialized data) but in a relational data setting. The definition and estimation of quality metrics are based on the identity attribute and is achieved by tracking the category change of each type of tuple during each operation (i.e. selection, projection, Cartesian product) in a relational data model. However, in an RDF data model there is no notion of identity key for tuples. Thus applying that approach for Linked Data is not directly possible. We propose extending the statistics of cardinality estimation techniques to estimate the quality of a query response. To the best of our knowledge there has been no solution for the second sub-problem of estimating the quality of various query splitting even in the relational data model.

## 3    Problem Description

Materialization of existing data sources at a local cache is one of the main tricks leveraged by data integration techniques to eliminate the time of fetching data from sources and provide fast response accordingly. As mentioned before, since changes of the original data might not be reflected in the cache, it may get out of date easily. The view maintenance is responsible to keep the materialized data up-to-date. This is done either by processing updates, in the presence of an update stream, or by executing queries lively, when no update stream exists. However maintenance of views can be postponed under two circumstances: 1) No query is accessing the materialized data targeted by that update. 2) The current materialized data can provide the response with the requested quality. The incentive of the user to be satisfied with lower response quality is to get faster response and consume less resources (e.g. pay less).

The decision of adaptively deferring the live execution of a sub-query until it is necessary to fulfil a response quality requirements is a complicated task and requires accurate quality estimation. This problem actually breaks down to two sub-problems. First the quality of the response provided with the current materialized data need to be estimated. If the current materialized data couldn't provide the requested quality, quality of the response can be boosted by live execution of sub-queries. Various query execution plans can be achieved by choosing different queries for live execution. The critical decision of choosing the less time consuming query plan which satisfies quality requirements of the response requires accurate quality estimation for each query plan and forms the second sub-problem. In this paper we only discuss and propose solution for the first sub-problem. In order to estimate quality, we need to explicitly define quality dimensions of a response in a Linked Data query processing environment. Response quality is expressed based on the response time, freshness (i.e., the cardinality based notion) and completeness [14].

In Figure 1, the shaded circle represents the result of executing query on the materialized data and the transparent circle represents the query results of the live engine. Area "A", which represents query results in the materialized data that doesn't exist in the result provided with the live engine, could contain inferred results from materialized data, data from not available sources and results which has been removed from original sources. However we relax the problem by only considering the latter source which means we assume all sources to be available and no inferred data is added to the materialized data. Area "B" represent

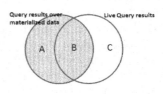

**Fig. 1.** freshness and completeness

the result set which exists both in live and local store. Area "C" represents the newly added facts to the live data which leads to changes in response of that particular query and still have not been reflected in the materialized data. We simplify the problem by assuming that live engine is able to cover all potential responses. With the above assumptions, *Freshness* quantifies the effect of the deletion on response quality and is defined as $B/(A+B)$. *Completeness* quantifies the effect of addition on response quality and is defined as $B/(B+C)$. In this paper we only consider the freshness as the quality metric of the response because in our synthetic data set we are assuming that data can only be removed from real world after materialization. However, the same technique can be applied to estimate the completeness of the response.

## 4    Proposed Method

Since our definition of freshness is based on the cardinality of fresh responses versus the total cardinality of the query response, we decided to extend the statistics of cardinality estimation techniques for estimating the freshness. This is achieved by storing the fraction of fresh responses per category and assuming that fresh entries have been uniformly distributed. In the following we explain how this applies to each type of cardinality estimation techniques.

**Indexing Based Approaches.** We extend two indexing approaches by storing both the total cardinality and the fresh cardinality per index entry. The higher the granularity level of indexing, the more accurate cardinality estimation, but it is more costly to build and maintain the index.

- **Simple Predicate Multiplication** In this approach we estimate join freshness by simply multiplying freshness of join's predicates. It requires indexing predicates along with their observed freshness. This approach works very well when join result is the Cartesian product of the result set of each predicate.

- **CS** Estimate join freshness by using the characteristic set [13]. It groups subjects with the same set of predicates together and index it as a "subject group". The whole dataset can be summarized into a set of "subject group"s with their associated predicates. Each predicate of each subject group store the cardinality of fresh and total entries. Analogously, the join's characteristic set(s) consist of individual subjects with their requested property(ies). For the freshness estimation of a join, we simply sum up the fresh cardinality and the total cardinality of characteristic sets that are super set of the join's characteristic set and divide the fresh cardinality by the total cardinality.

**Histogram Based Approaches.** Histogram and Qtree are among successful approaches for data summarization and estimation of query result size to compare query execution plans. The main appealing feature of histograms is having almost no run-time overhead [15]. The basic idea of histogram is to, first, group together attribute values of each dimension that have similar statistics. Having categorized entries per dimension, the histogram will consist of multi-dimensional buckets. Next step is to built the histogram by inserting entries into buckets. However, for the purpose of summarization, it doesn't keep all entries inside each bucket and only updates the statistics of the bucket on addition of new entries. Having the uniform distribution assumption per bucket, which is common among all histograms, it estimates cardinality and freshness of triple patterns based on the statistics and the ratio of overlap among their intersecting buckets.

Summarization in histogram-based approaches happens at two stages: hashing step and bucketing step. Histogram requires a hashing function to transfer string representation to numeric representation for processing data. A proper hashing mechanism is the heart of an efficient histogram-based approach since it will determine the uniformity of data distribution which is the main characteristics of histogram to summarize data. During this step, if some URIs happened to behave very similarly in terms of their joint freshness distribution, by hashing them to same value and treat them uniquely during summarization and query processing, histogram can still achieve a good freshness estimation.

Histogram buckets are determined based on a *partitioning rule*. For the sake of simplicity, we only implemented equi-width partitioning rule but the estimation rate can be improved by resorting to more advanced type of histograms introduced in [15]. According to the taxonomy proposed in [15], histogram based approaches require a sort and source parameter to define *partitioning rule* for specifying buckets. We tried sorting attribute values of dimensions based on their textual similarity and based on their freshness which is called *freshness hashing*. As mentioned before, we kept source parameter simple by dividing dimensions into buckets of equal length in histogram.

To adapt the histogram-based approaches for the freshness estimation problem, we proposed keeping two entries per bucket; the number of fresh and stale entries. In histogram-based approaches, join between triple patterns translates to intersection of buckets, assuming that entries are uniformly distributed all over each bucket. Interested readers are referred to [17] for more explanation on join processing using histograms. Table 2 summarizes how tuples should be character-ized in the result of joining two buckets. Thus, the cardinality of fresh result is the multiplication of the fresh cardinality in both buckets. Whatever that is left from the Cartesian product of both buckets is added to the stale result set.

Table 2. Tuple characterization for join operation

| B1 ⋈ B2 | Fresh | Stale |
|---------|-------|-------|
| Fresh   | Fresh | Stale |
| Stale   | Stale | Stale |

Qtree is an optimized histogram and its buckets are determined by identifying populated areas in the multi dimensional cube using a distance metric. Interested readers are referred to [17] for more detailed explanation.

## 5    Experiments

In order to measure the estimation accuracy of our proposed solution, we designed a synthetic dataset which consists of fresh and stale data. As pointed out before, to estimate the freshness of the materialized data for a query, the first step is to estimate its freshness for a join. We assumed each triple of the store is either fresh i.e. triple actually exists in live data sources or stale i.e triple doesn't exist in live data sources. We adapted selectivity estimation techniques in the DBMS for the join freshness estimation problem. That is, each bucket of our his-togram categorizes entries to fresh or stale instead of simply storing the count of bucket entries for summarization. After building the histogram by inserting all individual triples with their associated label to their corresponding bucket, we estimated the freshness of the materialized data for joins and compared it with its real freshness on the same data. Results are presented in Section 6.

### 5.1    Dataset Generation

We used the BSBM benchmark [2] (which is built for an e-commerce use case, where a set of products is offered by different vendors and different consumers have posted reviews about products) to generate a dataset as an snapshot of the materialized data in a data warehouse and a query set. We created N-TRIPLES benchmark dataset with 374,920 triples. In order to identify triples that actually exist in the current snapshot (i.e. fresh triples), we needed to assign true or false to triples. The generated dataset contains 40 predicates. We split existing predicates into 10 level of freshness(0-10%, 10-20%, ..., 90-100%) according to r-beta distribution of predicate freshness observed in [18]. Afterwards, we assign true or false to triples in dataset based on the freshness value of their predicate.

## 5.2   Query Set Generation

We used the BSBM query templates to generate queries then iterate over all query files and extract all triple patterns. We then replaced place holders with an arbitrary constant URI. Thereafter we extracted all pairs that have a common variable to make a join [s-s,o-o,s-o]. We separately investigated freshness prediction rate for each type of join because we expect different accuracy for each join category. In this paper we only presents results of s-s joins that have the best estimation performance. Finally, in the experiments we removed those queries that their real join cardinality was less than 5 because we don't expect our estimation to work well for queries with a few results. The final number of joins are 557, 64, 13 for S-S, S-O and O-O category respectively.

# 6   Result and Analysis

## 6.1   Estimating Freshness

We estimated the freshness of s-s join queries using the proposed indexing approaches. The Y axis shows the freshness values for queries of the X axis. We plot the real versus predicted freshness by each indexing approach in Figure 2. As depicted in Figure 2 predicted freshness of indexing based approaches started to disagree with real freshness after query 285 which is the point that bounded triple patterns started to appear in joins. This suggests that index-based approaches perform well for joins without bounded objects but don't perform well on joins having bounded object. This is because the freshness of bounded predicate is no longer similar to the freshness of the predicate while in fact indexing based approaches assume they are the same. By increasing the granularity level of the index i.e. storing the freshness per each bounded subject/object of triple pattern, we can increase the estimation performance but the process of building and maintaining such an index is a costly procedure. As indexing based approaches failed to provide good estimation for bounded triple patterns, which is due to both uniform distribution assumption and predicate independence assumption made by them, we asked if we can get good estimation for bounded triple patterns by extending the histogram approach for cardinality estimation.

By extending statistics stored in histogram's buckets with the fresh cardinality in addition to the total cardinality, we can estimate the freshness of a query response. However, it requires a proper hashing technique to transfer data from the string representation to the numerical representation. Thus we need to decide for a proper sort and source parameter for histogram to get good estimation [15].

## 6.2   Estimating Freshness Using Histogram

**Choose a Proper Hashing.** To summarize data with the extended histogram, we transfer dataset triples to their numeric representation using the similarity hashing and the freshness hashing proposed in Section 4. We plotted the estimated join freshness v.s. the real join freshness for all S-S joins using similarity hashing in Figure 3. Figure 3 depicts that the actual freshness of joins are

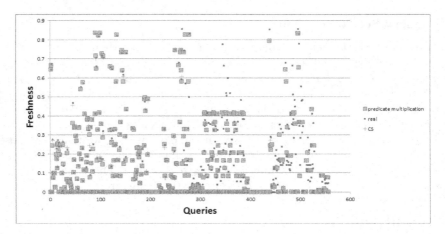

**Fig. 2.** freshness estimation in s-s joins using indexing approaches

extremely different than the estimated join freshness using the histogram with
mixed-hashing as a typical similarity-based hashing (Mixed hashing hashes the
subject and object values using prefix similarity hashing and hashes predicate
values using string hashing by checksums [17]).

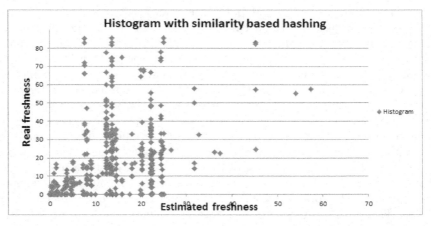

**Fig. 3.** predicted V.S. real freshness in histogram using similarity based hashing for
S-S joins

As shown in Figure 3, many joins with different actual freshness have been
predicted with similar values. This is because the BSBM dataset have a particular naming strategy in which many URIs share a long common prefix. Thus,
changing a subject or object constant in triple patterns, produces joins with various freshness values while similarity based hashing hashed them to same query
spaces (having similar hashed value). That leads to same freshness estimation

accordingly. In fact, URIs hashed to similar values will end-up in the same bucket and considered the same in summary data structure and join processing. Thus, changing them in bounded subject (object) will not make any difference while, in fact, its not true. Therefore, we proposed to sort dimension entries based on their freshness values and use it instead of the similarity hashing. It will keep entries with similar freshness values close together and will decrease the error in freshness estimation. Figure 4 shows that sorting histogram entries based on their observed freshness (freshness hashing) leads to better freshness estimates for joins.

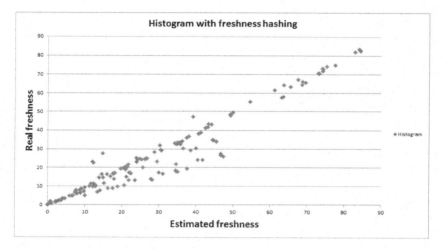

**Fig. 4.** predicted V.S. real freshness in histogram using freshness hashing for S-S joins

**Estimation Error.** The root-mean-square deviation (RMSD) is a frequently used measure for the differences between values estimated by an estimator and the values actually observed. We quantified the estimation error of different proposed techniques into one single value to get a sense of normalized error value and be able to compare the error over the course of storage space.

$$RSMD = \sqrt{\frac{\sum\limits_{i=1}^{n}(f_i - f_i')^2}{n}} \tag{1}$$

$f_i$ represents the actual freshness of a join and $f_i'$ represents the estimated freshness of that join. To normalize the RSMD, we divide it by the range of observed values, which is 100 in our case, because freshness values are all represented as a percentage.

The normalized RSMD of histograms using the freshness hashing are plotted in Figure 5 and it shows the estimation error of the histogram and the Qtree converged around 0.07 throughout various storage space while the simple predicate multiplication (an indexing based approach) consume very little space with a lower estimation error. However, the error of histogram based approaches can be further decreased by increasing the summary size or implementing more advanced type of histograms. In the future we are planning to achieve more accurate estimates using summaries that can capture more dependencies among join counterparts of queries such as probabilistic graphical models.

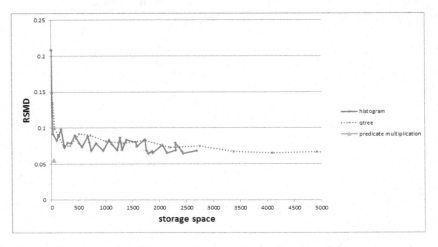

**Fig. 5.** consistency estimation error in s-s join in Qtree and histogram using sort hashing

## 7   Conclusion

In this paper, we study the problem of estimating the freshness of materialized data. This problem is one of the two sub-problems needed to be solved to optimally split queries among live and local processors based on response quality requirements. The existing approaches for quality estimation of materialized data for relational data model was heavily influenced by the existence of primary key and is not directly applicable for the RDF data model. We tackled this problem by extending the cardinality estimation techniques and proposed different approaches for the join processing phase on the extended cardinality estimation technique. Experimental results show that these approaches can estimate the freshness of the queries with a low error rate. We compared the performance of the proposed approaches for S-S queries on a synthetic dataset. In the future, we are planning explore more advanced type of summarization techniques to improve the freshness estimation of join.

**Acknowledgements.** I would like to thank Richard Cyganiak and Aidan Hogan for their great feedback during the experiment. This publication has emanated from research supported in part by a research grant from Science Foundation Ireland (SFI) under Grant

Number SFI/12/RC/2289, the European Commission EU-ICT VITAL project (Virtualized programmable InTerfAces for smart, secure and cost-effective IoT depLoyments in smart cities under contract number FP7-SMARTCITIES-608662 and the EU-ICT GAMBAS project (Generic Adaptive Middleware for Behaviour-driven Autonomous Services) contract number FP7-2011-7-287661.

# References

1. Bishop, B., Kiryakov, A., Ognyanov, D., Peikov, I., Tashev, Z., Velkov, R.: Factforge: A fast track to the web of data. Semantic Web 2(2), 157–166 (2011)
2. Bizer, C., Schultz, A.: The berlin sparql benchmark. International Journal on Semantic Web and Information Systems (IJSWIS) 5, 1–24 (2009)
3. Bouzeghoub, M.: A framework for analysis of data freshness. In: Proceedings of the 2004 International Workshop on Information Quality in Information Systems, pp. 59–67. ACM (2004)
4. Castillo, R., Rothe, C., Ulf, L.: Idexing RDF Data for SPARQL Queries. Professoren des Inst. für Informatik, RDFMatView (2010)
5. Dey, D., Kumar, S.: Data quality of query results with generalized selection conditions. Operations Research 61(1), 17–31 (2013)
6. Goasdoué, F., Karanasos, K., Leblay, J., Manolescu, I.: View selection in semantic web databases. Proceedings of the VLDB Endowment 5(2), 97–108 (2011)
7. Goldstein, J., Per-Åke, L.: Optimizing queries using materialized views: a practical, scalable solution. In: ACM SIGMOD Record 30, pp. 331–342. ACM (2001)
8. Hartig, O., Bizer, C., Freytag, J.-C.: Executing sparql queries over the web of linked data. In: Bernstein, A., Karger, D.R., Heath, T., Feigenbaum, L., Maynard, D., Motta, E., Thirunarayan, K. (eds.) ISWC 2009. LNCS, vol. 5823, pp. 293–309. Springer, Heidelberg (2009)
9. Kuno, H., Graefe, G.: Deferred maintenance of indexes and of materialized views. In: Kikuchi, S., Madaan, A., Sachdeva, S., Bhalla, S. (eds.) DNIS 2011. LNCS, vol. 7108, pp. 312–323. Springer, Heidelberg (2011)
10. Labrinidis, A., Qu, H., Xu, J.: Quality contracts for real-time enterprises. In: Bussler, C.J., Castellanos, M., Dayal, U., Navathe, S. (eds.) BIRTE 2006. LNCS, vol. 4365, pp. 143–156. Springer, Heidelberg (2007)
11. Labrinidis, A., Roussopoulos, N.: Exploring the tradeoff between performance and data freshness in database-driven web servers. The VLDB Journal 13(3), 240–255 (2004)
12. Lupei, D., Shaikhha, A., Koch, C., Nötzli, A.: Oliver Andrzej Kennedy, Milos Nikolic, and Yanif Ahmad. Higher-order delta processing for dynamic, frequently fresh views. Technical report, Dbtoaster (2013)
13. Neumann, T., Moerkotte, G.: Characteristic sets: Accurate cardinality estimation for rdf queries with multiple joins. In: 2011 IEEE 27th International Conference on Data Engineering (ICDE), pp. 984–994. IEEE (2011)
14. Parssian, A., Sarkar, S., Jacob, V.S.: Assessing information quality for the composite relational operation join. In: IQ, pp. 225–237 (2002)
15. Poosala, V., Haas, P.J., Loannidis, Y.E., Shekita, E.J.: Improved histograms for selectivity estimation of range predicates. ACM SIGMOD Record 25(2), 294–305 (1996)

16. Tummarello, G., Delbru, R., Oren, E.: Sindice.com: weaving the open linked data. In: Aberer, K., Choi, K.-S., Noy, N., Allemang, D., Lee, K.-I., Nixon, L.J.B., Golbeck, J., Mika, P., Maynard, D., Mizoguchi, R., Schreiber, G., Cudré-Mauroux, P. (eds.) ASWC 2007 and ISWC 2007. LNCS, vol. 4825, pp. 552–565. Springer, Heidelberg (2007)

17. Umbrich, J., Hose, K., Karnstedt, M., Harth, A., Polleres, A.: Comparing data summaries for processing live queries over linked data. World Wide Web **14**(5–6), 495–544 (2011)

18. Parreira, J.X., Umbrich, J., Karnstedt, M., Hogan, A.: Hybrid sparql queries: fresh vs. fast results. In: Cudré-Mauroux, P., Heflin, J., Sirin, E., Tudorache, T., Euzenat, J., Hauswirth, M., Parreira, J.X., Hendler, J., Schreiber, G., Bernstein, A., Blomqvist, E. (eds.) ISWC 2012, Part I. LNCS, vol. 7649, pp. 608–624. Springer, Heidelberg (2012)

# RDFa Parser for Hybrid Resources

Akihiro Fujii$^{(\boxtimes)}$ and Hiroyasu Shimizu

Faculty of Science and Engineering, Hosei University, 3-7-2Kajino-cho,
Koganei-shi, Tokyo 184-8485, Japan
fujii@hosei.ac.jp

**Abstract.** In this paper, we propose several mash-up design methodologies that are intended for enhancing human-readable aspects of semantically annotated data resources such as LOD. Proposals are based on the features provided by RDFa format which is an extension of RDF format. So called "semantic parser" have been developed which is an application software to provide data-conversion/transformation service through web API. The parser provides several hybrid features for both human-readable (HR) and machine-readable (MR) resources by using RDFa format.

**Keywords:** LOD · Semantic web · RESTful Web API · Mashup · Machine-Readable · Human-Readable

## 1 Introduction

Recent years, potentiality of semantic information annotated information resources, such as LOD (Linked Open Data) in web design is highly recognized. For the design paradigm of such system, light-weight software development is necessary aspect in terms of end-user computing and development. In the other words encouraging user-side mash-up with variety of resources is important for development of new services. The proposal of this paper is based on a hybrid approach. So called "semantic parser" have been developed which is an application software to provide data-conversion/transformation service through web API. Throughout this paper, we define two types of services of providing information resources through web API's (Application Programming Interface), human-readable (HR) and machine-readable (MR) services. For both HR and MR resources, the parser provides several features by using RDFa format. There are mainly two features in the system. One is to attach semantic annotation to ordinary web pages that does not have semantic information. Another is to provide pure semantic information through a certain XML-based format. At the beginning overall structure of the developed system is shown. Then the details of the provided features as well as potential applications and advanced utilizations of them are explained. The evaluations of proposed system are given in the later section.

## 2 Related Researches

In paper [3], semantic data authoring system is introduced. Based on MVP software model, existing web site is enhanced with semantic data and become machine-readable

© Springer International Publishing Switzerland 2015
T. Supnithi et al. (Eds.): JIST 2014, LNCS 8943, pp. 271–278, 2015.
DOI: 10.1007/978-3-319-15615-6_20

resources.   There is a comprehensive survey in [4], about mash-up methods over semantic annotated information resources.   In [5], they focused on RDFa formant and provide
additional features in order to enhance human-readable aspect of web sites.

# 3     Services Based on Semantic Annotation

Ordinary web sites consist of human-readable (HR) data and information. The amount
of such resource in the Internet is continuously increasing, or even exploding.   This
situation leads difficulties to extract necessary information through bare eyes of us.
Machine-readable (MR) data, on the other hand, is able to be processed with computer software or programs.   Semantic web technology, in a sense, is to cope with this
by broadening availability of resources.   W3C standard features such as RDF (Resource Description Format), OWL (Web Ontology Language) are powerful tools for
treating semantic information processing.   In this paper we are going to treat both
types of resource and flexible conversion of each other.   Services assumed in this
paper is based on RESTful Web API's.

When we talk about "Web Service", it may be taken as a rich, well-formatted configurations standardized by W3C such as SOAP or WDSL.   In recent years,   however, RESTful Web API is considered as common practice in order to provide "Web
Service" in larger context other than strictly defined configurations.   Throughout this
article, we assume that providing RESTful Web API in the format of RDFa, is meant
for wider definition of "Web Service".

# 4     Application Pattern of RDF

In this section, we will explain the overview of semantic resource management system over RESTful Web APIs.   We call such an architecture "semantic REST" for
short.   In 4.1 parser of semantic resource over semantic REST is introduced.   Then
in 4.2, means to produce MR resource from HR web sites or raw data resources are
explained.   In addition to them, how to export MR resource in HR format in order to
provide human-readable interface will be explained.

## 4.1     Parser over Semantic REST

In this sub-section, the design issue, such as structure and implementation, of Semantic REST is explained.

### Layered Structure for Implementation of Semantic REST
In recent years, there have been large amount of research results and proposals are
published in the field of semantic REST.   Examples are seen in [4][5].   Commonly
such systems take triple layered structure shown in Fig. 1.

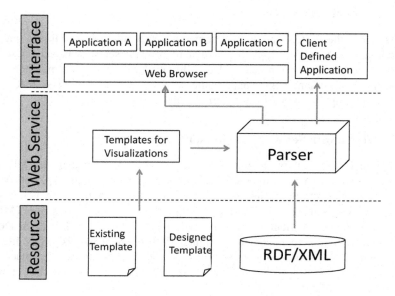

**Fig. 1.** Triple Layered Structure of Web Service

In the base layer, so called "Resource Layer", CRUD (Create, Retrieve, Update, Delete) feature is assigned to data and semantic information is prepared for the data through RESTful Web API's. The interface layer above it provides variety of Web interfaces depending on the types of devices from ordinary PC to tablet terminals. The middle layer placed in-between those two layers of the architecture is the main interest of this article. The functions of this layer are semantic data collection and processing's of the data. Some data may be collected through crawling job and that will be processed into, for example, a single data format. In the other example, data is filtered in order to match semantically intended applications. Later we will take an example that collect text data by crawling software. The resource is attached with additional semantic information by a portal web site. Consequently the semantic parser collects resources with or without semantic information and processes them either from HR to MR or vice versa.

**The Implementation of Semantic REST**

Generally client-server applications implemented over RESTful Web APIs, actions of clients are determined by state transitions obtained by API's. With semantic information, raw data becomes more useful resources with meta-data that explain the characteristics of the resources. Meta-data consists of collection of "name" and "value". The combinations of each item in the collection are regulated by "ontology" associated with them. The ontology is supposed to be designed a priori. The GET request is initiated upon the client's request that results in servers' control over the client statuses. The representation of state transition, the origin of acronym of REST, transferred to clients. There is some resource state management necessary in a system based on REST.

**RDFa Utilization over Semantic REST**

RDFa is one of the W3C's standard for semantic representations and this is one of the serialization format. We adopt this format to provide services based on semantic REST. Meta-data annotations are added and placed with XML tag members. With this placement if semantic information, the web site can be treated as MR data resource, even though it is originally designed only for MR usage.

## 4.2     Mash-Up with Semantic Annotation

The main feature of our proposing semantic parser is explained in the followings. MR data is annotated with semantic information by means of mash-up, both HR and MR data is provided from the single data resource. The operation flow is the followings. Firstly JSON format configure the relation between each items in HTML documents. Then RDF elements aiming for MR feature will be add. Each DOM in HTML document is associated with assigned RDF element in this JSON file. The semantic parser will process those two files then RDF elements will be placed in to HTML documents based on the indication from JSON file. The process completes with extracting RDFa format.

This sequence is explained from application developer's stand point where he or she will provide web service by using semantic parser. At the beginning, the developer uploads HTML and JSON formats on the server machine. The parser processes data and returns RDFa file. Finally the developer places RDFa and publishes data into appropriate position of HTML document which can be accessed from both HR and MR way.

Running this parsing features, there are several cases in providing services. One is that JSON data is placed into a prepared HTML document. The other case is that external HTML document or part of it is used for the purpose. In that case a group of DOM is collected from several web sites and configured into a new resource. This created resource can be used as RDFa format since JSON data is in-planted into RDF tags by the functions of parser.

## 4.3     Mash-Up Service for Visualizing the Semantic Data

The next important feature of proposing semantic parser is to create HR web site according to MR resources such as LOD (Linked Open Data). Templates for visualization are prepared depending on how to provide MR data, variety of visualization format are possibly used. Templates are based on XHTML (extensible Hyper Text Markup Language) and CSS (Cascading Style Sheets). For the visualization, XHTML template represents the features of human interfaces and CSS configure the relation to RDF/XML tags. Currently we have implemented some templates for the purpose. They process MR data resource as input of the templates and produce HR web sites based on the design of human interfaces. The parser produce output from RDF/XML based on visual template as indicated representation and structure. In order to use this feature, user indicates a specific URI as well as upload the file directly. In the next sub-sections, those two different realizations are explained in detail.

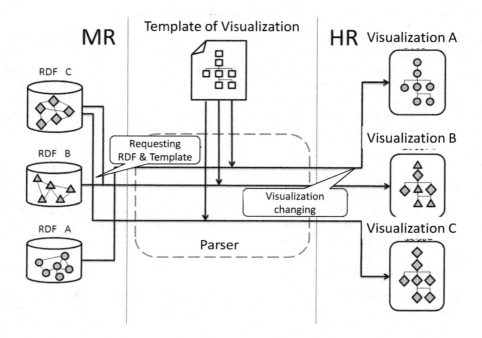

**Fig. 2.** Multiple-Data and Common Template

## Case Study 1 : Multiple-Templates for a Single Data Source

The situation is described in Fig.2.  Suppose there are 3 types of templates A, B, C, for one RDF data source as an example.  In this case there is a single MR data and is supposed to be used in several HR site through different types of templates.  Templates are variety in design and structure coming out from the parser.  This case is useful for providing variations in services which are tightly coupled with interface design.  The contents of them are coherent such as web site for a public relation of companies or introduction of a personal and organization.  In designing this type of web sites, there are cases that site designer needs to publish web sites for both desk-top PC's and tablets at the same time.  The process of designing this kind of web site usually follows the sequence described below:

1. RDF/XML data is created that contains core information for the intended service
2. Templates for different types of interface, desk-top and tablet are prepared.
3. They are sent to parser and processed there.

As the output of this process, two types of representational formats, both for desk-top PC and Tablets, in which contents are identical, will be obtained.  Modifications of the file are easily done by over-writing file through the parser.

**Case Study 2:    Multiple-Data Sources with a Single Template**

In this situation, multiple MR data sources are collected and published in a single representational format.    The relationship of data resources and parser is as the followings.    Suppose there are several RDF data sets which are different in content, and are represented by a common ontology.    In this case, the parser works according to class name or identifiers defined by the ontology so that some template file can be used for all those data resources.    Feasible applications to this feature are introduction of personal profiles, product catalogs, company profiles, and illustrated dictionaries, so forth.

## 4.4    System Implementation

In this section, we will explain how above mentioned features are implemented with examples.    Firstly, RDFa generation into HTML document is explained.    Then how the parser explained in 4.3.

**Translation from HTML to RDFa**

In this section, the implementation of feature to translate ordinary HTML document into RDFa file will be explained.    JSON data indicate the place and content for RDFa generations.    By selector value in JSON format the place of annotations are indicated.    The process is similar to design annotations of CSS.

**Implementation of Visualization Format of Semantic Data**

In this section, we will show how to generate RDFa implementation with visualization format from RDF/XML data resources that was explained in 3.3.    Here the case in 3.3.1 is taken as an example and explained.    We assume that there is a collection of company data base that is prepared in RDF/XML format.    Now the visualization format is suitable for this resource and publishing is prepared.    The format is based on HTML document that generate name and introduction of each company.    In processing the parser takes template inputs, and generates RDFa as visualization format.    The text in-between the <name> </name> tags are placed into {company: name} element.    If there is more than one candidate for transformation, every related data will be generated.

## 5    Case Study of Implementation in Realistic Business Web Site

In this chapter, implementation of parsing system is explained.    We take a realistic business application as an example.    "Makers-Inn" is an existing business information providing site, which provide public company profiling service as well as introductions of their main products and businesses in mechanical manufacturing and distributions.    For the data resources of the parsing system takes data items such as company name, address, and text sentences of introductions.    In 5.1 RDFa implementation of in 4.2 is explained.

## 5.1    RDF Annotation to HTML

Transformation to RDFa format from HTML with RDF resource is discussed here. In other words, this process convert HR web site with MR information and provide hybrid web site with both HR and MR feature.  The HTML and RFD resources associated with JSON code which creates the RDFa based web site.  This example indicate that with applying JSON format parsing RDFa is successfully done, and this RDFa web site works both HR and MR way.

## 5.2    RDF/XML Conversion of Resources to RDFa

In this section RDF/XML data as well as MR type resource can be converted into RDFa format according to visualization template which is based on HTML format. Let us explain two patterns of implementations.  Firstly, company name and introduction of the company profile is shown.  Input of this process is RDF/XML data source which consists of lists of company information and the template.  The RDFa output of those resources and file format will be created.  There is another example that web site publishes of names of companies and address of them.  Company RDF/XML provides name and address data which will be generated and published.

These two examples utilize ordinary company information as resource data and produces variety of output as web sites.  The examples indicated that visualization of MR resources could be flexibly managed by the proposed system.  In addition to this, if ontological information is prepared for different resource other than the example, it is possible to produce coherent RDFa format.  That means we have wider range of potentials in terms of MR resource utilization such as LOD.

# 6    Conclusion

In this paper, web services based on RESTful Web API's are proposed.  Applications are based on RDF/XML data and HTML representations and are to produce RDFa format that can be utilized both MR and HR way.  In the examples shown in this article, resource is limited to a simple existing data set.  Proposed parser can produce variety of possible applications.

We believe in the next generation web services require light weight semantic annotation.  In this view point, there are eight possible combinations in terms of resource provider and service consumer, HR/MR.  The proposal of this article is aiming to cope with the situation in which mash-up of web API and processing by parser is provided.

# References

1. An XML Pipeline Language. http://www.w3.org/TR/XPROC/
2. Heath, T., Bizer, C.: (Translation in Japanese, Hideaki Takeda, et al) Linked Data: Evolving the Web into a Global Data Space. Kindai-Kagakusya Publishing (2013)

3. Nitto, E.D., et al.: At Your Service: Service-Oriented Computing from an EU perspective. MIT Press (2009)
4. Ngu, A.H.H.: Semantic-Based Mashup o Composite Applications. In: IEEE Trans. on Service Computing **3**(1) (2010)
5. Khalili, A., et.al.: The RDFa Content Editor-from WYSIWYG to WYSIWYM. In: IEEE 36th Int. Conf. on Computer Software and Application (2012)
6. Louvel, J., Templier, T., Boileau, T.: REST in Action. Manning Publications (2013)
7. Next Generation Electronic Commerce Promotion Council of Japan (ECOM). http://www.ecom.or.jp/
8. Sherif, M.H.: standardization of business-to-business electronic exchanges. In: IEEE, Standardization and Innovation in Information Technology: SIIT 2007 Proceedings (2007)
9. Japan Information Processing Development Corporation (JIPDEC). An investigation of actual conditions of EDI/electronic tag application in Japanese industries (March 2010)
10. "GCommerce,": A case study of Microsoft Corporation (December 2010). http://www.microsoft.com/casestudies/
11. Fujii, A.: Standardization of electronic commerce in the cloud environment and its future evolution. In: Proceedings of PICMET 2012 Technology Management for Emerging Technologies (PICMET), pp. 2207–2213 (2012)

# Ontological Engineering

# Ontology Construction and Schema Merging Using an Application-Independent Ontology for Humanitarian Aid in Disaster Management

Pasinee Apisakmontri[1,3]([✉]), Ekawit Nantajeewarawat[1],
Marut Buranarach[2], and Mitsuru Ikeda[3]

[1] School of Information, Computer, and Communication Technology, Sirindhorn
International Institute of Technology, Thammasat University, Bangkok, Thailand
{pasinee.a,ikeda}@jaist.ac.jp, ekawit@siit.tu.ac.th
[2] National Electronics and Computer Technology Center, Bangkok, Thailand
marut.buranarach@nectec.or.th
[3] School of Knowledge Science, Japan Advanced
Institute of Science and Technology, Nomi, Japan

**Abstract.** Humanitarian aid information, e.g., information on the occurrences of disaster situations, victims, shelters, resources, and facilities, is usually rapidly dynamic, ambiguous, and huge. A system of humanitarian aid often involves data items from multidisciplinary environments, some of which have similar meanings but appear structurally different in various data sources. To achieve semantic interoperability among humanitarian aid information systems to be exchanged meaningful information, this paper contributes a methodology for construction of an application-independent ontology and proposes a guideline for merging information from different databases through the application-independent ontology that helps people to integrate systems with minimal modification. We demonstrate how to develop an ontology for *Humanitarian Aid for Refugee in Emergencies (HARE)* as a common ontology for sharing and re-use the current knowledge bases. We discuss an approach to merging *Relational Databases (RDBs)* to heterogeneous hierarchical ontology.

**Keywords:** Interoperability · Ontology construction · Merging resources · Humanitarian aid · Disaster management

## 1 Introduction

The frequency and severity of disasters have increased noticeably all over the world, causing impacts on societies, national economics, and environment. When a natural or human-made disaster occurs, a large amount of information is spread over, necessitating development of supporting information systems. A disaster management system is usually developed by many experts having a variety of backgrounds. Success of disaster information management depends on finding and effectively integrating related information to take decisions during information distribution [23], with the primary aim of getting the right resources to the

© Springer International Publishing Switzerland 2015
T. Supnithi et al.(Eds.): JIST 2014, LNCS 8943, pp. 281–296, 2015.
DOI: 10.1007/978-3-319-15615-6_21

right places at the right time and providing the right information to the right people to make the right decisions at the right levels at the right time [1].

Combining information from various sources is often problematic due to difficulties arising from interoperabilities of three basic types, i.e., system, syntax, and semantic interoperability. Interoperability at the system and syntax levels can be achieved by hardware improvement and a syntax standard, respectively. Semantic interoperability is more problematic. The use of different terms with different background knowledge causes confusion in the specification of semantics [4].

## 1.1   Ontology Construction and Merging Resources

### Strategy for Building Ontologies

- Application-dependent: the ontology is built based on an application knowledge base by a process of abstraction.
- Application-semi dependent: possible scenarios of ontology use are identified in the specification stage.
- Application-independent: the process is totally independent of the uses to which the ontology will be put in knowledge-based systems, agents, etc.

When establishing semantic interoperability of heterogeneous data, constructing the underlying ontology is a crucial part. Several publications about ontological development methodologies have been published, but still lack of widely accepted methodologies. The main reason is that mostly methodologies design for a particular project [9]. Our approach adopts the Uschold and King method [21], which a framework for enterprise modeling, to be a guideline. In highly dynamic environment, like a disaster management, to find a satisfactory mapping of Relational Databases (RDBs) onto a global model is a huge problem. The task of integrating heterogeneous information sources has several approaches. We consider a semantic mapping between an application-independent ontology and RDBs.

## 1.2   Objectives and Organization

Semantic Interoperability is challenging. If information systems are integrated with a common ontology, they will share a familiar underlying structure and then knowledge can be more easily shared. Our research explores the semantic interoperability on humanitarian aid and proposes a methodology for heterogeneous systems to be more readily shared with other different databases. An application-independence ontology in the humanitarian aid domain acts as a link between any data sources in the domain. We develop an application-independence common ontology for Humanitarian Aid for Refugee in Emergencies (HARE) that it can support interoperation among heterogeneous systems and provide the guideline for semantic interoperability among existing RDB of humanitarian aid in emergencies systems. The paper is organized as follows:

Section 2 provides background and related works. Section 3 describes our ontology development methodology. Section 4 and 5 describe how HARE ontology can be used to merging RDBs and explain the correspondence analysis among existing disaster management systems through HARE ontology. Section 6 concludes the paper.

## 2    Background

### 2.1    Semantic Interoperability Through Ontologies

A survey in [8] reported that ontologies and current technologies could be used to identify and associate semantically corresponding concepts in related disaster information in order that heterogeneous data can be integrated. An ontology is a conceptualization of data domain. It provides formal explicit definitions of concepts (known as classes), properties (known as attributes) describing various features of the concepts, restrictions on these properties and well-defined relationships between concepts. The target of ontology in Information Science is to define a common vocabulary for knowledge representation and specifies constraints on the relationships between objects in an application domain. The basis of an ontology can cope with the confusion and better enabling computers and people to work in cooperation [3, 12, 23]. Because of the potential of an ontology, one data source can be combined with another. Semantic interoperability allows the applications to cooperate with minimal modifications [13]. Thus, the need for ontologies to establish a common specification to deal with the entire disaster management cycle is necessary.

### 2.2    Related Works

Ontology has been used in several ways of the disaster management. [10] is finding suitable information in the open and distributed environment of current geographic information web services in order to overcome semantic heterogeneity during a keyword-based search. The strategy they used for building ontology is an application-semi dependent ontology. [19] models the conceptual ontologies that use the application-independent ontology strategy and this paper proposes service for disaster querying system. [11] proposes a practical emergency response ontology for collaborative crisis information management systems by using the application-dependent ontology strategy. [24] constructs emergency ontology by using the application-independent ontology strategy to support decision-making directly from the emergency documentation. [14] proposes an emergency ontology model using a hierarchical architecture, including upper ontology and application ontology by using the application-independent ontology strategy.

Ontology has been applied in many research areas. The above researches relate to ontology construction in different strategy. The ontology construction without an application-dependence, which is necessary for the common ontology construction, and the mapping between OWL and Relational Databases are challenging.

# 3    Construction of the HARE Ontology

The construction of the HARE ontology reflects the need to design a global ontology-based approach which is capable of dealing with an RDB. Our approach to ontology construction has been started with information exploring from a knowledge based of UNHCR handbook[2,15,16], which [15] contains 27 chapters of a standard knowledge of international organization in 576 pages, [16] contains 25 chapters in 325 pages, and [2] contains six chapters in 77 pages. Humanitarian aid in emergencies is part of a complex system in which different organization processes collaborate in sharing information on refugee's profiles, donations, rescue activities, and actors. The establishment of an ontology for the HARE domain is a pre-requisite to developing an application.

## 3.1    Ontology Building Methodology

There are several methods for building ontologies. Most of them focus on domain conceptualization and ontology implementation, but do not pay enough attention for integration of ontologies [4,18]. In additional, ontology design must consider the integration between ontologies and data sources. We adopt the basic phases of the Uschold and King method [21] as the skeleton of our ontology construction method.

The Uschold and King method consists of the following four main phases:

1. Identifying a purpose and scope: Defining the scope and granularity of the target ontology.
2. Building the ontology
   (a) Ontology capture: Identification of the key concepts and relationships in the domain of interest.
   (b) Ontology coding: Structuring the domain knowledge in a conceptual model.
   (c) Integrating existing ontologies: Reuse of existing ontologies to speed up ontology development process.
3. Evaluation: Making a technical judgement of the ontologies.
4. Documentation: Guidelines for documenting ontologies.

The Uschold and King method does not precisely describe the techniques and activities. It lacks the appropriate depth of methodological description and an iteration back and forth, which gradually refine the ontology. We need a common ontologies with the capability to intensely share information between relational databases. In order to make the ontology corresponding with RDB, we use hypernym, hyponym, and synonym in WordNet for ontology construction and finding the relations of matching between ontology and RDB.

The WordNet is similar to a thesaurus, but nouns, verbs, adjectives and adverbs in WordNet are distinguished into sets of cognitive synonyms (synsets). The WordNet database contains 155,287 words organized in 117,659 synsets. With this objective, the first three phases are extended and tailored for the construction of the HARE ontology as follows:

1. Identifying a purpose and scope
   (a) Getting requirements of refugee in emergencies
   (b) Creating the use case diagrams and use cases descriptions
2. Building the ontology
   (a) Ontology capture
       i. Considering knowledge models from the use case diagrams
   (b) Ontology coding and integrating
       i. Finding hypernyms of each concept to create hierarchical ontology
       ii. Integrating with upper ontologies
3. Evaluation
   (a) Validating ontology with the different databases from the humanitarian aid domain
   (b) Ontology modification

These three phases are detailed below.

## 3.2 Identify Purpose and Scope

**Getting Requirements of Refugees in Emergencies.** The operations of UNHCR cover many areas in refugee emergencies, including health, food, sanitation and water, as well as key field activities corroborate the operations such as logistics, community services, and registration. Such operations must be managed and controlled by many associate organizations. In the identify purpose step, information should be extracted carefully from documentations. In order to establish the ontology requirements, we have read the overall operations of the HARE from relevant chapters in the on Handbook for Emergencies [15] and related documents [2,16] to undertake the abstraction and processes of refugee emergencies from UNHCR. The HARE consists of five subsystems of UNHCR, i.e., Registration, Identification of Persons of Concern, Emergency Planning, Distribution of Assistance, and Donation.

**Creating the Use Case Diagrams and Use Case Descriptions.** We determined the domain, scope and purpose of the operations and developed the Unified Modeling Language (UML) Use Case Diagram that is used to create the functionality for the system to graphically represent and envision the concepts and relationships between elements. They are described below:

*Refugee Registration* [1]: Refugees should be registered as fast as possible after reach to a refugee center. Refugee profile must be the first information that organizations would like to know. The following information is recorded against a person of concern individual verification: name, unique identifying registration number, date and place of birth, sex, existing identity documents, marital status, special protection and assistance needs, level of education, occupational skills, religion, language, household and family composition, date of arrival, current location and address, place of origin, and photograph. This information will be collected to be the properties of refugee profile concept.

---

[1] Extracted from chapter 10 in [15] and all chapters in [16]

*Identification of Persons of Concern* [2]*:* After the refugee registration process, if time permits, a pre-screening should take place at this stage to identify those who may not be of concern to UNHCR. The refugee profile will be analyzed for refugees' needs assessment. An accurate estimate of numbers of refugees is a prerequisite for effective protection and assistance, and identification of beneficiaries, including persons with special needs.

*Emergency Planning* [3]*:* The planning process is very important. Efforts should be made to design and implement a shelter as soon as possible. Several organizations must rely on this planning system. The project is the structural planning for such as the training, logistics, telecom, security, sites(camp, shelter), etc. This will enable better management such as shelter management, non-food and food items' distribution.

*Distribution of Assistance* [4]*:* This is a simple system to handle the distribution of assistance and provision of service to refugees, including emergency health care, distribution of food and non-food items. Many staffs from organizations participate in this system for sharing help to refugees.

*Donation* [5]*:* The Donation system is designed to receive, manage, and distribute a mass of donated goods and services. With the help of refugee communities, they identify refugee individuals and groups with their needs, especially unaccompanied and separated children. This system operates with the distribution through an assistance system to support needs of refugees.

### 3.3    Ontology Capture

**Considering Knowledge Models From the Use Case Diagrams.** In the second step, the refugee emergencies use cases from the previous step will set the initiation to establish ontological conceptualization, which is the key to the HARE ontologies. The core concepts and the existing relations from use case diagrams help to build the scope and role of the ontology. Each use case consists of detail to extract relevant classes, their scopes and the required detail.

Firstly, a first set of terms is gathered by analyzing use cases. The selected terms are considered as the core concepts of HARE ontologies. That core concepts in the operations of the HARE, such as Commodity, Distribution Cycle, Family, Household, Head of Family, Head of Household, Ration Shop, Refugee, Registration Card, Scooping, Tipping, Vulnerable, RefugeeActivity, RefugeeNeed, Person, Plan, Project, Organization, Staff.

After the core concepts are defined, subclasses and disjoint decompositions are also identified, such as a food product is a particular type of Commodity. To

---

[2] Extracted from chapter 2 in [15]

[3] Extracted from chapter 3-9 in [15]

[4] Extracted from chapter 13 in [15] and all chapters in [2]

[5] Extracted from chapter 16 and 20 in [15]

get core concepts, the relations between concepts and axioms are important to define implicit meanings. The implementation of the HARE ontology required to choose an appropriate ontology editor and development environment. The Protégé development platform which contains the Protégé-OWL ontology editor for the Semantic Web is chosen for using in this research.

### 3.4   Coding and Integrating

**Finding Hypernyms of Each Concept to Create Hierarchical Ontology.**
Hypernym is a word or phase whose meaning includes the meaning of other words. A broad meaning of hypernym constitutes a category into which words with more specific meanings fall. Those core concepts are easiest to understand and interoperate when a hierarchy is diagrammed. In this paper, the core concepts are organized into hierarchies by hypernym of word from WordNet. For instance, there is a core concept 'RefugeeActivity'. The sense of the word 'RefugeeActivity' would have the hypernym hierarchy. The WordNet shows that a 'RefugeeActivity' is a kind of an 'HumanActivity'. After finding hypernyms of the entire concept, the concepts with the same a broad meaning are grouped into a same category in hierarchy as a Fig 1. An 'Action', 'Communication', 'HumanNeed', 'RefugeeActivity', 'Resonsibility', 'Role' and 'Service' are also a kind of a 'HumanActivity'.

**Fig. 1.** HARE hierarchical ontology

**Integrating with Upper Ontologies.** After creating ontology, we would notice that there are some classes in the ontology that can be hierarchical implementation on the upper ontologies. An upper ontology can be also called as a top-level ontology or foundation ontology, which describes the general concepts for all knowledge domains. As general concepts, several ontologies can be hierarchical implementation on the upper ontologies. Although upper ontology has several advantages, such as integrating with existing ontologies, providing a predefined set of ontological entities, turning away from conceptual ambiguity, however, semantic interoperability between ontologies is the main advantage of upper ontology [6,7,20]. Based on reviews of the characteristics of ontologies to be used as upper ontology, we decide that the below ontologies meet most of our requirements.

As the related upper ontology review, we found that the concepts of the upper ontologies, which are matched with the HARE ontology, are in the DOLCE,

**Table 1.** Related Upper Ontologies

|  | DOLCE | SUMO | SWEET | FOAF |
|---|---|---|---|---|
| Full Name | Descriptive Ontology for Linguistic and Cognitive Engineering | Suggested Upper Merged Ontology | Semantic Web for Earth and Environmental Terminology | Friend of a Friend |
| Language | First-Order Logic, KIF, and OWL | SUO-KIF, and OWL | OWL | RDF, and OWL |
| #Classes | 80 | 1000 | 6000 | 13 |
| #Properties | 80 | 4000 | 300 | 62 |

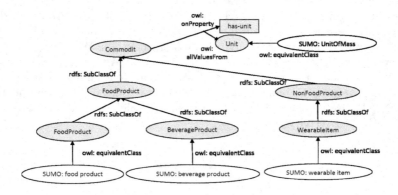

**Fig. 2.** SUMO upper ontology and HARE ontology

SWEET, SUMO, and FOAF (Table 1). They can provide the basis for common understanding in HARE. Fig.2 depicts an example of SUMO upper ontology to facilitate the semantic integration of HARE ontology. A blue oval means the core concepts of HARE ontology and a white oval means the concepts from upper ontology.

## 3.5   Evaluation the Ontology

As the result, our HARE ontology has 271 concepts, 67 object properties, 57 data properties. The information is structured in a common formal model including domain rules. The ontology we created can be used to be a common conceptualization of HARE that other related systems would be integrated with in this domain. As this section, we test the HARE ontology with two existing systems in order to present the example of the interoperability between two systems with the proposed ontology.

**Validating Ontology with the Different Databases From the Humanitarian Aid Domain.** In this paper, we consider the models specified in the five use cases in humanitarian aid, which are Refugee Registration, Identification of Persons of Concern System, Emergency Planning System, Distribution of Assistance, and Donation System as a good starting point of knowledge for our purpose of conceptualising a common understanding of disaster management model, for facilitating information exchange amongst applications that use different views of humanitarian aid. The entire procedures are shown in the section 4.

**Ontology Modification.** The HARE ontology is independent application ontology which intends to cover the knowledge in disaster management domain. So, an iterative process is important process to fulfill the domain knowledge. At each iteration, a decision is made on which new relation to add to the domain.

## 4  Semantic Interoperability with the HARE Ontology

### 4.1  Element Correspondences

A correspondence between an element of the HARE ontology and that of a RDB is a 4-tuple

$$\langle id, e_1, e_2, CT \rangle,$$

where $id$ is a unique identifier of the given correspondence, $e_1$ is an element, e.g., a class or a property, of HARE, $e_2$ is an element, e.g., a table or a column, of the RDB, and $CT$ is a correspondence type, which is one of the following: equality ($=$), overlapping ($\cap$), mismatch ($\bot$), more general ($\supseteq$), and more specific ($\subseteq$) [5]. Methods for determining class-table correspondences, property-column correspondences, and property-table correspondences are described below.

**Class-Table Correspondences.** We adopt a linguistic approach to determine correspondences between classes and tables from schema information. The approach exploits linguistic properties of schema elements. We compare name strings for syntactical name matching and compare their meanings for semantic name matching. As a preparation step, names are cleaned by (i) changing uppercase letters to lowercase letters, (ii) removing special symbols, (iii) expanding abbreviations to full forms, (iv) replacing punctuations with spaces. Let $C$ be a concept in HARE and $T$ a table in a RDB. Let $name(C)$ and $name(T)$ denote the names of $C$ and $T$, respectively, after they are preprocessed. A correspondence between $C$ and $T$ is determined by comparing $name(C)$ and $name(T)$ as follows:

1. *Syntactical name matching:* Construct a correspondence $\langle id, C, T, = \rangle$ if a word in $name(C)$ is the same as at least one word in $name(T)$.
2. *Semantic name matching:* Class-table correspondences are also determined based on semantic relationships such as synonym, hypernym, and hyponym relationships, given by WordNet.

- Construct $\langle id, C, T, = \rangle$ if a word in $name(C)$ is a synonym of at least one word in $name(T)$.
- Construct $\langle id, C, T, \cap \rangle$ if a word in $name(C)$ and a word in $name(T)$ have a common hypernym.
- Construct $\langle id, C, T, \supseteq \rangle$ if a word in $name(C)$ is hypernym of at least one word in $name(T)$.
- Construct $\langle id, C, T, \subseteq \rangle$ if a word in $name(C)$ is hypornym of at least one word in $name(T)$.

**Property-Column Correspondences.** The linguistic approach used earlier for determining class-table correspondences is also applied for determining correspondences between properties and columns. In addition, constraints on data types are also used; i.e., a non-foreign key column may correspond to only a data type property, and a foreign key column may correspond to only an object property.

**Property-Table Correspondences.** A correspondence may exist between a property and a junction table. A junction table is a database table that contains common columns from two or more other tables. A correspondence $\langle id, prop, T_J, = \rangle$ between a property $prop$ and a junction table $T_J$ is constructed if $T_J$ is a bridge between a table $T_1$ and a table $T_2$, $prop$ is an object property of a class $C_1$ with $C_2$ being its range, and $C_1$ and $C_2$ correspond to $T_1$ and $T_2$, respectively.

### 4.2   Mapping a RDB to the HARE Ontology

Based on the correspondences between their elements, a given RDB $\mathcal{D}$ is mapped into HARE with the assistance of a domain expert using the following steps (Table. 2):

**Table 2.** Techniques and Steps for mapping RDB to HARE ontology

| Step | Techniques for finding correspondences | | |
|---|---|---|---|
| | Class-table | Property-column | Property-table |
| 1. Select eligible tables from database | | | |
| 2. Find candidate class-table correspondences | / | | |
| 3. Filter out candidates by a domain expert | | / | |
| 4. Add new correspondences by a domain expert | | / | / |

1. *Select eligible tables from $\mathcal{D}$:* Tables of $\mathcal{D}$ that are relevant to the use cases under consideration are selected. More specifically, a table is selected if its name is related to some keyword in the description of some identified use case.
2. *Find candidate class-table correspondences:* Determine the correspondences between classes in HARE and tables in $\mathcal{D}$, using the method given in section 4.1.

3. *Filter out candidates by a domain expert:* An obtained correspondence between a class $C$ and a table $T$ is then examined as follows: Let $Prop(C)$ denote the set of all properties of $C$ and $Col(T)$ the set of all columns of $T$. To determine whether $C$ and $T$ describe entities of the same type, correspondences between the properties in $Prop(C)$ and the columns in $Col(T)$ are considered by a domain expert. If it is unlikely that $C$ should be mapped to $T$, then the domain expert recommends the removal of the correspondence between $C$ and $T$.

4. *Add new correspondences by a domain expert:* After filtering some candidate class-table correspondences out, a table with no corresponding class is further investigated. Let $T$ be a table currently having no corresponding class and $Col(T)$ the set of all columns of $T$. Based on property-column correspondences and property-table correspondences, a new correspondence is created using the following criteria:

    (a) Let $C$ be class in HARE and $Prop(C)$ the set of all properties of $C$. A new correspondence between $C$ and $T$ is created if a domain expert recommends that the properties in $Prop(C)$ and the columns in $Col(T)$ describe entities of same type.

    (b) If $T$ is a junction table and there is a property-table correspondence between a property *prop* of a concept in HARE and $T$, then add a correspondence between $T$ and *prop*.

## 5    Evaluation Results

For considering to ontology mapping with existing systems, two RDBs have been evaluated. The first existing system is called Sahana Eden (Emergency Development ENvironment for Rapid Deployment Humanitarian Response Management) [17] and the second one is the Ushahidi platform [22].

Sahana Eden is an open source disaster management software platform which has been built specifically to help in disaster management, whose mission is to help alleviate human suffering by giving emergency managers, disaster response professionals and communities access to the information that they need to better prepare for and respond to disasters. Sahana Eden provide a number of different modules, such as organization registry, Project Tracking, Messaging, Scenarios & Events, Human Resources, Inventory, Assets, Assessment, and Map. The database of Sahana has 245 tables to find the correspondences to HARE ontology.

Ushahidi is an open source to visualize information on a map in disaster domain, track reports on the map, filter data by time, and see when things happened and where. It easily collects information via text messages, email, twitter and web-forms. The Ushahidi has 53 tables to find the correspondences to HARE ontology. In session 5, we demonstrate an example of mapping between the two RDBs through HARE ontology.

### 5.1    Mapping the Sahana Eden Database to HARE Ontology

From 245 tables of Sahana Eden database, 56 tables are selected as in-scope database that relevant to use case description. 57, 4, 1, and 1 class-table

**Table 3.** Correspondences of the relevant information

| HARE and Sahana | HARE and Ushahidi |
|---|---|
| $\langle m_{17}, Location, gis\_location, =\rangle$ <br> $\langle m_{18}, Location, gis\_location\_name, =\rangle$ | $\langle m_3, Location, location, =\rangle$ |
| $\langle m_{37}, Person, pr\_person, =\rangle$ <br> $\langle m_{40}, Person, pr\_physical\_description, =\rangle$ <br> $\langle m_{67}, has - skill(Person), hrm\_competency, =\rangle$ | $\langle m_4, Person, incident\_person, =\rangle$ |
| $\langle m_{10}, Event, event\_event, =\rangle$ <br> $\langle m_{13}, Event, event\_activity, =\rangle$ <br> $\langle m_{14}, Event, event\_asset, =\rangle$ <br> $\langle m_{15}, Event, event\_human\_resource, =\rangle$ <br> $\langle m_{16}, Event, event\_site, =\rangle$ <br> $\langle m_{58}, Event, event\_incident, \supseteq\rangle$ | $\langle m_5, Event, incident, \supseteq\rangle$ |

correspondences of type $=, \cap, \supseteq, and \subseteq$, respectively, are identified. Fig 3 illustrates some resulting of class-table correspondences. Consider example, the correspondence

$$\langle m_{60}, Responsibility, req\_commit, \cap\rangle.$$

As shown in Fig. 4a "Responsibility" is a concept in HARE. According to the part of WordNet shown in Fig. 4b, "Responsibility" and "Commitment" has "Group Action" as a common hypernym. The correspondence $m_60$ is therefore constructed. With the assistance of a domain expert, 34 correspondences are removed and 10 new correspondences are added in step 3 and 4, respectively, of section 4.2.

$\langle m_1, Assets, asset\_asset, =\rangle$
$\langle m_2, Shelter, cr\_shelter, =\rangle$
$\langle m_{58}, Event, event\_incident, \supseteq\rangle$
$\langle m_{59}, Skill, hrm\_competency, \cap\rangle$
$\langle m_{60}, Responsibility, req\_commit, \cap\rangle$
$\langle m_{61}, Responsibility, req\_commit\_item, \cap\rangle$
$\langle m_{62}, Responsibility, req\_commit\_person, \cap\rangle$
$\langle m_{63}, Need, req\_req, \subseteq\rangle$

**Fig. 3.** Comparing with structure-level matching

Altogether 39 tables of 56 tables in Sahana database that is 64.28 are mapped to concepts in HARE ontology.

## 5.2   Mapping the Ushahidi Database to HARE Ontology

From 53 tables of Ushahidi database, 7 tables are selected as in-scope database that relevant to use case description. 4, and 2 class-table correspondences of type $=, and \supseteq$, respectively are identified. For example,

$$\langle m_2, Event, Incident, \supseteq\rangle$$

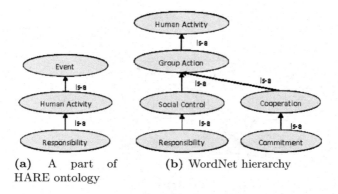

(a)  A  part  of          (b)  WordNet  hierarchy
HARE ontology

**Fig. 4.** Finding relation by WordNet

$\langle m_{64}, need(Need), req\_req\_item, = \rangle$
$\langle m_{65}, need(Need), req\_req\_skill, = \rangle$
$\langle m_{66}, has - document(Need), req\_document, = \rangle$
$\langle m_{67}, has - skill(Person), hrm\_competency, = \rangle$
$\langle m_{68}, is - responsed - by(Assistance), req\_commit\_person, = \rangle$
$\langle m_{69}, response - item(Assistance), req\_commit\_item, = \rangle$

$\langle m_{70}, Commodity\_type, supply\_item\_category, = \rangle$
$\langle m_{71}, Staff, hrm\_human\_resource, = \rangle$
$\langle m_{72}, Assistance, req\_commit, = \rangle$
$\langle m_{73}, Commodity, supply\_item, = \rangle$

**Fig. 5.** Add new corresponding relationship

$\langle m_1, Category, category, = \rangle$
$\langle m_2, Country, country, = \rangle$
$\langle m_3, Location, location, = \rangle$
$\langle m_4, Person, incident\_person, = \rangle$
$\langle m_5, Event, incident, \supseteq \rangle$
$\langle m_6, Event\_type, incident\_category, \supseteq \rangle$

**Fig. 6.** All corresponding relationships between HARE and Ushahidi

'Incident' from Ushahidi database is mapped with 'Event' from HARE ontology
by more specific relation. Because the WordNet show that the hypernym of 'Incident' is 'Event'. With the assistance of a domain expert, no correspondence pairs
is removed and no new correspondence is added in steps 3 and 4, respectively,
of section 4.2.

Altogether 6 tables of 7 tables in Ushahidi database that is 85.71 are mapped
to concepts in HARE ontology (fig.6).

### 5.3  Merging Resources: Emergency Response Information

Natural disasters affect people worldwide. Having access to relevant information
is a key for humanitarians to be prepared and effective in response. An emergency
response system has several processes of the humanitarian aid in emergencies.
The HARE ontology aims to interoperate the disaster information systems to
be useful and relevant to the humanitarian community. The section 4 and 5

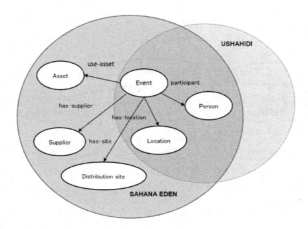

**Fig. 7.** Relations between Sahana Eden and Ushahidi after matching with HARE

explain the steps for finding correspondence between HARE ontology and the database of existing systems,i.e., Sahana Eden, and Ushahidi. The result shows that the relevant information between Sahana Eden and Ushahidi are location, person, and event information. The relevant information is derived from the correspondences (Table 3).

Fig. 7 illustrates the relations between Sahana Eden and Ushahidi database that are connected through HARE ontology. Event, Person, and Location are the common correspondences between Sahana Eden database and HARE, and Ushahidi database and HARE. The common correspondences are used to be a bridge over the gap between systems. Missing information can be shared. Ushahidi system would get additional information from Sahana Eden system. For example, Ushahidi system do not know the information about asset, supplier, and distribution site, but those information is shared through HARE ontology. Several existing systems can be linked by HARE ontology. The more linking systems, the more completed sharing information.

## 6   Conclusion

We have described a method for humanitarian aid in emergencies interoperability between an ontology and RDB. Our approach is based on ontology engineering and hypernym, hyponym, and synonym in WordNet to model the common ontology in humanitarian aid domain without dependent application. The contribution in this paper include the HARE ontology that is a common ontology in humanitarian aid in emergencies domain with the capability to intensely share information between relational databases, and a methodology for developing common ontology from documents to interoperate with existing systems. We have presented the mapping approach between an HARE ontology and two relational databases that are Sahana Eden and Ushahidi system. The results can reflect that HARE ontology can be a common ontology and the humanitarian aid RDBs systems can be shared information through the HARE ontology.

# References

1. Geomatics Solutions for Disaster Management (2007). http://books.google.co.jp/ books?id=eoB6nTkhLqkC
2. Division, O.S.: Commodity Distribution. United Nations High Commissioner for Refugees (1997)
3. Fan, Z., Zlatanova, S.: Exploring ontologies for semantic interoperability of data in emergency response. Applied Geomatics 3(2), 109–122 (2011)
4. Garrido, J., Requena, I.: Proposal of ontology for environmental impact assessment: An application with knowledge mobilization. Expert Systems with Applications 38(3), 2462–2472 (2011)
5. Giunchiglia, F., Shvaiko, P., Yatskevich, M.: Semantic matching. In: Encyclopedia of Database Systems, pp. 2561–2566. Springer (2009)
6. Gómez-Pérez, A., Fernández-López, M.: Ontological engineering : with examples from the areas of knowledge management, e-commerce and the semantic web. Springer (2010)
7. Hoehndorf, R.: What is an upper level ontology? (2010). http://ontogenesis. knowledgeblog.org/740
8. Hristidis, V., Chen, S., Li, T., Luis, S., Deng, Y.: Survey of data management and analysis in disaster situations. Journal of Systems and Software 83(10), 1701–1714 (2010)
9. Iqbal, R., Murad, M.A.A., Mustapha, A.: Sharef, Mohd, N.: An analysis of ontology engineering methodologies: A literature review. Research Journal of Applied Sciences, Engineering and Technology 6(16), 2993–3000 (2013)
10. Klien, E., Lutz, M., Kuhn, W.: Ontology-based discovery of geographic information services–an application in disaster management. Computers, environment and urban systems 30(1), 102–123 (2006)
11. Li, X., Liu, G., Ling, A., Zhan, J., An, N., Li, L., Sha, Y.: Building a practical ontology for emergency response systems. In: 2008 International Conference on Computer Science and Software Engineering, vol. 4, pp. 222–225. IEEE (2008)
12. Miller, E.: W3c technology and society domain: Semantic web points. www.w3. org/2001/sw/EO/points (accessed Septermber 04, 2012)
13. Missikoff, M., Taglino, F.: An ontology-based platform for semantic interoperability. In: Handbook on Ontologies pp. 617–634 (2004)
14. Peng, Y., Wang, W., Cunxiang, D.: Application of emergency case ontology model in earthquake. In: International Conference on Management and Service Science, MASS 2009, pp. 1–5. IEEE (2009)
15. Preparedness, T.E., Section, R.: Handbook for Emergencies. United Nations High Commissioner for Refugees, UNHCR Headquarters, Case Postale 2500, Switzerland, 3rd edn. (2007)
16. Profile, P., the Population, Section, G.D.: United Nations High Commissioner for Refugees: UNHCR Handbook for Registration. http://www.unher.org/3f8e93e9a.pdf (accessed September 09, 2003)
17. SAHANA: Eden:sahana software foundation. http://sahanafoundation.org/ products/eden/ (accessed September 09, 2013)

18. Scheuer, S., Haase, D., Meyer, V.: Towards a flood risk assessment ontology knowledge integration into a multi-criteria risk assessment approach. Computers, Environment and Urban Systems **37**, 82–94 (2013). http://www.sciencedirect.com/science/article/pii/S0198971512000762
19. Sotoodeh, M., Kruchten, P.: An ontological approach to conceptual modeling of disaster management. In: 2nd Annual IEEE Systems Conference 2008, pp. 1–4. IEEE (2008)
20. Tripathi, A., Babaie, H.A.: Developing a modular hydrogeology ontology by extending the SWEET upper-level ontologies. Computers & Geosciences **34**(9), 1022–1033 (2008). http://www.sciencedirect.com/science/article/pii/S009830040800085X
21. Uschold, M., Gruninger, M., Uschold, M., Gruninger, M.: Ontologies: principles, methods and applications. Knowledge Engineering Review **11**, 93–136 (1996)
22. Ushahidi: Open source, global impact, freedom of information: Ushahidi. http://www.ushahidi.com/, (accessed July 12, 2014)
23. Xu, W., Zlatanova, S.: Ontologies for disaster management response. Geomatics Solutions for Disaster Management, pp. 185–200 (2007)
24. Yu, K., Wang, Q., Rong, L.: Emergency ontology construction in emergency decision support system. In: IEEE International Conference on Service Operations and Logistics, and Informatics, IEEE/SOLI 2008, vol. 1, pp. 801–805. IEEE (2008)

# Development of the Belief Culture Ontology and Its Application: Case Study of the GreaterMekong Subregion

Wirapong Chansanam[1], Kulthida Tuamsuk[2], Kanyarat Kwiecien[2],
Taneth Ruangrajitpakorn[3(✉)], and Thepchai Supnithi[3]

[1] Chaiyaphum Rajabhat University, Chaiyaphum, Thailand
`wirapongc@kkumail.com`
[2] Information and Communication Department, KhonKaen University,
KhonKaen, Thailand
`{kultua,kandad}@kku.ac.th`
[3] Language and Semantic Technology Laboratory,
National Electronics and Computer Technology Center, Bangkok, Thailand
`{taneth.rua,thepchai.supnithi}@nectec.or.th`

**Abstract.** In this paper, the development of an ontology that represents the knowledge of belief culture in the Greater Mekong subregion(GMS) is presented. The ontology was carefully designed to specify the concepts relevant to intangible and tangible cultural heritage and the relations among them. The knowledge domain in this work focuses on the cultural context and implicit attributes of the GMS as an initial case study. To further illustrate the potential of the developed ontology, a semantic search application was implemented and then evaluated by experts. On the evaluation processes, several complicated queries were designed in order to fully utilize the relations among ontological-classes, and the results were returned accurately. The evaluation proved that the ontology was well defined in aspects of its hierarchical structure and relations from intermediate concept layers.

**Keywords:** Belief culture · Intangible attributes · Ontology · Semantic search · Greater Mekong subregion

## 1    Introduction

Belief and culture are a major part of the human way of life. Beliefs have critical impacts on human behaviors, values, characteristics, mindsets, and manners. Although belief and culture exist in every society, the scope and meaning of the term are still arguable. Several works have attempted to define the scope and meaning of belief - culture, but scope and meaning vary based on the understanding of the people in each society and the context that surrounds them, such as religion, ethnic background, and geographic settings.

© Springer International Publishing Switzerland 2015
T. Supnithi et al. (Eds.): JIST 2014, LNCS 8943, pp. 297–310, 2015.
DOI: 10.1007/978-3-319-15615-6_22

With a population of 326 million people, the Greater Mekong subregion(GMS) is the world's third most populous. In terms of land area, its 2.6 million square kilometers rank it as the 10th largest country on earth. The GMS is home to resilient economies, a wealth of natural and human resources, pristine environments, and a rich cultural heritage that attracts tourists from around the world. These powerful attributes have been recognized by country members of the GMS: Cambodia, the People's Republic of China, the Lao People's Democratic Republic, Myanmar, Thailand, and Viet Nam [1].Because the Mekong River Basin is a fertile land and the home of different ethnic groups with a long history, it can be said that the GMS is one of the origins of world civilizations and cultural heritage. As part of this cultural heritage, belief culture has been an interesting issue for regional development because the beliefs of people in the GMS can be both obstacles and driving forces for social and economic development. Beliefs have significant impacts on the way of life as well as on people's values, attitudes, and behaviors, which are all based on their ethnicity, religion, and the surrounding environment. This has made the belief culture of the people in the GMS one of the most interesting areas of research in this region.

Ontology is a widely used tool for describing knowledge representation[2, 3].The benefits of ontology include[4] its interoperability to share common understanding among people or software agents, it enables the re-use of domain knowledge, and it enables explicit assumptions. Some ontologies of cultural knowledge have been developed by During, Eide & Ore, Signore, and Szasz, et al. [5-8].However, those ontologies focus on tangible cultural heritage and are intended to be used as information resources rather than as tools for information retrieval. More importantly, the ontology of the knowledge of belief and culture has not been developed.

In this research, we developed ontology ofthe belief culture in the GMS and constructed the knowledge domain by using a thesaurus approach. The terms in the context of belief culture were extracted from two existing thesauri, a database of Human Relations Area Files [eHRAF] [9] and UNESCO's Intangible Cultural Heritage (ICH)thesaurus [10],and the list of terms and their relations were input, defined, and confirmed by experts. The thesaurus was used as a source for ontological development using the Hozo-ontology editor. To further illustrate the potential of the developed ontology, a semantic search application was implemented and then evaluated by experts in ontology and the GMS belief culture.

## 2    The Ontology Design Framework

Figure 1 shows the design framework for constructing the ontology to represent the knowledge domain of belief culture in the GMS. The ontological construction process consisted of several steps: 1) building a thesaurus of belief cultural knowledge; 2) annotating the facts to incorporate their semantic content and to build semantically enhanced facts using domain-related concepts based on class hierarchies. These steps were performed by using semantic similarities based on the principle of ontological engineering[11, 12],evaluating completeness by ontology experts, and visualizing the knowledge in the ontological vision.

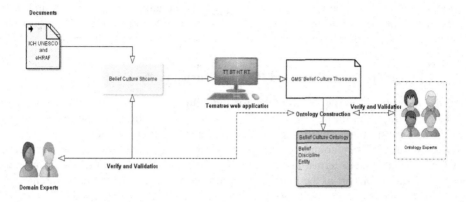

**Fig. 1.** The design framework for belief culture ontology construction

# 3    Ontology Development

Many of the domain-based initiatives recommend the use of closed vocabularies, such as eHRAF[9] and the ICH thesaurus[10] but do not associate particular parts of a thesaurus with a domain. We first discuss what the knowledge requirements are based on the existing thesaurus. This will enable us to create an accurate knowledge foundation for intangible cultural descriptions.

## 3.1    Setting Objectives of the Thesaurus Based on Previous Work

After studies on the existing thesauri, the objectives of this research thesaurus construction were developed. The first objective was to construct a useful thesaurus for the domain by using a structure-based approach to provide a hierarchical structure with explicit interpretation. Some hierarchically organized thesauri, such as eHRAF and the ICH thesaurus, do not have a clear definition on the sub/super class relations and partial relations but use a strict sub/super class relation in a single inheritance hierarchy. Single inheritance limits the amount of information about a term that can be derived from its position in the hierarchy, despite the fact that the terms can be classified in multiple ways. For example, the concept "belief" is represented by two terms: belief (religion) and belief (culture). Unfortunately, the difference between the two qualified terms may not be clear to a user, and it is difficult to decide where subclasses of the concept belong. The second objective was to link the fields in each description to particular parts of that thesaurus. For example, the field "religion" should be linked to that part of the thesaurus that contains a hierarchy of religion types. In some cases this is straightforward, such as a hierarchy of belief can be clearly defined with terms to be of value for the religion field; however, there are many cases where values are assigned to fields in several parts of the thesaurus. The third objective was to manage the indexing space. A human indexer, using the structure-based approach, was confronted with large sets of possible values to choose from. Belief hierarchy in the eHRAF and ICH thesaurus contains several hundreds of terms. A solution to this problem is to constrain the value sets for a particular field, based on a partial description of the entity.

## 3.2 Expanding the eHRAF and ICH Thesaurus for the GMS

As the basis for building an ontology for indexing entities, we used the knowledge structure based on a thesaurus. The eHRAF and ICH thesaurus are the most complicated and most standardized bodies of knowledge concerning the classification of culture heritage. They contain a large number of main and diverse terms, including synonyms and related terms, as well as scope notes—textual definitions of the concepts—for a major part of the main terms, with concepts represented in hierarchies. A particular concept occurs only once in a thesaurus hierarchy. Intermediate concepts (guide terms) are used to group the concepts at a lower level in the hierarchy. We handled the main terms as concept names in the knowledge base. Although this is possible because each main term in the eHRAF and ICH thesaurus is unique, it causes problems when a concept can also be identified by its synonyms, as is the case in WordNet synsets[13]. Searching both thesauri for the terms *culture* and *belief* returns the concept belief (area with intangible cultures)first. It was decided to represent each concept in the knowledge base by a unique identifier derived from their record number. The hierarchy of both thesauri was converted into a hierarchy of concepts, where each concept has a label slot correlated with the main term in the two thesauri and a synonyms slot where alternate terms are represented. The values for these slots were partly derived using explicit tables of periods and partly found using the intermediate concepts in both databases. Finally, we added knowledge about the relation between possible values of fields and nodes in the knowledge base.

## 3.3 Ontology Design and Development Process

In our case study, ontology is a knowledge structure containing concepts and relations regarding the belief culture in the GSM. The ontology is organized as directed cyclic graphs. Each node represents a concept, and there is relational link between them and a relation between node and data.

We designed and developed the belief culture ontology for a subset of culture, entity, and disciplines. Our ontology was developed by applying the following steps:

**Step #1.**To correct and reorganize the knowledge contained in an initial conceptual framework and to detect missing knowledge.

**Step 1.1:** Establish the properties list of an initial set of concepts. First, the hierarchies and taxonomic relations were drawn between concepts and instances, ad hoc relations between concepts and instances of the same or another hierarchy. We identified the functions and axioms of the ontology. The initial set was either a source model from direct interpretation of an existing data structure or a collection of "base level" concepts in the sense of cognitive studies in the eHRAF and ICH thesaurus of the domain and its instinctive list of properties. One important aspect is that concepts have to be sufficiently real and concrete to have well-defined properties. Finally, we generated a draft document reflecting the preliminary conceptual framework from this step.

**Step 1.2:** Determine new concepts from attribute values. Attribute values, particularly literals, actually describe many concepts. If the attribute value designates a universal, for example, *has role: God*, the attribute should be transformed to representation in the respective concepts. If it is not clear, we return to step 1.1 to describe the properties of the additional concepts.

**Step 1.3:** Determine entities vague in attributes. The meaning of already identified concepts may be involved in attributes, e.g., *Imaginary Being: Animals Liked Imaginary Being*, hides *Imaginary Being*. If it is not clear, we return to step 1.1 to describe the properties of the additional concepts.

**Step 1.4:** Examine property consistency. Visual representations are useful for consistency control to investigate distinct viewpoints and reasoning scenarios. The framework under construction needed to describe well the world seen from the point of view of a domain or range concept, e.g., an entity description or an intangible cultural description. This can lead to the detection of new properties among the existing concepts. It may also motivate a change in the domain and range, e.g., when alternative paths lead to different ranges. Similarly, completeness of reasoning, e.g., about time period, place, or location, leads to the detection of additional properties. If it is not clear, we return to step 1.2 to redetect hidden concepts.

**Step 1.5:** Establish the concepts hierarchy. Specify and merge equivalent properties using multiple inheritances for domain and/or range. Identify the lowest domains and ranges over the complete scope. This process leads to the detection of the more abstract concepts. Our proposed work will not start with instinctive abstract concepts. If it is not clear, we return to step 1.1 to describe the properties of the additional concepts or to step 1.4 for the merged properties.

**Step 1.6:** Check the correctness and completeness of each hierarchy by analyzing the following three points: (a) whether the taxonomic relations between concepts are correct, (b) whether the concepts present in the original hierarchy should be further specified or generalized, and(c) whether all the concepts or instances required emerge in the original hierarchy. If necessary, the fourth point is whether to add/delete from the original ontology any concepts or instances that have to be processed.

**Step #2.**To link culture properties to particular subsets of the belief culture thesaurus. Establish property hierarchies. Having checked that the hierarchies are correct, analyze the correctness and completeness of the definitions of concepts, instances, properties, relations, functions, and axioms. We analyzed the initial conceptual framework attached to the code in which the ontology was implemented. Specialized resources for this purpose (such as books, dictionaries, research papers)are required as isthe assistance of specialists in the belief cultures domain defined in the ontology. This step may lead to the detection of more properties and incompatibilities. Therefore recheck step 1.4, else end with step # 3.

**Step #3.** To describe additional domain knowledge, in particular, the constraints between culture and property values.

**Step 3.1:** To review and correct an original thesaurus, design a new ontology including all the above-mentioned changes, build correct and complete hierarchies, and output the correct and complete definitions for their later implementation. We defined a synthesis document identifying the actions carried out and the design criteria using a bottom-up approach.

**Step 3.2:** To implement the belief culture ontology on the basis of the new conceptual framework, including all the recorded changes. This will output a document containing the code of the new ontology implementation.

**Step 3.3:** To check and remove useful concepts. We need to avoid a situation where steps #1 through #2 develop a boundless ontology. In practice, people have difficulty in knowing when to stop designing more and more formal properties. The conceptual framework "ends" naturally at primitive values—numbers, time-span, free text. Other concepts must be explicitly stated as intermediate concepts, i.e., properties that would introduce range concepts out of scope are considered deleted. In an application database, respective intermediate concept information may be kept in free text. Extensions may continue there, adding formal properties.

We highlight that the driving force is the properties and functions rather than the concepts, which is in contrast to the well-established Enterprise ontology, Ontology Development 101[12], and other design methodologies. We do not explicitly state below the application of all principles described in the previous sections. Each step can be understood as incremental.

This work aims at proposing a Hozo-ontology editor tool[11] that allows the construction and development of a belief culture ontology. This tool enables the creation of an ontology using rule-based representation of belief culture knowledge. The knowledge, that is, facts and rules, is represented first in an ontology vision format. The proposed tool will enable domain experts to perform verification and validation of created belief culture knowledge.

In ontology construction, concept and property relationships in the belief culture domain are defined and domain ontology is constructed based on the eHRAF and ICH thesaurus according to the examples of belief culture concepts, instances, and values of domain ontology, as shown in Figure 2.

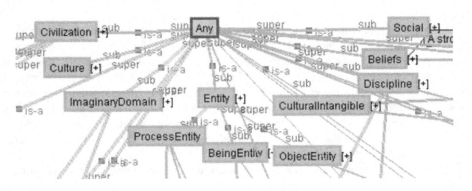

**Fig. 2.** An example of belief culture ontology

There are categories in belief culture domains determined as concepts, such as beliefs, culture, entity, discipline, and cultural intangible. These categories include several subcategories or sub-concepts. For example, entity has subcategories of object entity, being entity, and process entity. Each instance has values. Ontology is applied not only in the process of query classification to obtain the concepts of each term but also to match target category.

The proposed work methodology is a sound initial approach to carrying out the above-mentioned process, although it could be improved in future studies using a more complex knowledge structure. In order to extend the reusability of the ontology to be rebuilt, guidelines and criteria to achieve a higher degree of reusability are needed in the construct process. Another open issue regards the relationship between the ontology that is being recreated and top-level ontologies, if any.

# 4    Ontology-Based Application Framework

## 4.1    OAM Architecture

The ontology-based application management (OAM) framework[14] is a development platform for simplifying the creation and adoption of a semantic web application. OAM allows the user to interactively define mapping between a database schema and ontology's OWL to produce the RDF data. It also provides several application templates that support RDF data processing, such as semantic search and recommender system applications. The OAM framework is implemented to support ontology developers in implementing a prototype for easily testing their created ontology in practical application since it requires less technical programming skill in implementing a semantic web application[14].The overview architecture of OAM is illustrated in Fig. 3.

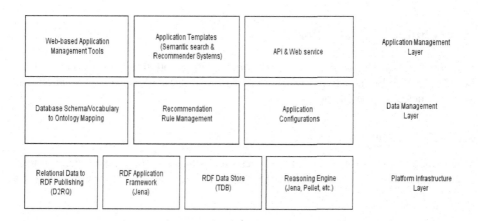

**Fig. 3.** A layered architecture of the OAM framework

## 4.2     Semantic Search for Belief Culture in the GMS

Semantic search is an improved version of contextual search in terms of ability to understand the relation of the contextual meaning based on ontology schema. In this work, belief culture ontology of the GMS culture was employed as a knowledge base for the semantic search system. The system was implemented as a web-based application for easy and free access. The system considers various aspects, including context information of search, location, pictures, variation of words, synonyms, generalized and specialized queries, concept matching, and natural language queries, to provide relevant search results. The application is able to understand the nuances of culture heritage and ensure the most relevant results. These are advantages that only a semantic approach can guarantee because it has a contextual understanding of the meaning of content.

**Fig. 4.** A snapshot of the GMS belief culture knowledge-based systems

# 5     Ontology Evaluation

In this section, we design the framework for evaluating both the ontology and application. Evaluation by the experts was applied for both cases and system evaluation was also conducted to show its potential.

## 5.1     Evaluation by the Experts

This evaluation focuses on assigning experts and specialists to evaluate the ontology and its application. A set of criteria for the evaluation and for the experts was designed separately.

### 5.1.1 Evaluation of the GMS Ontology

Based on the requirements outlined in the ontology evaluation, a generic evaluation framework was created. In this stage, experts were separated into two groups for evaluation:

(1) experts in GMS culture and belief domain

(2) specialists in ontology development.

To seek both knowledge domain experts and ontology experts, we applied a snowball technique [15], which first identified one or more experts and second used these experts to find further experts until the criteria has been achieved. With this method, we guaranteed no bias from the expert selection process.

The set of criteria involved two aspects. The first was to evaluate formal features of the ontology, e.g., its consistency, well-formedness, and completeness of definitions, according to the standards given in Gomez-Perez [16, 17]. The second aspect was the utility and usability of the ontology. This aspect aimedto address "how suitable it is for the task it was created for" and "how well it represents the domain of interest". The methods and criteria on the belief culture ontology evaluation were as follows:

- *Provide a questionnaire for the evaluation of ontology by humans.* It is necessary to verify and validate ontology with a human evaluation method. We designed a questionnaire by considering three aspects:
  1. assessors' basic information
  2. knowledge representation and knowledge structure in a variety of dimensions, such as scope, concept class, properties, instance, and reusability potential for future development and application
  3. open-ended questions for suggestion.

- *Provide the assessors' evaluation by assigning responsibility to each group.*

  1. Ontology specialists—to prove the completeness ontology structure for software agents and ontological components.
  2. Experts in GMS culture and belief domain—to verify and validate correctness, consistency, properties, and hierarchical relations among concepts in ontology.

The evaluation assessed the quality of the ontology by drawing upon semiotic theory [18], taking several metrics into consideration for assessing the syntactic, semantic, and pragmatic aspects of ontology quality. With criteria and setting, the ontology was evaluated by three experts in the GMS culture and belief domain and four ontology specialists, as shown in Table 1.

**Table 1.** Evaluation of knowledge structure for system results

| Aspects | Average | Meaning (Level) |
|---|---|---|
| Determine Scope | 4.00 | High |
| Define Classes | 4.00 | High |
| Define Properties | 3.80 | High |
| Define Instance | 3.50 | High |
| Future Development and Application | 4.25 | High |
| Total | 3.91 | High |

### 5.1.2 Evaluation of the GMS-Based Application

To prove the practical usefulness of the developed ontology, a prototype of the semantic search system was implemented. The aspects to evaluate included:

- Performance—This is specific to the type of ontology. For knowledge retrieval performance in discovery activities, measurements, such as Precision, Recall and F-Measure, are usually used [19]. More generic performance measures are execution time and throughput.
- Scalability—Scalability of a semantic search system is associated with the ability to perform an activity involving an increasing amount of semantic explanations. This can be measured together with performance; however, this is also related to the scalability of repositories.
- Correctness—This is related to the ability of a semantic search system to respond correctly to different inputs, contents, or changes in the application problem by changing the instance. This criterion is related to mediation and request of the semantic search system. Information resulting from the request or interaction of services should be checked against a reference set.
- Usability—It might be useful to know which semantic search system has an easy to use graphic user interface or environment. We consider that due to the paucity of frontends for semantic search system development, a comparison would be more easily done using feedback forms.

To evaluate the semantic search application, the experts were separated into three groups based on their specialist duty:

1. Semantic web expert—to test defining a query and approve the answer based on a given ontology
2. Domain experts—to define a query based on their knowledge and to provide a score to retrieved results
3. Information retrieval specialist—to calculate the efficiency of the semantic search system.

With given criteria and duties, experts gave satisfaction scores based on semantic search performance and results, as shown in Table 2. Eight experts were involved in this evaluation.

**Table 2.** Semantic search system evaluation results by humans

| Semantic Search System | Average | Meaning (Level) |
|---|---|---|
| Content | 5.00 | High |
| Accuracy | 5.00 | High |
| Format | 4.50 | High |
| Ease to Use | 4.50 | High |
| Timeliness | 5.00 | High |
| Perceived Usefulness | 5.00 | High |
| Total | 4.83 | High |

## 5.2  System Performance Evaluation

The standard evaluation approach to a knowledge retrieval system revolves around the notion of relevant and non-relevant documents. With respect to a user-context information need, a document in the test collection is given a binary classification as either relevant or non-relevant. This decision is referred to as the gold standard or ground truth judgment of relevance. The test document collection and suite of information need to be of reasonable size; they need to average performance over fairly large test sets as results are highly variable over different documents and context information needs. As a rule of thumb, 80% of context information needs has usually been found to be a sufficient minimum. The accuracy of application using the developed ontology is shown in Table 3.

**Table 3.** Knowledgeretrieval efficiency results

| | | Relevant | | |
|---|---|---|---|---|
| | Semantic searching | Relevant meaning | Non-relevant meaning | Overall meaning |
| Retrieved | Able | 52 | 0 | 56 |
| | Unable | 0 | 12 | |

| | | |
|---|---|---|
| Recall | = | 0.7647 |
| Precision | = | 0.9286 |
| F-Measure | = | 0.8387 |

The results of the knowledge retrieval showed that the semantic search application was effective regarding values of Precision, Recall, and F-measure.

# 6    Discussion

In this paper, an ontology representing tacit knowledge of belief, culture, cultural heritages, and human thoughts was presented. The satisfaction scores for the developed ontology satisfied the domain experts, and the ontology worked well as a knowledge resource in a semantic search application.

From our observations, this ontology is the first abstract ontology for the cultural heritage domain. It gives explicit and explainable relations of beliefs to culture that leads to cultural heritage within culture. By applying GSM cultural data, the relation of the imaginary entity "Nāga" stands out as the core of the culture. The ontology schema provides a flexible but well-defined scope to help users understand that Nāga is the race of a serpent being worshiped as a deity by the people living in the area. The individuals of Nāga by their name were mentioned with story detail in many intangible cultural heritages, such as folklore, music lyrics, and poems. Those individuals can be related to physical cultural objects, such as sculptures or paintings, by the well-defined relations of the developed ontology schema. It is not only the text data that can be instantiated; the ontology was designed to represent related details, such as geographic information and historical information. These data are proof that the ontology was well designed to show interoperability of abstract knowledge hidden in the human way of living.

As used in the semantic search application, the ontology shows its best potential as it cannot only be queried for direct questions about simple hierarchy and property but can also perform lengthy questions via intermediate concepts. Those intermediate concepts were carefully tailored to link concepts based on facts and logic. Moreover, the hierarchy of the ontology was considered and designed to prevent conflicts of ambiguity and overlapping concepts, hence there were multiple hierarchical layers. With those attributes of the developed ontology, it noticeably outran the existing thesauri for its clearness and ease to understand by both humans and machine. Furthermore, the ontology was designed to be extendable with global belief cultural data; thus this ontology can be used as an intermediate schema of linked open data to collect world-wide instances.

# 7    Conclusions and Future Work

In this paper, we presented a newly developed ontology focused on representing tacit knowledge of belief and culture via cultural heritage. This work employed vocabularies and a rough hierarchy from eHRAF and UNESCOICH thesaurus as a base. The ontology consistedof409conceptsfrom those thesauri and was expanded within formation specific to the GMS cultural resource, with 344concepts and 290 instances. We also implemented a web-based semantic search application to serve as a semantic browser for the ontology.

To extend the content in ontology, we plan to add instances starting from the localized area and expanding the content until it reaches a global scale. More applications based on the developed ontology will be implemented, such as a recommendation system and a question–answer system.

**Acknowledgments.** This research is supported by the Office of the Higher Education Commission (OHEC).

# References

1. Asian Development Bank: Greater Mekong Subregion: Twenty years of partnership. Mandaluyong, The Philippines, ADB (2012)
2. Gruber, T.: What is an ontoloty. In: Liu, L., Özsu, M.T. (eds.) Encyclopedia of Database Systems. Springer (2008)
3. Chandrasekaran, B., Josephson, J.R.: The ontology of tasks and methods. In: AAAI 1997 Spring Symposium on Ontological Engineering, March 24-26. Stanford University, CA (1997)
4. Milton, S., Keen, C., Kurnia, S.: Understanding the benefits of ontology use for australian industry: a conceptual study. In: 21st Australasian Conference on Information Systems (2010)
5. During, R.: Cultural Heritage Discourses and Europeanisation: Discursive Embedding of Cultural Heritage in Europe of the Regions [n.p.]. Wageningen University (2010)
6. Eide, Ø., Ore, C.: TEI and Cultural Heritage Ontologies: Interchange of information ? Literary and Linguistic Computing **24**(2), 161–172 (2009)
7. Signore, O.: The semantic web and cultural heritage: ontologies and technologies help in accessing museum information. In: Information Technoloty for the Virtual Museum, December 6-7, 2006 (2008)
8. Szász, B., Saraniva, A., Bognár, K., Unzeitig, M., Karjalainen, M.: Cultural Heritage on the Semantic Web – the Museum24 project. Digital Semantic Content across Cultures conference Paris, the Louvre May 4-5 (2006)
9. Subjects, Cultures, and Traditions Covered in eHRAF World Cultures & eHRAF Archaeology Outline of Cultural Materials (OCM) Subject List. http://ehrafworldcultures.yale.edu/ (accessed July 7, 2014)
10. The UNESCO Thesaurus. http://databases.unesco.org/thesaurus/ (accessed July 5, 2014)
11. Kozaki, K., et al.: Hozo: An environment for building/using ontologies based on a fundamental consideration of "role" and "relationship". In: Gómez-Pérez, A., Richard Benjamins, V. (eds.) EKAW 2002. LNCS, vol. 2473, pp. 213–218. Springer, Heidelberg (2002)
12. Noy, N.F., McGuinness, D.: Ontology Development 101: A Guide to creating your first Ontology. Stanford Knowledge Systems Laboratory Technical Report KSL-01-05 and Stanford Medical Informatics Technical Report SMI-2001-0880 (March 2001)
13. George, A.M.: WordNet: A Lexical Database for English. Communications of the ACM **38**(11), 39–41 (1995)
14. Buranarach, M., Thein, Y.M., Supnithi, T.: A community-driven approach to development of an ontology-based application management framework. In: Takeda, H., Qu, Y., Mizoguchi, R., Kitamura, Y. (eds.) JIST 2013. LNCS (LNAI), vol. 7774, pp. 306–312. Springer, Heidelberg (2013)
15. Handcock, M.S., Gile, K.J.: On the Concept of Snowball Sampling arXiv: 1108. 0301v1 [stat.AP], 1554, pp. 1–5 (August 1, 2011)
16. Gómez-pérez, A., Fernández-López, M., Vicente, A.J.: Towards a method to conceptualize domain ontologies. In: Working Notes of the Workshop on Ontological Engineering, vol. 46, p. 41 (2003)

17. Gómez-pérez, A., Rojas-amaya, M.D.: Ontological reengineering for reuse. In: Proceedings of the 11th European Workshop on Knowledge Acquisition, Modeling and Management, EKAW 1999, pp. 139–156 (1999)
18. Stamper, R., Liu, K., Hafkamp, M., Ades, Y.: Understanding the roles of signs and norms in organizations - a semiotic approach to information systems design. Behaviour & Information Technology **19**, 15–27 (2000)
19. Belew, R.K.: Finding out about: A cognitive perspective on search engine technology and the www. Cambridge University Press, Cambridge (2000)

# Representations of Psychological Functions
## of Peer Support Services for Diabetic Patients

Ikue Osawa[1(✉)], Mitsuru Ikeda[1], Takashi Yoneda[2], Yoshiyu Takeda[2],
Masuo Nakai[3], Mikiya Usukura[3], and Kiwamu Abe[3]

[1] School of Knowledge Science, Japan Advanced Institute of Science and Technology,
1-1 Asahidai, Nomi, Ishikawa 923-1211, Japan
{osawa,ikeda}@jaist.ac.jp
[2] Department of Endocrinology and Metabolism, Kanazawa University,
13-1 Takara, Kanazawa, Ishikawa 920-8641, Japan
{endocrin,takeday}@med.kanazawa-u.ac.jp
[3] Houju Memorial Hospital, 11-71 Midorigaoka, Nomi-shi, Ishikawa 923-1226, Japan
{masuo,uscratch,k-abe}@houju.jp

**Abstract.** One of the functions of peer support services for diabetic patients is psychological changes through communications among patients. Medical professionals in the practice of this research request peer support services through a web system. The design of the psychological functions requires tailoring depending on contexts. An important thing for the adaptive design is to discuss the needed psychological functions among designers such as medical professionals, patients, and researchers. However, since the designers tend to set intentions of psychological functions by taking a seat-of-the-pants approach, even for the designers, describing their intentions of psychological functions is not easy. In this paper, we propose a framework to represent psychological functions for the designers to share and discuss intentions of psychological functions in web systems.

**Keywords:** Peer support service · Definition of service · Diabetes education

## 1 Introduction

One of the functions of peer support services for diabetic patients is psychological changes through communications among patients. Medical professionals in the practice of this research request the peer support services through a web system for those who are too busy to attend face-to- face peer support services held in hospitals.

The design of the psychological functions requires tailoring depending on contexts such as participants' lifestyles, life-cycle, locations, customs and other factors [1]. Important processes for the adaptive design of the psychological functions are to discuss the needed psychological functions according to each context among designers such as medical professionals, patients, and researchers. However, since even for the designers themselves, it is not easy to describe their intentions of psychological functions, the designers are not able to have same understanding of the intentions.

© Springer International Publishing Switzerland 2015
T. Supnithi et al. (Eds.): JIST 2014, LNCS 8943, pp. 311–318, 2015.
DOI: 10.1007/978-3-319-15615-6_23

In this paper, we suggest a framework to represent psychological functions for the designers to discuss and share intentions of psychological functions in web systems. The design objects of the peer support services are as follows.

*Patient community: realized community in practice*
*Online activities: online activities in peer support services*
*Web system: means to realize the online activities*
*Intentions of web system functions: the intentions of web system behavior*
*Intentions of psychological functions: the intentions of psychological changes and communications among patients*
*Principles: mechanisms behind psychological functions*

In service engineering, general services are designed through phases of "development of functional structures" and "embodiment of functions" [2]. In the "development of functional structures", an assembly of functions to satisfy customers is modeled. The modeling is associated with behaviors to realize each function in the phase of embodiment of functions. In this research, following this view, we set the view of the design phase. The phases are intentions of psychological functions and that of web system functions. The intentions of web systems have been often taken for the intentions of psychological functions, and this is one of the factors that impede sharing and discussing psychological functions.

The principles offer assumed mechanisms concerning psychological functions as grounds to intend psychological functions. On the basis of the principles, designers intend psychological functions with considering each context. The intentions of the psychological functions play a purpose role in designing web systems, and designers intend functions of web systems. Consistent with the intentions of the web systems, web systems are implemented. On the system, patients communicate with one another, and the communities are realized.

Necessity for the discussion among the designers is to make the intentions in each phase clear, and to show the connection between the phases. In this research, we suggest a framework to explicate the intentions of psychological functions, and explain the way to relate intentions of psychological functions and that of web system functions.

In the next section, we discuss the construct of the psychological functions, and in section 3, we examine relations between intentions of psychological functions and that of web system functions. As space is limited, for the detail of principles, please refer to proceedings [3] [4].

## 2      Intentions of Psychological Functions

Conceptual frameworks are necessary to represent intentions of psychological functions. Compared with well-defined machinery functions, intentions of ill-defined functions like psychological changes functions in peer support services are hard for the designers to explain. This is because the view of the psychological functions is ambiguous. There is almost neither a single view of the functions everyone agrees with, nor obvious criteria.

One of the approaches to address this issue is ontology-based approach. Ontology is an explicit specification of conceptualization, clearly represents how the target world is captured by people, and particularly plays an important role for comprehensive understanding of a complex domain. In order to define notions of psychological functions to represent the intentions of the psychological functions, we have developed ontology of the psychological functions from service ontology [5]. In this section, after explaining definitions of the functions, we show the construct of psychological functions with figures [6].

## 2.1    Definitions of Psychological Functions

Sumita, Kitamura, Sasajima and Mizoguchi [5] who research on ontology-based description of functional design of services state "The concepts of services and product functions have the same conceptual structure". The definition of functions of artifacts by Kitamura and Mizoguchi [7] is that "a result of teleological interpretation of a behavior (e.g. state-changes) of the operand(s) under an intended goal" [7]. On the basis of this definition, Sumita, Kitamura, Sasajima and Mizoguchi [5] describe that a service function is a state-change of operand as effects of act by agent [8], and that "the function performer is an agent to perform functions" [5].

In accordance with this definition, we consider agent, act (performance), its operand, and a state-change (effect) as major factors in concept of psychological functions. In this research, we define psychological functions as follows.

**Definition**
*A state-change of psychological aspects of both performers and receivers influenced by communication acts which are carried out by diabetic patients in the aim of overcoming psychological problems associated with diabetes.*

  *Agent : Diabetic patients*
  *Act : Communication acts*
  *Operand : Psychological aspects*
  *Effect (State-change) : Improvement of psychological aspects*

## 2.2    Framework of the Psychological Functions

We configured construct of the definitions of the psychological functions, using ontology editor named "Hozo" [9]. Fig. 1 shows the overall of the ontology. The ontology mainly provides the construct of peer support service functions, and parts to make up the functions such as "self-efficacy" or "helping others". The construct and the parts are the bases of the framework to explicate psychological functions. Fig. 2 simply shows the construct of the peer support service functions with major components in an understandable way. We outline the summary of the diagram first, and mention the specific example in the next chapter.

The diagram in Fig. 2 shows the major structure to represent the intentions of the psychological functions. The structures consists of three kinds of parts; main body (A), main parts (B), (C), (D), (E), (F), and the other parts. Each main part has a role in the psychological function (A). The role of (B), (C), (D), (E), and (F) are act, principle, operand, and Effect respectively.

The principle (D) is important to express principles behind intentions. For the detail of principles, please refer to proceedings [3] [4]. The diagram is to represent how the act (C) of agent (B) causes operand (E) a favorable effect (F) with what kind of principles (D).

To be more precise, act (C) does not directly cause the effect (E), but (I). The straightforward operand of the act (C) is operand (G). Attribute (H) of the operand (G) changes from previous state (J) to next state (K) as an effect (I). The principle (D) relates effect (I) to effect (F). In the effect (F), operand (E) changes from previous state (L) to next state (M).

This construct is to represent the intentions of the psychological functions behind the web system for the designers to share and discuss the intentions of the functions.

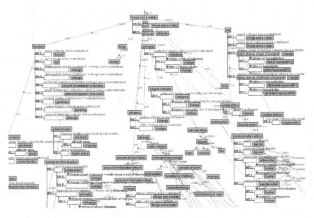

**Fig. 1.** Overview of ontology

## 2.3    An Example

Fig. 3 shows an example of intentions of the psychological functions. The intentions of this example are the psychological function of improving self-efficacy by helping others (A). Diabetic patients play an agent role (B) in this function, act is helping others (C), operand is information (G) of act, attribute (H) of the operand (G) is amount of information, effect(I) by act (C) is changes of amount of information, the previous state (J) of effect (I) is little information, the next state (K) of effect (I) is much information, the operand (E) of this function (A) is self-efficacy, the effect(F) of this function(A) is psychological changes(F), previous state(L) of effect(F) is low self-efficacy is, and the next state (M) of the effect (F) is high self-efficacy. That is the intention of this examples is letting diabetic patients (A) help others (B), as its effect (I), the amount (H) of information (G) changes (I) from little (J) to much (K). The change (I) of the amount of information causes self-efficacy (E) to change (F) from low (L) to high (M). The principles (D) explain designers' assumption of how helping others influence the state change of self-efficacy.

Without the framework based on the ontology, different views of the psychological functions bring about confusions in discussion. For instance, functions can be considered is "helping others" with the focus of the acts, "improving self-efficacy", and "emotional functions". They possibly mention the same function, or different functions. These intentions of psychological functions behind web systems remain ambiguous.

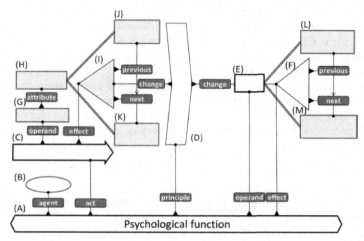

**Fig. 2.** Structure of function

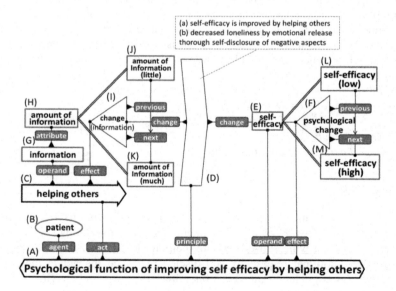

**Fig. 3.** An Example of intentions of a psychological function

# 3      Relation between Psychological Intentions and Web Systems

## 3.1      Intentions of Psychological Functions and of Web System Function

Designers translate intentions of psychological functions into intentions of web systems to realize the intentions. In this sub-section, we explain how to relate intentions of psychological changes and that of web systems.

Shimomura et al [10], in general service design, discuss relations between means for realizations and intentions of general services. Shimomura et al [10] deliver their view that "A service is defined within a framework consisting of service provider, service receiver, service contents, and service channel", and that "A service is an act by which a provider causes... a receiver to change into the state..., where both contents and a channel are means to realize the service". The contents causes state changes of receivers, like information, and is transferred, amplified, and controlled by the channel [10]. Converting this definition to peer support services, both the service providers and the service receivers are diabetic patients, the service channel are web systems, contents are communication contents conveyed by the system to patients. The communication contents cause state changes of psychological aspects.

On the basis of Shimomura et al [10], we contemplate how designers set intentions of web systems from intentions of psychological functions. We explained major concepts of psychological functions in 2.1: agent, act, its operand, and a state-change of the operand. Among these concepts, the concepts related to intentions of web-system are the act of the communications, and the operand of act. As mentioned above, the communication contents change psychological state. From the view of the psychological functions, since the psychological changes are caused by the operand of act, the operand of act correspond to communication content (e.g. information - the messages of encouragement). In addition, communication contents are conveyed by the web-system behavior, and the operand of act is conveyed by act of patients. Accordingly, act of patients correspond to web-system behavior (e.g. helping others - showing messages on a board).

The followings show the connections between intentions of the psychological functions and that of web system functions.

*Act: web system behavior*
*Operand of act: Communication contents*

## 3.2      Examples

The table 1 shows an example of relations between intentions of psychological functions and that of web systems. The intentions of the psychological functions are two: <agent 1> improving self-efficacy by helping others, and <agent 2> improving loneliness by emotional release by self-disclosure. Since the act is the subject to translate into system behavior, the each intention of the system behavior is <agent 1> show message posted on message board, <agent 2> also show message posted on message board. Operand of act is subject to translate into communications contents. The intentions of communication contents are <agent 1> encouragement, and <agent

2> negative emotion. Fig. 4. Realized web system based on the intention shows web systems and its intentions of psychological functions.

This is how designers set intention of psychological functions and intentions of web-system functions.

**Table 1.** Intentions of psychological functions and web system function

| Function | Subject | agent 1 | agent 2 |
|---|---|---|---|
| psychological function | act | helping others | emotional release by self disclosure |
| | effect | improve self-efficacy | imrpove loneliness |
| web system functions | system behavior | show message posetd on MB | show message posetd on MB |
| | communication content | encouragement | negative emotion |

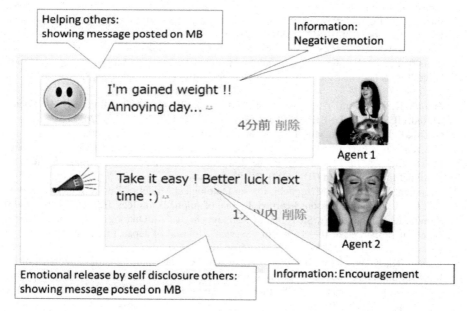

**Fig. 4.** Realized web system based on the intention

## 4    Conclusion

In this paper, we suggest the framework to represent the psychological functions for the designers to discuss intentions of psychological functions in web systems, and relations between the intentions of the psychological functions and that of the system functions.

We have developed the web system for online peer support services in the practice of this research with making the intentions of the psychological functions explicit. In the next step, on the basis of representations of our intentions, we share and discuss the intentions of psychological functions and improve the peer support services..

**Acknowledgement.** This research was supported by the Japan Society for the Promotion of Science (JSPS) KAKENHI Grant Number 2510574.

# References

1. Fisher, E.B., Boothroyd, R.I., Coufal, M.M., Baumann, L.C., Mbanya, J.C., RotheramBorus, M.J., Sanguanprasit, B., Tanasugarn, C.: Peer support for self-management of diabetes improved outcomes in international settings. Health Affairs **31**(1), 130–139 (2012)
2. Akasaka, F., Nemoto, Y., Kmita, K., Shimomura, Y.: Service Design Knowledge Management based on Input-Output Function Representation. The Japan Society for Precision Engineering **77**(11), 1050–1056 (2011)
3. Osawa, I., Ikeda, M., Nabeta, T., Yoneda, T., Takeda, Y., Nakai, M., Usukura, M., Abe, K.: Incremental Design of Web community for Diabetics through Practice. The 3th Forum on Knowledge Co-Creation, II2-1–II2-10 (2013). (in Japanese)
4. Osawa, I., Ikeda, M., Nabeta, T., Yoneda, T., Takeda, Y., Nakai, M., Usukura, M., Abe, K.: Construction of Web Community Function for diabetics. Special Interest Group on Knowledge Based Systems, The Japanese Society for Artificial Intelligence, 98, 1–7 (2013). (in Japanese)
5. Sumita, K., Kitamura, Y., Sasajima, M., Mizoguchi, R.: Are services functions? In: Snene, M. (ed.) IESS 2012. LNBIP, vol. 103, pp. 58–72. Springer, Heidelberg (2012)
6. Osawa, I., Ikeda, M., Nabeta, T., Yoneda, T., Takeda, Y., Nakai, M., Usukura, M., Abe, K.: The relation between function of an online community for diabetic patients and psychological changes. In: Proceedings of the 26th Annual Conference of the Japan Society for Artificial Intelligence (in CD-ROM) (2012). (in Japanese)
7. Kitamura, Y., Koji, Y., Mizoguchi, R.: An Ontological Model of Device Function: Industrial Deployment and Lessons Learned. Journal of Applied Ontology **1**(34), 237–262 (2006)
8. Sumita, K., Kitamura, Y., Sasajima, M., Mizoguchi, R.: An Ontological Consideration on Essential Properties of the Notion of "Service". Journal of Japan Industrial Management Association **63**(3), 138–153 (2012). (in Japanese)
9. Kozaki, K., Kitamura, Y., Ikeda, M., Mizoguchi, R.: Development of an environment for building ontologies which is based on a fundamental consideration of "relationship" and "role". In: Proceedings of the Sixth Pacific Knowledge Acquisition Workshop (PKAW 2000), pp. 205–221 (2000)
10. Shimomura, Y., Hara, T., Watanabe, K., Sakao, T., Arai, T., Tomiyama, T.: Proposal of Service Engineering (1st Report, Service Modeling Technique for service engineering). Transactions of the JSME (The Japan Society for Mechanical Engineering) **71**(702), 669–676 (2005). (in Japanese)

# Semantic Social Web

# Personalized Model Combination for News Recommendation in Microblogs

Rémi Lam, Lei Hou[(✉)], and Juanzi Li

Knowledge Engineering Group, Department of Computer Science and Technology,
Tsinghua University, Beijing 100084, People's Republic of China
{remilami92,greener2009}@gmail.com,
lijuanzi@tsinghua.edu.cn

**Abstract.** Facing large amount of accessible data everyday on the Web,
it is difficult for people to find relevant news articles, hence the importance of news recommendation. Focused on the information to be used
and the way to model it, each of the existing models proposes its own
algorithm to recommend different news to different users. For these models, personalization is only done at the recommendation level. But if the
user chooses a model that is not appropriate for him, the recommendation may fail to work accurately. Therefore, personalization should also
be done at the model level. In our proposed model, the first level is defining four atomic recommendation models that make fully use of the social
and content information of users and the second level is adapting to each
user that atomic models effectively used. Experiments conducted on two
real datasets built from *Twitter* and *Tencent Weibo* give evidence that
this double level of personalization boosts the recommendation.

**Keywords:** Recommender systems · User-generated content · Personalization · Social network · Self-adaptive

## 1 Introduction

In modern era, people face the problem of information exploitation: finding relevant pieces of information among large volume of data is not a trivial task. To
overcome this issue referred as *information overload*, many companies get interested in designing recommender systems whose aim would be to provide relevant
information to the user. As most of them deal with punctual events, the interest
created by a piece of news decreases over time [12]. Because of this ephemeral
nature, users should read a news article quickly after its publication. However,
it often suffers from a data sparsity problem, namely it is hard to make recommendations for users who only provide little available information since there is
little or no feedback provided within such short time.

Fortunately, the booming online social networking applications have become
the dominant information acquisition and dissemination systems. Microblogging, as a light-weight media, enables the users to post short messages for daily

© Springer International Publishing Switzerland 2015
T. Supnithi et al.(Eds.): JIST 2014, LNCS 8943, pp. 321–334, 2015.
DOI: 10.1007/978-3-319-15615-6_24

chatter, conversation, sharing information, and maybe reporting news. Choosing microblogging platforms like *Twitter* or *Tencent Weibo* to provide news recommendation has two benefits: as a real-time media, it is easier to recommend fresh news (news often appear in microblogging platforms before breaking out in traditional media [5]); and as a social media, based on the principle of *homophily*, we can leverage the social network of a user to recommend news.

Researchers have proposed many recommendation models which have proven being efficient in many applications. We will refer these models as the *existing* models. However, the problem is that each model ignores the fact that different users follow different patterns to get informed. For example, some users might only read articles that match their personal interests while others might be influenced by their social groups and prefer to read articles that their friends recommend. Ignoring the user behavior differences is a problem because if we indifferently use the same recommendation model to provide all the users with news recommendation, we may have applied a model that is not adapted to some of the users because the personalization would only be done at the recommendation level. We propose to add one more level of personalization by combining, for each user, the existing models. The personalization would also be done at the model level. We will refer our model with second level of personalization as the *global* model. The challenge here remains in finding the most appropriate way to combine the existing models.

The main contributions of this paper can be summarized as follows:

1. We focus on the problem of news recommendation in microblogging platforms by paying particular attention to the fact that different users may have their personal preferred recommendation models, fact that is not considered in the present models.
2. We propose a global recommendation model with two levels of personalization. The first level naturally comes from the existing recommendation models. Particularly, we propose four atomic recommendation methods to make fully use of all the information that can be used in recommendation, including both content and social information. For the second level, we define a global model that, depending on the user, combines differently the existing recommendation models.
3. Our experiments conducted on real datasets built from Twitter and Tencent Weibo prove that adding social information as a parameter of the recommender system improves the accuracy of the recommendation, and our global model achieves better performance than single level of personalization existing models.

The rest of the paper is organized as follows. In Section 2, we discuss the related work. In Section 3, we formally define the problem of personalized news recommendation before describing our global model in Section 4. Our experimental results are presented in Section 5 and Section 6 concludes our work.

## 2    Related Work

Recommender systems follow at least one of these two paradigms: content-based filtering and collaborative filtering [10].

Content-based recommender systems provide item recommendations based on the construction of item and user profiles where the item profile (respectively the user profile) only represents the intrinsic properties of the corresponding item (resp. the corresponding user) [14]. Item- and user-profiles are computed independently and are domain-dependant. The key step of content-based recommendation paradigm is the question of profiling which is not a trivial problem [7]. Generally, the users and the news to be recommended can be characterized through a vector space model or topic distributions obtained by statistical models (e.g. *Latent Dirichlet Allocation* [1]). After this profiling step, content-based models usually provide the matching by computing the similarity (e.g. cosine similarity) between the profiles of the users and the profiles of the news to be recommended. In the system proposed by Liang et.al, users' profiles are computed as a ponderation of the profile of the browsed news (computed through keyword extraction) with the news browsing duration and recency. The longer the duration of news browsing by a user is, the more the news would count in building this user's profile [11]. Careira et al. used a Bayesian classifier to select the interesting news based on the users' profiles. A user's profile is first computed by asking directly what are his interests and then, after the user has read a news article, updated manually (the system asks the user if he is interested by the news) or automatically (using duration of reading and number of read lines) [2].

The paradigm of collaborative filtering is a good alternative and has actually been the most successful user and news filtering paradigm in recent years [8,9]. Indeed, the features that are used to compute the profiles are completely domain-free and are based on the feedbacks given by the users to the recommender system: for example, the widely used *Google News* [4] relies on users' click-through behavior to provide the recommendation. However, we can notice that pure collaborative filtering recommendation fails to recommend articles to a new user because this user would not have interacted with the system; it would also fails to recommend a new article to users because no user would have interacted with it in the system. This problem is referred as the *cold-start* problem. One possible solution to this problem is to include some content information in the recommendation process. The *hybrid models* successfully implement both of the two content-based and collaborative filtering paradigms [2,16].

Recently, the proliferation of the social media (e.g. the most popular micro-blogging platform, *Twitter*[1] with 650 millions users in early 2014[2]) making the social news recommendation an active area of research [3,6,17]. Firstly, users can publish content, generally called *posts*, up to 140 characters, which can improve the user profiling process. Secondly, as they enable interaction between the users, microblogging platforms naturally contain a social network structure. In social

---

[1] Twitter: http://twitter.com
[2] Twitter statistics: http://www.statisticbrain.com/twitter-statistics/

networks, the *homophily* principle stipulates that users with common interests tend to be friends and friends tend to have common interests. Social-based recommender systems use this principle for interest extraction and recommendation [5, 6].

However, we can notice that most of these recommender systems force the usage of one recommendation method for all the users and thus, don't personalize the recommendation method. Our model not only incorporates the social information as a parameter of the recommendation but also personalizes the recommendation method to each user.

# 3   Problem Definition

In this section, we formalize the problems of news recommendation in microblogging platforms (Section 3.1) and model combination (Section 3.2).

## 3.1   The Problem of News Recommendation

**Definition 1.** *Let $\mathcal{U} = \{u_0, u_1, ...\}$ be a set of users registered in the microblogging platform, let $\mathcal{P} = \{p_0, p_1, ...\}$ be a set of posts published in the microblogging platform and let $\mathcal{N} = \{n_0, n_1, ...\}$ be a set of news articles arriving from the news portals.*

**Definition 2.** *Let $\mathcal{P}_i = \{p_{i0}, p_{i1}, ...\} \subset \mathcal{P}$ be the set of posts in $\mathcal{P}$ published by the user $u_i$, let $\mathcal{F}_i = \{u_{i0}, u_{i1}, ...\} \subset \mathcal{U}$ be the set of friends in $\mathcal{U}$ of the user $u_i$ [3], let $\mathcal{N}_i = \{n_{i0}, n_{i1}, ...\} \subset \mathcal{N}$ be the set of articles in $\mathcal{N}$ relevant to the user $u_i$, and let $\mathcal{U}_j = \{u_i \in \mathcal{U} | n_j \in \mathcal{N}_i\}$ be the set of users in $\mathcal{U}$ that find the article $n_j \in \mathcal{N}$ relevant.*

**Definition 3.** *We are given a set of articles $\mathcal{N}$, a set of users $\mathcal{U}$ and for each user $u_i \in \mathcal{U}$, the list of his posts $\mathcal{P}_i$, his friends $\mathcal{F}_i$ and the articles $\mathcal{N}_i$ that are relevant to him. The problem of personalized news recommendation is given a user $u_i \in \mathcal{U}$, to find the top-k relevant articles in $\mathcal{N}$ without using the set of articles $\mathcal{N}_i$ relevant to that particular user.*

We raise the following question: for a given user $u_i \in \mathcal{U}$, what is concretely $\mathcal{N}_i$ (the list of news articles that are relevant to $u_i$)? Since we cannot get the exact list of news articles relevant to each user, we make the following strong assumption which prevails in our model and our experiments.

**Assumption.** A user $u_i$ finds an article $n_j$ relevant (1) if and only if $u_i$ forwards the news article $n_j$ through a post (2).

The implication (2) $\Rightarrow$ (1) is natural contrary to its converse (1) $\Rightarrow$ (2) which is an approximation. Indeed, if a user $u_i$ reads an article $n_j$, he might find the news article relevant without forwarding it through a microblog post. The user might also not have read all the news articles relevant to him (and he would not forward them).

---

[3] The term *friends* denotes the users that $u_i$ follows.

## 3.2   The Problem of Model Combination

**Definition 4.** *Let* $\mathcal{M} = \{M_1, M_2, ..., M_m\}$ *be the set of existing recommendation models. We suppose that each of these models* $M_1, M_2, ..., M_m$ *respectively has* $I_1, I_2, ..., I_m$ *interest score functions. The interest score between a user and an article represents to what extent the user would like the article to be recommended to him. Then, by ranking the articles based on the scores with one user, each of the models recommend* $k$ *articles to each of the users.*

To the best of our knowledge, each of the existing models has its own recommendation algorithm and forces the user to use it to have news recommended to him. In other words, considering one recommendation model, when recommending news, the same method would be used for all the users. This is what we call personalization at the recommendation level. The contribution of this paper is adding one more level of personalization: at the model level. It means that, depending on the user, the model recommendation would be different. Concretely, we define a global model that for each user, first, computes a combination of different existing models and then, uses this combination to provide the recommendation. We still keep the first level of personalization since we use the existing models. These definitions enable the formalization of the problem, drawn in Fig. 1.

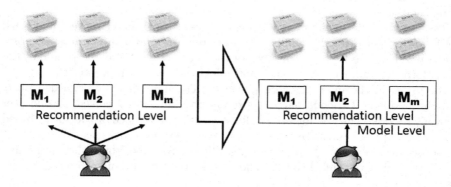

**Fig. 1.** Our idea (Left: Before - One level of personalization, Right: After - Two levels of personalization)

**Definition 5.** *The problem of model combination consists in, given a user* $u_i \in \mathcal{U}$ *and a set of existing recommendation models* $\mathcal{M}$, *combining the existing recommendation models before using the combination to provide the recommendation.*

## 4   The Proposed Models

We name our global model as the *Mixture* model. This model has two versions (presented in Fig. 2).

- *Mixture-S* (Mixture-**S**elect): this version *selects* and applies one atomic recommendation model to provide the recommendation. The selection depends on the user.
- *Mixture-C* (Mixture-**C**ombine): this version *combines* the atomic recommendation models to provide the recommendation. The combination depends on the user.

**Fig. 2.** *Mixture* model (Left: *Mixture-S*, Right: *Mixture-C*)

### 4.1    *Mixture-S* Version

The *Mixture-S* version consists in two steps: use a logistic regression classifier to select the model $M_i \in \mathcal{M}$ and use the selected model to compute the recommendation.

As input of the logistic regression classifier, the user is represented with the following features: the number of posts published and the number of posts reforwarded by the user, the number of the user's friends and the profile of the user. Both the number of published and reforwarded posts represent the activity of the users on the microblogging platform. The number of reforwarded posts and the number of friends partly represent the influence received by one user from his friends, which is intuitively linked to the preference of the social-based models. The profile of the user represents the interests of the user. We use such signal because in different domains, the influence received from other users may be different. The output of the classifier is the preferred atomic recommendation model.

The training data of the classifier is composed by the users respecting the following condition: at least for one of the existing recommendation model, one of the $k = 50$[4] recommended articles using that model is relevant to the user. Our training data is thus exclusively composed of users satisfied by at least one of the recommendation models. To label each user of the learning set with the preferred recommendation model, for each user, we choose the one for which when providing the recommendation of articles, the first relevant news article is the best ranked.

---

[4] Such value of $k$ is evaluated experimentally to guarantee an adequate number of users in the training set.

## 4.2  *Mixture-C* Version

The *Mixture-C* model combines the existing recommendation models. For each user $u_i$, we define the the $m$ coefficients $\alpha_1$, $\alpha_2$, ..., $\alpha_m$ (computation detailed later).

First, the *Mixture-C* computes the interest scores between each user and each news article. Then, the *Mixture-C* global model recommends $k$ articles to each of the users.

$$I_{mixtC}(u_i, n_j) = \alpha_1 \times I_1(u_i, n_j) + \alpha_2 \times I_2(u_i, n_j) + ... + \alpha_m \times I_m(u_i, n_j) \quad (1)$$

To compute the coefficients, we use $m$ linear regression models: one for each coefficient. The input is the user represented with the same features as for the Logistic Regression classifier in the *Mixture-S* version. The output is the real value of the coefficient ($0 \leq$ coefficient $\leq 1$). When building the training sets (one for each coefficient), the user selection process is the same as for the Logistic Regression classifier of the *Mixture-S* model. To label the users belonging to the training set, the coefficients are simultaneously computed as follows.

$$\alpha_1 = \frac{1}{r_1(u_i)} \qquad \alpha_2 = \frac{1}{r_2(u_i)} \qquad ... \qquad \alpha_m = \frac{1}{r_m(u_i)} \quad (2)$$

where $r_1(u_i)$, $r_2(u_i)$, ..., $r_m(u_i)$ are respectively the ranks of the first article relevant to the user $u_i$ using $M_1, M_2, ..., M_m$ in the recommendation provided by respectively the models 1, 2, ..., $m$. These coefficients are then normalized so that $\sum_{i=1}^{m} \alpha_i = 1$. In each training set, the data is labeled with the value of the corresponding coefficient. Such definition is justified by the fact that the better the rank is (i.e. $r$ becomes lower (closer to 1)), the better the weight of the corresponding recommendation algorithm is (i.e. the coefficient is bigger).

## 4.3  Atomic Recommendation Models

We would like to evaluate the performance of our *Mixture* global model, compared to the existing recommendation models. Many recommender systems already exist, comparing our model to all these systems would be too fastidious. Therefore, we choose to define four recommendation models that are representative of the different recommendation paradigms introduced in Section 2. We will refer these recommendation models as *atomic* recommendation models. They all compute interests scores between each user and each article in order to compute the recommendation.

We compute the user's profile from the content of his own published posts (aggregated as a unique document for each user) and we compute the profile of a news article from its content. We use the *Latent Dirichlet Allocation* (LDA) topic model [1] with Gibbs Sampling inference technique, to extract the topics $\mathcal{Z} = \{z_0, z_1, ...\}$ from the corpus. As for the hyperparameters, we choose $K = 50$ topics, $\alpha = 50/K$ and $\beta = 0.01$ as recommended in [15], and perform 1500 iterations of sampling. After that, each user $u_i$ (resp. news article $n_j$) is represented as a vector $\overline{u_i}$ (resp. $\overline{n_j}$) where each dimension is a topic $z$ and the corresponding weight represents the probability that $u_i$'s posts (resp. $n_j$) refer to the topic $z$. We have similar representation for news articles.

Once the profiles obtained, for each of these models (summarized in Fig. 3), we describe below the interest score computation. We are given one user $u_i \in \mathcal{U}$ (represented by $\overline{u_i}$) and one article $n_j \in \mathcal{N}$ (represented by $\overline{n_j}$).

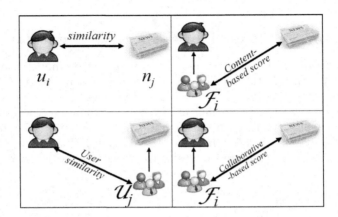

**Fig. 3.** Atomic models. From top to bottom, left to right: content-based, social content-based, collaborative, social collaborative-based.

**Content-based Model.** In the content-based atomic model, we use the content information of users' posts and news articles by computing the similarity between them. The interest score $I_{ct}$ is the cosine similarity ($cs$) between the user $u_i$ and the article $n_j$.

$$I_{ct}(u_i, n_j) = cs(\overline{u_i}, \overline{n_j}) \quad \text{where} \quad cs(\overline{u_i}, \overline{n_j}) = \frac{\overline{u_i} \cdot \overline{n_j}}{\|\overline{u_i}\| \|\overline{n_j}\|} \tag{3}$$

**Collaborative-based Model.** In the collaborative-based atomic model, we use the profile of the users that are the most similar to the user (user-oriented neighborhood model). The interest score $I_{cl}$ is computed as the average of the cosine similarities between the user $u_i \in \mathcal{U}$ and all the users except himself that forwarded the article $n_j$ (users of the set $\mathcal{U}_j \setminus \{u_i\}$).

$$I_{cl}(u_i, n_j) = \frac{\sum_{u \in \mathcal{U}_j \setminus \{u_i\}} cs(\overline{u_i}, \overline{u})}{|\mathcal{U}_j \setminus \{u_i\}|} \quad \text{where} \quad cs(\overline{u_i}, \overline{u}) = \frac{\overline{u_i} \cdot \overline{u}}{\|\overline{u_i}\| \|\overline{u}\|} \tag{4}$$

**Two Social-based Models.** We define two social-based atomic models: social content-based and social collaborative-based. These two models leverage the social network of the users, through the *following* relationships. As different friends have different influence on one user, we define the influence $\text{infl}(u_i, u_j)$ that a user $u_i$ has on a user $u_j$ as follows.

$$\text{infl}(u_i, u_j) = 3 \times \text{nReforward}(u_i, u_j) + 2 \times \text{nMention}(u_i, u_j) + 1 \tag{5}$$

where nReforward($u_i, u_j$) is the number of $u_i$'s posts reforwarded by $u_j$ and nMention($u_i, u_j$) is the number of mentions of $u_i$ in the $u_j$'s posts.

Such definition comes from the fact that if a user $u_j$ establishes friend relationships with another user $u_i$ then $u_j$ is probably influenced by $u_i$ (term 1). Furthermore, if the user $u_j$ often mentions the user $u_i$ then $u_j$ is probably very influenced by $u_i$ (coefficient 1: strong signal). Last but not least, if the user $u_j$ often reforwards user $u_i$'s posts then $u_j$ is certainly very influenced by $u_i$ (coefficient 2: very strong signal). We validated the values of these coefficients through experiments.

The interest score $I_{sct}$ (resp. $I_{scl}$) is computed as the weighted average of the interest scores of the content-based model (resp. collaborative-based model) between the user itself, all the users that the user $u_i$ follows (users of the set $\mathcal{F}_i$) and the article $n_j$.

$$I_{sct}(u_i, n_j) = \frac{\sum_{u \in \mathcal{F}_i \cup \{u_i\}} \mathrm{infl}(u, u_i) I_{ct}(u, n_j)}{\sum_{u \in \mathcal{F}_i \cup \{u_i\}} \mathrm{infl}(u, u_i)} \tag{6}$$

$$I_{scl}(u_i, n_j) = \frac{\sum_{u \in \mathcal{F}_i \cup \{u_i\}} \mathrm{infl}(u, u_i) I_{cl}(u, n_j)}{\sum_{u \in \mathcal{F}_i \cup \{u_i\}} \mathrm{infl}(u, u_i)} \tag{7}$$

# 5   Experiments

In this section, we provide experimental evaluations to show the effectiveness of the proposed models in previous sections. First we introduce the data collection and description, then present the evaluation metric and experiment setup, and finally report the performance results for different approaches in different settings as well as analysis and discussions.

## 5.1   Data Collection

To the best of our knowledge, no public existing benchmark dataset is available for news recommendation in microblog platform. Therefore, we need a standard dataset to evaluate our recommendation algorithm. To build our dataset, we need two streams of microblog and news stream, the social network of users, and user-news relationships (i.e. the news articles forwarded by the users). We build our datasets from two microblogging services *Twitter* and *Tencent Weibo* and news portals. The statistics related to these datasets are summarized in Table 1.

**Twitter.** To build the *Twitter* dataset, we use Twitter's API[5] which enables to listen to a stream of a sample of the tweets. This stream of tweets is actually estimated as 1% of the total tweets. Morstatter et al. study the implication of using this API instead of the *Firehose* which is a stream of all the tweets [13].

---

[5] Twitter API: http://api.twitter.com/1.1/

We extract the posts published by 5,000 users from 2013/12/24 to 2014/01/13 [6]. During this posts crawling period, 2,563 users actually publish at least one post including links to news portals. From all the extracted posts, 1,845,345 of them are published by these users. Some of these posts include links to the news portals. From these links, we extract 10,869 unique news articles and we have 28,923 user-news relationships. We also extract the 40,549 users friendship relationships among these users.

**Tencent.** We first use Tencent's API and crawl the microblogs from the public timelines. By identifying the most active users (i.e. the users who most frequently post or forward the news articles), we obtain 568,365 microblog posts posted by 4,940 users. Then, we select the posts that contain links to the news articles, access the latter, and collect 7,722 news articles associated to 1,856 users. We have 22,285 user-news relationships. Finally, to obtain a microblog posts corpus, we employ the API to get the users' timelines and their social relations. We obtain 3,593,781 microblog posts and 3,470 friendship relationships.

**Table 1.** Datasets statistics

|                                    | Twitter   | Tencent   |
|------------------------------------|-----------|-----------|
| Number of users                    | 2,563     | 1,856     |
| Number of microblogging posts      | 1,845,345 | 3,593,781 |
| Number of news articles            | 10,869    | 7,722     |
| Number of friendship relationships | 40,549    | 3,470     |
| Number of user-news relationships  | 28,923    | 22,285    |

As we collect the microblogging posts and the news on two different systems (respectively the microblogging platform and the news portals), we emphasize the fact that we don't have for each user the exact list of news articles relevant to him. This is why we use the links to news articles that are present in the microblogging posts of the user as a signal of interest. Such approximation will considerably have negative impact on the precision of our evaluation and make our news recommendation task particularly challenging.

## 5.2 Evaluation Metrics

To evaluate the recommendation performance of the proposed models, we use three types of metrics, including precision at k, mean reciprocal rank and the average discounted cumulative gain at k, which are defined as follows:

---

[6] These users were selected because they were those who published the most English posts including links to news portals (*Yahoo News*, *Huffington Post*, *New York Times*, *CNN* and *Fox News*) during 24h crawling period.

*Precision-at-k.* For each user $u$, we define the condition $c_u^k$ which is *true* if and only if among the $k$ articles recommended to the user, at least one of them was forwarded by the user. The precision-at-k ($p@k$) is computed as follows:

$$p@k = \frac{\sum_{u \in U} I_{|c_u^k}}{|U|} \tag{8}$$

where $I_{|c_u^k}$ is the indicator variable equal to 1, if $c_u^k$ is true, and 0 otherwise. $p@k$ is higher when the system provides the recommendation of at least one relevant article among the $k$ recommended articles.

*Mean reciprocal rank.* For each user $u$, we define $r(u)$ as the rank of the first article relevant to the user $u$ in the recommendation (supposing we are recommending all the articles to the user, what counts here is the ranking of the recommended articles). The mean reciprocal rank ($MRR$) is computed as follows:

$$MRR = \frac{1}{|U|} \sum_{u \in U} \frac{1}{r(u)} \tag{9}$$

$MRR$ is higher when the recommendation puts the first relevant article into the top ranks.

*Average discounted cumulative gain-at-k.* For each user $u$, we define the discounted cumulative gain-at-k ($DCG_u@k$) as follows:

$$DCG_u@k = G(1) + \sum_{i=2}^{k} \frac{G(i)}{\log_2(i)} \tag{10}$$

where $G(i)$ is the relevance of the article at rank $i$. The average discounted cumulative gain-at-k ($ADCG@k$) is the average over all the users:

$$ADCG@k = \frac{\sum_{u \in U} DCG_u@k}{|U|} \tag{11}$$

$ADCG@k$ is higher when the recommendation puts relevant articles into the top ranks.

### 5.3   Evaluation of Our *Mixture* Model

Through a 4-fold cross-validation scheme, we compute the average precision of our Logistic Regression classifier used in the *Mixture-S* version and the average squared mean errors of our Linear Regression classifiers used in the *Mixture-C* version. These results are summarized in Table 2.

We also evaluate our global model compared to the atomic models. The results are summarized in Table 3. The $p@k$ ($k = 10, 50$) values are all relatively small. For example, the highest value of $p@10$ is 8.71% which can be interpreted

**Table 2.** Evaluation of the learning models: the *LR precision* represent the average of our Logistic Regression classifier used in the Mixture-S version, and $\alpha \sim \delta$ denote the squared mean errors of the linear regression classifiers in Mixture-C

| Dataset | LR Precision | Squared Mean Errors | | | |
|---------|--------------|-------|-------|-------|-------|
| | | $\alpha$ | $\beta$ | $\gamma$ | $\delta$ |
| *Twitter* | 39.2 % | 0.063 | 0.085 | 0.064 | 0.093 |
| *Tencent* | 48.6 % | 0.030 | 0.065 | 0.093 | 0.056 |

as: for only 8.71% of the users, the recommendation of 10 news articles is satisfying (at least one of the recommended articles is relevant to the user). Small values are mainly due to the imperfection of our data as we approximate the news relevant to a user with the news he forwards. However, we should focus more on the comparison between the values.

**Table 3.** Evaluation of the *Mixture* model

| Models | *Twitter* dataset | | | | *Tencent* dataset | | | |
|--------|------|------|------|---------|------|------|------|---------|
| | P@10 | P@50 | MRR | ADCG@20 | P@10 | P@50 | MRR | ADCG@20 |
| Cont | 8.49% | 24.0% | 0.038 | 0.10 | 3.33% | 11.9% | 0.019 | 0.03 |
| Coll | 6.46% | 16.4% | 0.034 | 0.07 | 4.84% | 14.1% | **0.032** | **0.07** |
| SCont | 7.04% | 20.1% | 0.034 | 0.08 | 3.73% | 12.8% | 0.020 | 0.03 |
| SColl | 6.90% | 18.3% | 0.040 | **0.10** | 3.91% | 12.0% | 0.024 | 0.06 |
| Mixt-S | **8.71%** | 22.6% | **0.043** | **0.10** | 4.14% | 13.6% | 0.028 | 0.06 |
| Mixt-C | 8.70% | **25.3%** | 0.041 | **0.10** | **5.54%** | **16.2%** | **0.032** | **0.07** |

The performance of at least one version of our *Mixture* model is better than the performance of each of the atomic models. It shows that adding one level of personalization in the global model is beneficial to the recommendation. Furthermore, as our *Mixture* global model uses the social information and performs better than content-based and collaborative-based models, it gives evidence that considering the social network of a user improves the quality of the recommendation.

## 5.4    Comparison between the Two Versions of the *Mixture* Model

In the previous experiment, we show that adding one level of personalization is beneficial to the recommendation rather than using independently each of the atomic models. Now, we would like to know firstly, if the features mentioned earlier have an impact on the preference towards one of the two versions of the *Mixture* model and secondly, if we can increase the performance of the recommendation by automatically choosing between the two versions of the model. For that purpose, we learn a Logistic Regression classifier to automatically select between the *Mixture-S* and the *Mixture-C* versions of the *Mixture* global model.

The input is the user represented with the same features as for the classifier of the *Mixture-S* model.

For the *Twitter* dataset, we evaluate the performance of *Mixture* with classifier on 1365 users. Other users belong to the training set used for training the classifier. Among these 1,365 users, 789 are classified as preferring the *Mixture-S* version and 576 as preferring the *Mixture-C* version. The precision of the classifier is 57.5 %. We summarize the results in Table 4.

**Table 4.** Evaluation of the *Mixture* model with classifier

|  | **P@10** | **P@50** | **MRR** | **ADCG@20** |
|---|---|---|---|---|
| Mixture-S | 9.08 % | 22.6 % | 0.047 | **0.11** |
| Mixture-C | 8.86 % | 23.6 % | 0.042 | 0.10 |
| **Mixture with classifier** | **9.74 %** | **24.2 %** | **0.048** | **0.11** |

We observe that the performance of our *Mixture* model with classifier is better than using independently the two versions of our *Mixture* model. Therefore, we conclude that the chosen features have influence on the preference to one of the two versions.

## 6    Conclusions

Our basic intuition is that there is not a unique news recommendation method that suits all the users. But the existing recommender systems usually constrain every user to one recommendation method. This is why, we build an original global model for news recommendation that personalizes for each user the recommendation method. Therefore, beyond the recommendation level (different news are recommended to different users), we also add personalization in the model level (different recommendation methods are used for different users). The experiments that we conduct on real datasets built from *Twitter* and *Tencent Weibo* validate our choices.

**Acknowledgement.** The work is supported by 973 Program (No. 2014CB34-break0504), NSFC (No. 61035004), NSFC-ANR (No. 61261130588), European Union 7th Framework Project FP7-288342 and THU-NUS NExT Co-Lab.

## References

1. Blei, D.M., Ng, A.Y., Jordan, M.I.: Latent dirichlet allocation. The Journal of Machine Learning Research **3**, 993–1022 (2003)
2. Carreira, R., Crato, J.M., Gonçalves, D., Jorge, J.A.: Evaluating adaptive user profiles for news classification. In: Proceedings of the 9th International Conference on Intelligent User Interfaces, pp. 206–212. ACM (2004)

3. Chen, J., Nairn, R., Nelson, L., Bernstein, M., Chi, E.: Short and tweet: experiments on recommending content from information streams. In: Proceedings of the SIGCHI Conference on Human Factors in Computing Systems, pp. 1185–1194. ACM (2010)

4. Das, A., Datar, M., Garg, A., Rajaram, S.: Google news personalization: scalable online collaborative filtering. In: Proceedings of the 16th International Conference on World Wide Web, pp. 271–280. ACM (2007)

5. De Francisci Morales, G., Gionis, A., Lucchese, C.: From chatter to headlines: harnessing the real-time web for personalized news recommendation. In: Proceedings of the Fifth ACM International Conference on Web Search and Data Mining, pp. 153–162. ACM (2012)

6. Gartrell, M., Han, R., Lv, Q., Mishra, S.: Socialnews: Enhancing online news recommendations by leveraging social network information. Tech. rep., Technical Report CU-CS-1084-11, Dept. of Computer Science, University of Colorado at Boulder (2011)

7. Gauch, S., Speretta, M., Chandramouli, A., Micarelli, A.: User profiles for personalized information access. In: Brusilovsky, P., Kobsa, A., Nejdl, W. (eds.) The Adaptive Web. LNCS, vol. 4321, pp. 54–89. Springer, Heidelberg (2007)

8. Hofmann, T.: Probabilistic latent semantic indexing. In: Proceedings of the 22nd Annual International ACM SIGIR Conference on Research and Development in Information Retrieval, pp. 50–57. ACM (1999)

9. Koren, Y., Bell, R., Volinsky, C.: Matrix factorization techniques for recommender systems. Computer 42(8), 30–37 (2009)

10. Li, L., Wang, D.D., Zhu, S.Z., Li, T.: Personalized news recommendation: a review and an experimental investigation. Journal of Computer Science and Technology 26(5), 754–766 (2011)

11. Liang, T.P., Lai, H.J.: Discovering user interests from web browsing behavior: an application to internet news services. In: Proceedings of the 35th Annual Hawaii International Conference on System Sciences, HICSS 2002, pp. 2718–2727. IEEE (2002)

12. Matsubara, Y., Sakurai, Y., Prakash, B.A., Li, L., Faloutsos, C.: Rise and fall patterns of information diffusion: model and implications. In: Proceedings of the 18th ACM SIGKDD International Conference on Knowledge Discovery and Data Mining, pp. 6–14. ACM (2012)

13. Morstatter, F., Pfeffer, J., Liu, H., Carley, K.M.: Is the sample good enough? comparing data from twitters streaming api with twitters firehose. In: Proceedings of ICWSM (2013)

14. Phelan, O., McCarthy, K., Bennett, M., Smyth, B.: Terms of a feather: content-based news recommendation and discovery using twitter. In: Clough, P., Foley, C., Gurrin, C., Jones, G.J.F., Kraaij, W., Lee, H., Mudoch, V. (eds.) ECIR 2011. LNCS, vol. 6611, pp. 448–459. Springer, Heidelberg (2011)

15. Rosen-Zvi, M., Griffiths, T.L., Steyvers, M., Smyth, P.: The author-topic model for authors and documents. In: Proceedings of the 20th Conference on Uncertainty in Artificial Intelligence, pp. 487–494. ACM (2004)

16. Shmueli, E., Kagian, A., Koren, Y., Lempel, R.: Care to comment?: recommendations for commenting on news stories. In: Proceedings of the 21st International Conference on World Wide Web, pp. 429–438. ACM (2012)

17. Wang, Y., Zhang, J., Vassileva, J.: Personalized recommendation of integrated social data across social networking sites. In: Proceedings of the Workshop on Adaptation in Social and Semantic Web (SASWeb 2010). CEUR Workshop Proceedings, vol. 590, pp. 19–30. Citeseer (2010)

# A Model for Enriching Multilingual Wikipedias Using Infobox and Wikidata Property Alignment

Thang Hoang Ta[1](✉) and Chutiporn Anutariya[2]

[1] Shinawatra University, Bangkok, Thailand
tahoangthang@gmail.com
[2] Asian University, Chonburi, Thailand
chutiporna@asianust.ac.th

**Abstract.** Wikipedia supports a large converged data with millions of contributions in more than 287 languages currently. Its content changes rapidly and continuously every hour with thousands of edits which trigger many challenges for Wikipedia in controlling, associating and balancing article content among language editions. This paper provides some processes to enrich Wikipedia content, which will retrieve semantic relations based on alignment between infobox properties and Wikidata properties in various languages. Then, the outcomes mainly contribute these semantic relations back to Wikidata and Wikipedias, especially ones are based on the Latin alphabet. The case study will offer a specific case about aligning biological infoboxes and detecting missing interwiki links of biological species at Vietnamese Wikipedia and English Wikipedia.

**Keywords:** Multilingual wikipedias · Wikidata · Semantic relations

## 1 Introduction

Wikipedia is an encyclopedia that allows the public community to develop content voluntarily in numerous languages. It covers the content differentiation which is from the differences of language structure and editor contributions. Wikipedia must face with some difficulties, such as content management, anti-vandalism, data number values verification and content synchronization among its projects. Some Wikipedias have a high collaborative quality, such as English Wikipedia consists of more than 4.57 million articles with highest depth of 874[1]; in contrast, there are still many Wikipedias with modest quality and a small number of editors. To enrich for these poor Wikipedias, semantic relations will be chosen as an approach consistently that machine can comprehend in inserting content. To extract semantic relations from various languages of Wikipedia, DBPedia [1] is a proper and prominent project that is popularly examined by many researchers. DBPedia supports many linguistic tasks such as Entity Linking, Question Answering and Relationship Extraction [1,2]. DBPedia uses English with its rich content as the starting point when extracting semantic relations. The result relations are also applied to other language articles which have interwiki

---

[1] https://meta.wikimedia.org/wiki/List_of_Wikipedias.

© Springer International Publishing Switzerland 2015
T. Supnithi et al. (Eds.): JIST 2014, LNCS 8943, pp. 335–350, 2015.
DOI: 10.1007/978-3-319-15615-6_25

links with English articles instead of executing semantic extraction independently. For non-English articles that have no interwiki links with English, their semantic relations may be missing or limited because of the insufficient support of DBpedia's extraction mechanism for all languages. With low data update frequency and contribution prohibition from public community broadly, this shortage becomes larger when DBPedia offers not enough semantic relations to enrich the content for small-scale Wikipedias effectively. Another component of DBPedia, DBpedia Live Extraction Manager could solve the low data update frequency problem, but unfortunately, it mainly supports English that depends on update threads. [3,4] Besides, Wikidata is a sister project of Wikipedia, which allows editors contribute semantic relations (Wikidata statements[2]) on it freely and separately. This project also supports multilingual database [5] for reducing the complication of interwiki links among Wikipedias. However, Wikidata is still in the development process so it comprises not much semantic content.

We apparently see the problem that it is difficult to enrich articles which lack of interwiki links with English or any languages have a high collaborative quality. Therefore, we firstly research about the interlink connection. Infobox is a fixed-format table which includes properties (parameters) and its values to describe basic information related to articles. Infobox has a structured metadata which we can retrieve and assess semantic relations after aligning its properties and Wikidata properties semi-automatically. Next, based on the assessments, we can conclude which articles have interwiki links with English and make the connections. After all these, we contribute gathered relations to articles which have interwiki links with English.

This paper aims to create a proposed model which contains some uncomplicated processes for aligning infobox properties and Wikidata, extracting semantic relations from infoboxes (or navigation templates) of multilingual Wikipedias and enriching these gathered results for some Wikipedias in order to improve the overall quality. We focus on the semantic relations among several Latin languages or languages based Latin alphabet, in particular between Vietnamese Wikipedia and English Wikipedia. Wikidata is used as a data server to compare and analyze the interwiki links between Wikipedias. We expect to retrieve as many semantic relations as possible from Wikipedia articles, Wikidata and DBPedia. We also avoid using translation tasks and complex NLP algorithms in recognizing and retrieving semantic relations to make implementation faster. At last, the accuracy of article content and errors are made from editors will not be mentioned in this paper.

The paper structure contains reviewing related works in Section 2; a detailed description of the proposed model is described in Section 3; Section 4 is a case study about biological species of English Wikipedia and Vietnamese Wikipedia; Section 5 lists some conclusions and recommendations for next future works.

## 2     Related Works

One research used the technical core of DBPedia to form a new extraction framework. [7] From that, we can inherit all the valuable mechanisms of DBPedia and add more

---

[2] https://www.wikidata.org/wiki/Help:Statements

custom methods to enhance the productivity of extracting semantic relations from Wikipedia. The difficulty is we should have a very deep understanding about DBPedia architecture as well as robust servers to run the extraction rapidly and smoothly. We prefer to find a simpler solution for extracting semantic relations. Other research of Thanh Nguyen and co-authors deployed WikiMatch as a new approach for aligning infoboxes in different languages with its case study aligns infoboxes in Vietnamese, Portuguese and English. [8] WikiMatch can be good for high cross-language heterogeneity and few data instances. We have to investigate the use of a fixed point-based matching strategy to improve the effectiveness. This research excludes machine translation and dictionaries which we also apply to our research in the property alignment. Research of Eytan A., Michael S. and Daniel S. W offered Ziggurat, a system retrieves infobox properties automatically in English, Spanish, French, and German. [12] Some others researches mapped Wikipedia properties to DBPedia properties in many languages. [3][9] However, with all researches above, we could not do this mapping for all languages, especially many rare languages such as Waray-Waray[3] and Cebuano[4] languages. Many researchers used DBPedia which inhibited editors contribute semantic relations and extraction mechanisms. In contrast, Wikidata opens widely for all editors, utilizes human in editing its content and has a strong connection with Wikipedia. [4] Therefore, Wikidata will be a promised semantic source in future with more languages. The current semantic relations of Wikidata have conflicted with DBPedia when its developers specify these relations in some distinctive ways. However, we can make some alignments to compare the similarity of semantic relations between two sources and update each other data.

The researchers pointed out some outstanding tendencies in studying about enriching Wikipedia content between languages. There are also many areas of Wikipedia, which can enhance its performance by applying a framework for multilingual wikis. The rapid development of Wikidata may create many opportunities for new researches and applications in the future.

Besides, category classification is also an important part of enriching Wikipedias content. For small Wikipedias, automatic classification can lessen the efforts of editors who must classify articles into proper categories manually. In some small-scale Wikipedias, many articles followed the English classification system when editors used tools to convert English categories to their language categories. Before executing this task, we need to restructure category system where is still exist many loops and redundant categories. Research of Simone and Roberto (2009) [10] solved this case throughout using a method of comparing English with WordNet. When English category system is aligned with a good structure, other Wikipedias can follow on its category system. Then, we create new categories with NLP patterns which showed that categories and their network can generate concept relations and semantic knowledge [6][11].

---

[3] https://war.wikipedia.org/wiki/Winaray
[4] https://ceb.wikipedia.org/wiki/Sinugboanon

# 3    The Proposed Model

Figure 1 points out the proposed model with three vital processes:

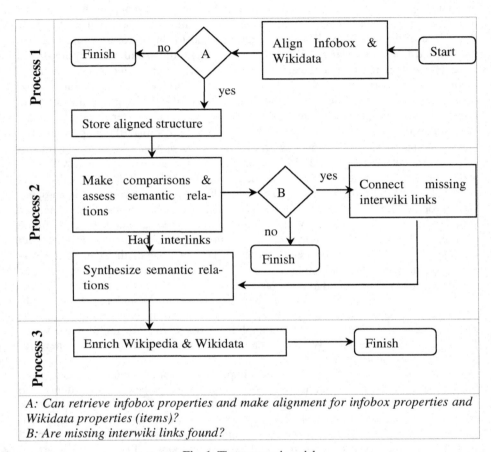

**Fig. 1.** The proposed model

- **Process 1:** This process will make property alignment between infobox and Wiki-data. If gathering inadequate results, DBPedia may be replaced for Wikidata. DBPedia may change the alignment more differences compared with Wikidata, therefore to comply with Phase 2 of Wikidata, we should use Wikidata. Then, the aligned structure is stored hidden inside infoboxes to avoid affecting their usage. Editors can modify this structure appropriately so it will support for next researches publicly. Moreover, we can make alignment between navigation templates and Wikidata to extend the aligned structure. The significant advantage of this process is to create aligned structure that can support for retrieving semantic relations easily with an uncomplicated mechanism among Wikipedias and also support for the Phase 2 development of Wikidata. The outcome will directly affect to Process 2 because it cannot operate if no aligned structures are established.

- *Process 2*: Comparisons of semantic relations and assess their correlation will be executed to detect missing interwiki links. Depend on some assessments or matching algorithms, we can conclude the interwiki links between articles of different languages and then update sitelinks at Wikidata. For articles had interwiki links, we don't have to do anything above. Next, we synthesize all semantic relations and other optional relations (categories, images, geographic coordinates, etc.) to prepare for next process.

- *Process 3*: This last process will enrich article content and Wikidata statements after implementing the comparisons of gathered semantic relations from Process 2. We also can crosscheck data (semantic relations which are mainly from infobox properties) among Wikipedia's to enrich so the anti-vandalism may be detected and prevented. We expect to enrich more data for articles which have new interwiki links in Process 2.

### 3.1   Align Infobox Parameters with Wikidata Properties

A semi-automated tool will be created to support searching and aligning the semantic equivalence between Wikidata properties (or items which has no "relevant property") and infobox properties. First of all, we choose infoboxes of non-English Wikipedias which have interwiki links with English Wikipedia. The reason why we do this because these infoboxes will tend to have more similar properties, even in different languages. Then, we get all the properties from these infoboxes. Next, with each property we search it on Wikidata to find the corresponding property or item. If we can not find anything on Wikidata, we will pass this property and mark it as a specific label "unknown". We check the alignment between Template:Infobox school in English and Wikidata which is shown in Table 1.

**Table 1.** Alignment between Template:Infobox school and Wikidata

| Properties of Template:Infobox school | Corresponding property at Wikidata |
|---|---|
| image | Property:P18 - *Image*: a relevant illustration |
| name | unknown |
| location | unknown |
| country | Property:P1 - *Country*: sovereign state of this item |
| coordinates | Property:P625 - *Coordinate location*: geocoordinates |

We can put more information in alignment process such as redirects and related templates which help in detecting missing interwiki links more effectiveness in Table 2.

**Table 2.** Infobox structure alignment of templates Bản mẫu:Trường học in Vietnamese and Template: Infobox school in English with Wikidata properties

|  | **Vietnamese** | **English** | **Wikidata** |
|---|---|---|---|
| *Template name* | Trường học | Infobox school | Q5618975 |
| *Properties* | Hình | image | Property:P18 |
|  | Tên | name | - |
|  | - | location | - |
|  | Nước | country | Property:P17 |
|  | Coor | coordinates | Property:P625 |
|  | Hiệu trưởng trường | principal | Q1056391 |
| *Redirects* | Bản mẫu:Đại học Bản mẫu:Infobox University Bản mẫu:Infobox university | Template:School Template:Infobox High-School Template:Infobox Other Education Template:Infobox Private School ... |  |
| *Related templates* | NA | NA |  |

*\* A tool will help to search similar properties on Wikidata and human decisions are made to assign which best property on Wikidata for every property of infobox.*

We mainly use human judgments in supervising the execution and making final decisions for the alignment. In this paper, we prefer to allow editors freely contribute to the aligned structure of infoboxes as the way that semantic relations are developed on Wikidata. Thus, we can utilize the community power to align more infoboxes that we are unable to implement by ourselves. Besides, the meaning of infobox properties is uncomplicated so linguistic experts are not really necessary to appraise this alignment. A problem of data exactness and data management may arise when editors contribute content but we would like to leave it for next research which offers some better solutions for improving the accuracy of alignment of properties with dictionary, WordNet, translation, NLP algorithms and assessments of linguistic experts.

If there is no alignment between Wikidata properties and infobox properties, we can use DBPedia as an optional source to make the alignment. Notwithstanding, this is not our recommendation because DBPedia will change aligned structure that is not matched with Wikidata metadata and Phase 2 of Wikidata plan. [5]

Aligned results can be stored as XML format in infoboxes between `include` tax, for example `<noinclude>aligned results</noinclude>`. This will not

---

[5] https://www.wikidata.org/w/index.php?title=Wikidata:Introduction&oldid=42871496

affect the infoboxes, which are embedded in Wikipedia articles. Like mentioned above in Process 1, these XML fragments can be reused for next research and help the infobox alignment of Wikidata at Phase 2.

**Table 3.** XML pattern of Template:Infobox school in English after alignment

| Template:Infobox school |
|---|

```
...  <noinclude><!--
<infobox lang="en" name="Template:Infobox school"
syno-nyms="" redirects="Template:School, Template:Infobox
High-School, ..." wikidata="Q5618975" relationship="">
      <properties>
          <property name="image" synonyms="portrait,
illustration, picture" wikidata="Property:P18"
descripttion="a relevant illustration" datatype="Commons
media file">
          </property>
          ...
      </properties>
</infobox>
--></noinclude>
```

### 3.2    Detect Missing Interwiki Links and Connect Them to Articles in Different Languages as well as Synthesize All Semantic Relations

We define two types of semantic relations:

- *Semantic Relations Based on Article Structure*: semantic relations are retrieved from redirects, categories, external links, internal links, images, videos, audios of articles. These semantic relations can be represented in RDF triples which are not always found in DBPedia because of its insufficient support and low update frequency. When there are no semantic relations on DBPedia, we will create these by ourselves. The simple solution is to use a bot to get semantic relations from article content throughout APIs of Wikipedia.[6]
- *Semantic Relations from Infobox and Navigation Templates*: We retrieve semantic relations from infoboxes or any templates that have structured metadata. RDF triples will be set up from these semantic relations. When infoboxes regularly summarize the information of articles, these semantic relations are helpful for detecting interwiki links.

---

[6] https://www.mediawiki.org/wiki/API:Main_page

The sample articles will be classified into two groups: non Latin-based alphabet Wikipedias and Latin-based alphabet Wikipedias. We prefer to focus on the latter. As stated in the introduction section, English Wikipedia has a high collaborative quality. It may be a valuable source for identifying interwiki links with other Wikipedias. Likewise, any Wikipedias have high collaborative quality such as German Wikipedia, French Wikipedia and Spanish Wikipedia will be considered as sources to find interlinks. In this paper, we want to compare articles of all Wikipedias with English articles to seek interwiki links.

Supposed that to detect interwiki link for an unlinked article in Vietnamese, firstly we should have a look at the article. We must understand the article content and search the relevant articles in English by some defined keywords. If we find a needed article, we will connect it to Vietnamese article. This task requires the understanding of English, Vietnamese and knowledge about that article. However, we try to make this task simpler that machine can comprehend when we exclude human, translation and NLP approaches to find the similarities among articles in various languages. Instead, we use article name and its redirects. There is a huge tendency to use the same or nearly same article name in the Latin-based alphabet Wikipedias. This case may only correct for article about cities, people, biological species, proper nouns, acronyms, etc. Additionally, there are a lot of articles being translated from English to non-English languages. This reduces user's efforts in building and developing articles from the beginning. Therefore, it is easier to identify a certain article name of Latin-based alphabet Wikipedias which has or does not have in the English. In contrast, it is totally difficult for recognizing an article of non Latin-based alphabet Wikipedias, which has its version in English or not because of the different alphabets.

**Comparison List**

The most difficult thing is to search an article A in language A has interwiki link with which article in language B. To do that, we have to create a comparison list (candidate articles) of language B to which the article A will compare. Supposed that language B has 4 million articles, it is not feasible to a execute linear algorithm to match A with 4 million articles of B.

From the difficulty above, we must reduce the size of comparison list. Normally, when we search for an object, we always use its name as the first criterion in searching. In this case, article name and its redirects can be used to define the comparison list.

**Table 4.** Article "Chó" in Vietnamese and "Dog" in English

| Vietnamese | English |
| --- | --- |
| Article name: **Chó** | Article name: **Dog** |
| Redirects: Con chó, *Canis lupus familiaris*, Cún, Chú chó, Chó nhà, cẩu | Redirects: Canis familiaris, Dogs, *Canis lupus familiaris,* Canis Canis, |

In Table 4, with the article name "Chó" in Vietnamese, we can never find any article which has the same name in English because of the language differences. However, if we search by redirects we realize that "Chó" article in Vietnamese may have a relationship with "Dog" article in English because the two contain redirect "Canis lupus familiaris" which is a dog's scientific name. Creating a comparison list from searching by name and its redirects can be used for Latin-based alphabet Wikipedias which have many resemblances of usage article name and redirects to get more benefits. This method typically provides one article in the comparison list. It can reduce the compared times, but may affect the outcome when there are no matching results are found or the comparison list is empty. Thus, in our future researches, we will apply many methods which can detect and compare the similarities of using images, videos, categories, internal links and semantic structure of certain articles.

**Semantic Relations**

*Semantic Relations Based on Article Structure:* Besides article name, an article must be built a semantic structure which machines can understand when they automatically execute matching processes. The simplest structure is to organize an article by its relationships of categories, images, terms, templates and others. For example, in Vietnamese Wikipedia, "Alcina" article does not have interwiki link. By reading its source code (Wiki markup), we form its structure. In Table 5, we use sitelinks of Wikidata to translate terms from Vietnam into English.

**Table 5.** The semantic structure of Alcina article on Vietnamese Wikipedia in English

| **Alcina** (Vietnamese Wikipedia) | |
|---|---|
| *Term:* link-to | [[opera]], [[composer]], [[George Frideric Handel]], [[human]], [[**anh hùng ca**]] (epic), [[Riccardo Broschi]], [[**Orlando Furioso**]], [[Ludovico Ariosto]], [[London]], [[England]], [[1735]] |
| *Category:* has-category | [[Category:Opera]] |
| *Template:* has-template | {{Reflist}} |
| *Image:* has-image | George Frideric Handel by Balthasar Denner.jpg |
| * Note: terms in bold do not have their own articles or interlinks | |

Then, from Table 5, we also form a graph in Fig. 2. This graph will remove all terms which could not be translated into English.

In Fig. 2, these semantic relations can be seen as weak relations because Wikipedia article content depends on user contributions. So, different articles in different language may form different semantic structures. That is a crucial reason why we can not use these structures for detecting interlinks. Our first idea is to compare these structures and make conclusions that interwiki links may exist among articles. However, we can use this structure to support the assessment of detecting missing interwiki links of articles in next section.

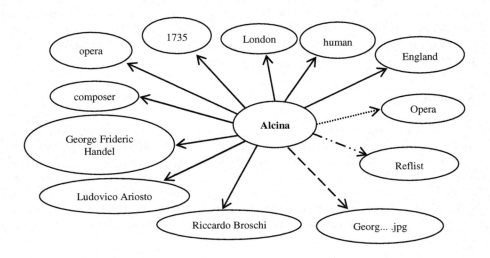

**Fig. 2.** The semantic structure of Alcina article on Vietnamese Wikipedia in English

*Semantic Relations from Infobox and Navigation Templates:* In this section, we will primarily retrieve semantic relations from infoboxes and navigation templates which were embed in articles. Other templates may be used if they serve some good semantic relations. All articles will be scanned in order to choose ones that contained infoboxes with their aligned structure in Process 1. For articles had interwiki links with English Wikipedia, we just retrieve the semantic relations from the infobox properties. For others, we detect the missing interwiki links, connect these links and then also synthesize the semantic relations. For example, Chó article in Vietnamese does not have interwiki link with English Wikipedia. After searching by its name, we can find the candidate Dog article in English (Table 4). Then, a bot will read the content of two articles and collect semantic relations from Template:Taxobox which is aligned as Section 3.1.

**Table 6.** Semantic relations of Chó (vi) & Dog (en) articles

| Chó (Vietnamese) | Dog (English) |
|---|---|
| *Language*: vi | *Language*: en |
| *Page_name*: Chó | *Page_name*: Dog |
| *Redirects*: Con chó, Canis lupus familiaris, Cún, Chú chó, Chó nhà, Cẩu, Chó | *Redirects*: Canis familiaris, Dogs, Canis lupus familiaris, Canis Canis, Domestic Dog, A man's best friend, Doggy, Dog (Domestic), Dog groups, Dogs as our pets, Dog |
| *Name*: | *Name*: Domestic dog |
| *Type*: species | *Type*: species |
| *Regnum*: Động vật | *Regnum*: Animalia |
| *Ordo*: Bộ Ăn thịt | *Ordo*: Carnivora |
| *Familia*: Họ Chó | *Familia*: Canidae |
| *Genus*: Chi Chó | *Genus*: Canis |
| *Species*: Sói xám | *Species*: Gray Wolf |
| *Binomial*: | *Binomial*: |
| *Binomial_authority*: | *Binomial_authority*: |
| *Synonyms*: | *Synonyms*: |
| *Wikidata: Q144* | *Wikidata: Q144* |

We can establish an assessment to compare semantic relations of two articles. In the Table 6, we will compare two articles, Chó in Vietnamese and Dog in English which are biological species. We set up our own assessment by comparing some semantic relations which are Regnum, Ordo, Familia, Genus and Species. The result details are shown in Table 7.

**Table 7.** The comparison result between Chó (Vietnamese) and Dog (English) is done by bot

```
PAGE      vi:Chó --- en:Dog
Type:     species
RESULT    Score: 5/5 --- Percentage: 100%
DETAIL
Species (OK)      vi:Sói xám      <-----> en:Gray Wolf
Genus   (OK)      vi:Chi Chó      <-----> en:Canis
Ordo    (OK)      vi:Bộ Ăn thịt <-----> en:Carnivora
Familia (OK)      vi:Họ Chó       <-----> en:Canidae
Regnum  (OK)      vi:Động vật   <-----> en:Animalia
```

If all these semantic relations are matched, we can hypothesize that these two articles may have an interwiki link. After that, a bot will automatically connect them together by adding sitelinks on Wikidata or sets an alert template which notices editors and let them make final decisions. This section does not mention about using a fixed assessment for all articles which are identified by short abstract (may refer to a type). Different articles can have different types and therefore they will have different assessments based on gathered semantic relations and how we apply the proper assessments. For example, there are two articles in two languages contain Template: Infobox person, we need an assessment with enough semantic relations to prove these two articles are talking about the same person. In some cases, two people have the same name, same birthday, same nationality, same sex, but we cannot conclude that they are the same person. Lastly, we will aggregate all semantic relations are found for the next process.

### 3.3    Enrich Semantic Relations for Articles and Wikidata Statements

With the semantic relations from Section 3.2, we will enrich article infoboxes and Wikidata statements by comparing semantic relations of different articles of Wikipedias. For example, in Table 6, supposed that we have `Binomial_authority` property has a value is Carl Linnæus in English, we can update the `Binomial_authority property value` in Vietnamese if it does not exist. Then, if Wikidata item `Q144` lacks statement `taxon author`, we can insert it with value Carl Linnæus.

We also can enrich other data, such as categories, external link section, gallery section, images, etc. For categories, we can create new ones from basic NLP patterns (Table 8) depend on existing English categories and classify the articles into them based on English classification if needed. Our purpose is to create a category taxonomy of small-scale Wikipedias and category classification system more fine-grained.

**Table 8.** Some Category NLP patterns in English [6] and Vietnamese

| English | Vietnamese |
|---------|------------|
| *[X] in [Y]* | *[X] ở [Y]* |
| Cities in France | Các thành phố ở Pháp |
| X= Cities (plural) | X = Các thành phố |
| X1 = City | X1 = Thành phố |
| Y = France | Y = Pháp |
| *XY* | *YX* |
| Information Technology | Công nghệ thông tin |
| X = Information | X = Thông tin |
| Y = Technology | Y = Công nghệ |
| *X by Y* | *X theo Y* |
| Culture by nationality | Văn hóa theo quốc tịch |
| X = Culture | X = Văn hóa |
| Y = Nationality | Y = Quốc tịch |

## 4    Case Study

Recently, Vietnamese Wikipedia crossed over 1 million articles with many thousand biological articles were mainly created by bot. These stub articles miss interlinks because bot generate them automatically from external databases. Furthermore, many local editors did not pay attention to enrich these boring articles. Therefore, we need to find a solution to solve the problem. One of the feasible solutions is we can enrich these articles from other Wikipedias, for example English Wikipedia. To do so, firstly, we need to connect these articles to English articles. This case study only applies this first step. We use Process 1 and some parts of Process 2 of the proposed model for the implementation. We do not enrich the article content, but try to link these articles to English to know how many interlinks we can connect and conjecture the content enrichment. We will choose biological articles which contain Template:Taxobox and have no interwiki links[7] to English Wikipedia.

The biological classification for infoboxes of Wikipedia articles is mainly complied to ICZN[8] and ICN[9] standards. Firstly, we align Bản mẫu:Bảng phân loại in Vietnamese and Template:Taxobox in English with Wikidata properties or items. Template:Taxobox has relationships with Template:Automatic taxobox and Template:Speciesbox so we also align these two templates as well.

**Table 9.** Alignment of Bản mẫu:Bảng phân loại (vi) and Template:Taxobox with Wikidata

| | **Vietnamese** | **English** | **Wikidata** |
|---|---|---|---|
| **Template name** | Bảng phân loại | Taxobox | Q52496 |
| **Properties** | status_system | status_system | Property:P141 |
| | image, hình | image | Property:P18 |
| | range_map | range_map | Property:P181 |
| | binomial | binomial | Property:P225 |
| | species, loài | species | Q7432 |
| | genus, chi | genus | Q34740 |
| | familia, họ | familia | Q35409 |
| | ordo, bộ | ordo | Q10861678 |
| | class, lớp | class | Q37517 |
| | regnum | regnum | Q36732 |
| | domain | domain | Q146481 |
| | … | … | … |
| **Redirects** | Bản mẫu:Phân loại khoa học, Bản mẫu:Taxobox | Wikipedia:TX, Wikipedia:TAXOBOX, … | |
| **Related templates** | NA | Template:Automatic taxobox Template:Speciesbox | |

Next, we search for English articles with Vietnamese articles corresponding by article name and its redirects and then make comparisons between them. With one Vietnamese article, we choose one English article that has the same article name or redirects and make comparisons of this couple. The result of each comparison is similar to Table 6 and Table 7 above. Our assessment likes *Semantic relations from infobox and navigation templates* part of Section 3.2. We choose five compared semantic relations: regnum, class, ordo, familia, genus for the alignment. Our mandatory conditions are two biological objects must be the same name (or matching any scientific name, binomial) and the same object type (species, genus, ordo, familia) before comparing. Then, we calculate the result of each couple by a matching percentage of

---

[7] https://vi.wikipedia.org/w/index.php?title=Đặc biệt:Không liên wiki_wiki&limit=500&offset=0

[8] http://iczn.org/iczn/index.jsp

[9] http://www.iapt-taxon.org/nomen/main.php

semantic relations. We will not solve any article which is related to Monospecificity (monospecies, monogenus, etc.). There are some cases that English articles have interwiki links with other Vietnamese articles, but not with the comparing articles. So we need to insert `Template: Merge` into these Vietnamese articles to merge these articles each other. Besides, we also have a bit problem with compare property values of binomial because of the plurality of binomial value and naming convention.

**Table 10.** Results of comparing article couples in Vietnamese and English

| No. | No. Random Couples | Manual Matching | >=80% Matching | =100% Matching | Merge needed | New interlinks |
|---|---|---|---|---|---|---|
| 1 | 100 | 80 | 77 | 64 | 37 | 40 |
| 2 | 100 | 84 | 83 | 67 | 40 | 43 |
| 3 | 100 | 77 | 76 | 67 | 31 | 45 |
| 4 | 100 | 78 | 76 | 58 | 32 | 44 |
| *Mean* | | *79.75%* | *78%* | *64%* | *35%* | *43%* |
| | | | | | | |
| **5** | 200 | 165 | 163 | 120 | 98 | 65 |
| 6 | 200 | 164 | 156 | 118 | 89 | 67 |
| 7 | 200 | 155 | 149 | 119 | 81 | 68 |
| 8 | 200 | 160 | 158 | 130 | 85 | 73 |
| *Mean* | | *80.5%* | *78.25%* | *60.88%* | *44.13%* | *34.13%* |
| | | | | | | |
| 9 | 1000 | 819 *(81.9%)* | 788 *(78.8%)* | 575 *(57.5%)* | 463 *(46.3%)* | 325 *(32.5%)* |

In Table 10, we executed the comparisons: 4 times with 100 random couples, 4 times with 200 random couples and 1 time with 1000 random couples. We received the result of higher-and-equal-80%-matching which is not much different from the manual method. The matching percent can be higher a bit because we removed the articles which are related to Monospecificity. We realized that a large number of couples need to be merged which could be from the mistakes of bots and editors. This helped to reduce the repetitive of articles. The new interlinks we found in this case study around 30%-40%, which showed that there are still many articles that lack of interlinks in biology articles. To connect the interwiki links of articles, we will set a suggested template into these articles and may let the judgments for the editor community. However, bot can automatically connect interwiki links for the articles which have higher-and-equal-80%-matching. The next step is to retrieve as much as possible semantic relations which can help to enrich the article content. In this case study, the machine can easily detect the missing interwiki links among articles because of the similarities of using infobox format and article names as well as redirects of Latin-based alphabet Wikipedias.

# 5    Conclusion

Our proposed model is a new approach which based on the alignment between Wikipedia infoboxes and Wikidata to enrich the articles of different Wikipedias as well as Wikidata. This model provides many aligned structures of infoboxes which can be as a good source for many researchers when retrieving the semantic relations openly and independently. According to Phase 2 of Wikidata plan, we believe that these aligned structures may help Wikidata developers in unifying the infoboxes of all languages. In this model, we can utilize the community power in property alignment which DBPedia inhibited. Nevertheless, our model is in the development stage which may not support the content enrichment process completely. In Process 1, we should use some translation tools and parsers to improve the property alignment. Furthermore, we need more algorithms to evaluate the correlation of properties more exactness and inherit other previous researches to widen the alignment property database. Creating a comparison list by searching the article name is still not the best solution to detect missing interlinks. Thus, we will compare semantic structures and other data of articles in this task. In the case study, we conclude that our model can work well with the biological articles of Latin-based Wikipedias. However, to apply to other domains effectively, many efforts needed to be made to improve our model. That is the reason why we will build more assessments for different article domains in Process 2. The enrichment process depends on the gathered semantic relations. We may use external links and article references to enrich more data to article content. In short, we expect this paper will open up to many researches about the correlation between Wikidata and Wikipedia.

# References

1. Mendes, P.N., Jakob, M., Bizer, C.: In: DBPedia: A Multilingual Cross-Domain Knowledge Base. LREC, pp. 1813–1817. European Language Resources Association ELRA (2012)
2. Cabrio, E., Cojan, J., Gandon, F., Hallili, A.: Querying multilingual DBpedia with QAKiS. In: Cimiano, P., Fernández, M., Lopez, V., Schlobach, S., Völker, J. (eds.) ESWC 2013. LNCS, vol. 7955, pp. 194–198. Springer, Heidelberg (2013)
3. Morsey, M., Lehmann, J., Auer, S., Stadler, C., Hellman, S.: DBPedia and the Live Extraction of Structured Data from Wikipedia. Electronic Library and Information Systems 46(2), 157–181 (2012)
4. Kim, E.-K., Weild, M., Choi, K.-S.: Metadata synchronization between bilingual resources: case study in wikipedia. In: International Workshop on the Multilingual Semantic Web (2010)
5. Vrandečić, D., Krötzsch, M.: Wikidata: A Free Collaborative Knowledge Base. Communications of the ACM (2014, to appear)
6. Nastase, V., Strube, M.: Decoding Wikipedia Categories for Knowledge Acquisition. AAAI (2008)
7. Lehmann, J., Isele, R., Jakob, M., Jentzsch, A., Kontokostas, D., Mendes, P.N., Hellmann, S., Morsey, M., van Kleef, P., Auer, S., Bizer, C.: DBpedia - A Large-scale, Multilingual Knowledge Base Extracted from Wikipedia. IOS Press (2012)

8. Nguyen, T., Moreira, V., Nguyen, H., Nguyen, H., Freire, J.: Multilingual schema matching for Wikipedia infoboxes. Proceedings of the VLDB Endowment **5**(2), 133–144 (2011)

9. Palmero Aprosio, A., Giuliano, C., Lavelli, A.: Towards an automatic creation of localized versions of DBpedia. In: Alani, H., Kagal, L., Fokoue, A., Groth, P., Biemann, C., Parreira, J.X., Aroyo, L., Noy, N., Welty, C., Janowicz, K. (eds.) ISWC 2013, Part I. LNCS, vol. 8218, pp. 494–509. Springer, Heidelberg (2013)

10. Ponzetto, S.P., Navigli, R.: Large scale taxonomy mapping for restructuring and integrating wikipedia. In: IJCAT 2009 (2009)

11. Chernov, S., Iofciu, T., Nejdl, W., Zhou, X.: Extracting semantic relationships between wikipedia categories. In: ESWC, June 2006

12. Adar, E., Skinner, M., Weld, D.S.: Information arbitrage across multi-lingual wikipedia. In: WSDM 2009 Proceedings of the Second ACM International Conference on Web Search and Data Mining, pp. 94–103 (2009)

# Template-Driven Semantic Parsing
# for Focused Web Crawler

Michał Blinkiewicz$^{(\boxtimes)}$, Mariusz Galler, and Andrzej Szwabe

Institute of Control and Information Engineering, Poznan University of Technology,
M. Sklodowska-Curie Sq. 5, 60-965 Poznan, Poland
{michal.blinkiewicz,mariusz.galler,andrzej.szwabe}@put.poznan.pl

**Abstract.** We present Template-Driven Semantic Parser (TDSP) capable to represent, at least to some degree, the semantics of Web pages being processed. Data extraction process realized by means of TDSP is driven by a set of instructions stored in an easily modifiable XML-based template. In order to enhance the precision of Web page data extraction, the TDSP template format allows to use a specialized Expression Language (EL). The template may be easily created and modified using a tool called Visual Template Designer. TDSP provides an output document containing an RDF graph composed of triples that represent the website resources under exploration. In accordance to the Semantic Web paradigm, each resource has its semantics assigned and is connected to other resources by means of one or many relations. The semantic types of the resources and the relations between them are predefined in an ontology of Web artifacts.

**Keywords:** Template · Parsing · Focused Web Crawler · Semantic Web · Expression Language

## 1 Introduction

The Web crawler is usually understood as an application that automatically browses WWW, typically for the purpose of Web content indexing. More specifically, a focused crawler visits only pages that contain information associated with a predefined set of subjects [5]. The subjects may be defined in many ways, e.g., using keywords, examples of relevant documents or even RDF triples [7].

The crawling process consists mainly of fetching Web pages, extracting relevant information and collecting links to other potentially relevant websites [8]. Extraction of information from Web pages is performed by a component referred to as a parser.

Very few data extraction tools maintain the semantics of the extracted data. This is an important issue considering the fact that most of existing Web pages do not contain any semantic metadata. Usually the raw data retrieved from Web pages cannot be easily accessed and manipulated by a data extracting machine, partly because modern Web pages contain many elements responsible for data presentation [10]. Taking into account the size of the Internet and the pace of

© Springer International Publishing Switzerland 2015
T. Supnithi et al.(Eds.): JIST 2014, LNCS 8943, pp. 351–358, 2015.
DOI: 10.1007/978-3-319-15615-6_26

changes in this area, one should take into consideration that such situation may not change in the near future.

This paper presents Template-Driven Semantic Parser (TDSP) which is capable to provide the semantics of extracted Web data in the RDF format. TDSP is the main parser of CAT Focused Crawler (CATFC) system. The TDSP is very versatile — it can process practically any Web page as long as it has the form of a well-formatted XML document and an appropriate template has been provided by the user. Based on the instructions from the template the parser is able to extract data from selected page elements together with the related URL string.

The word "semantic", appearing in the name of presented parser, should be understood in the reference to the parser's compatibility with the Semantic Web data representation standards. More specifically, TDSP allows to represent the semantics of extracted data at the lowest, technological-only level, by referring to classes and properties defined in a domain-specific ontology. The ontology that supports the process of Web data harvesting realized by means of TDSP includes definitions of Web artifacts such as a discussion list post, user nickname, list. The TDSP template enables the determination of the semantics for data that originally was not accompanied by any semantic metadata.

## 2   Related Work

One of the basic features of TDSP is its versatility understood as the ability to parse HTML pages of diverse structure. In order to let the parser "know" how to process the particular page, it has to be provided with an appropriate template. The template-based approach to the design of TDSP is in line with the state-of-the-art data extractors controlled by templates or configuration files.

The first of them is OXPath [4] which is an extension of XPath that facilitates data extraction from the Web. While XPath allows to extract data only from an HTML document, OXPath allows to navigate between several HTML pages, execute some actions and aggregate the extracted information. Furthermore, OXPath enables information selection in accordance to visual features, e.g., text color.

Another solution is Web-Harvest (WH) [1]. WH is primarily used for parsing HTML/XML based Web pages and it provides techniques and technologies for XML manipulation, i.e., XQuery and XSLT. Furthermore, WH allows to process and modify the extracted data by means of Java-based scripting languages and regular expressions.

Unfortunately none of the above-mentioned solutions meets all the requirements with regard to the main parser of CATFC. Moreover, as the decision of creating a new tool for data extraction was made by the authors of TDSP, the issue of an appropriate template format selection still remains open. As a consequence, similarly like in the case of the referred parsing tools, the template format of TDSP was created without using any ready-made solutions. The reasons of such decision are described in Section 4.1.

An interesting template format that served as an inspiration for the work on TDSP template format is Website Parse Template (WPT). WPT [3,9], developed by OMFICA, is an open format which provides an HTML Web pages structure and content description. WPT is based on an XML and is consistent with the Semantic Web concept specified by W3C.

## 3   TDSP Design

TDSP consists of several components. Modification, extension and maintenance of a multi-component parser is much easier than in the case of applications with a single component structure.

TDSP OSGi Service implements methods of the *ParserService* interface which is an entry point to the parser. One of its task is to inform the Parser Manager module of parsing capabilities with regard to particular Web pages. Moreover, the component is responsible for starting parsing process and passing the parsing result back to the Parser Manager.

Template Manager is responsible for the TDSP parsing templates management. When the parser starts, all correct templates contained in CATFC database are loaded into the *Template Manager*. The correctness of a template is verified using a Relax NG [11] schema. The main task of *Template Manager* during the parsing process is to provide the appropriate template for each particular Web page.

Template Processor is the core component of TDSP. Its input is the fetched Web page and the template. Based on the template and the results of the *Expression Language (EL) Interpreter* work, *Template Processor* extracts the selected data from the Web page content in the form of RDF triples and composes the RDF graph.

EL Interpreter, at the request of the *Template Processor* component, interprets the EL expressions occurring in the template instructions.

### 3.1   Template Structure

The TDSP template format [2] is based on the XML document format. Each TDSP template consists of multiple instructions. Every such instruction is represented by a single element of the XML document. Instructions comprise of a specific set of parameters that control the parser operation. An XML attribute of the XML element represents the single template instruction parameter. Every parameter value is in the form of an EL expression.

It follows that the template consists of two layers. The first are instructions while the second are EL expressions. Instructions are responsible for ensuring semantic information interconnections and template related moves. Moreover, there are few instructions responsible for looping and conditions. On the other hand expressions are used to Web page data manipulation and extraction [2].

## 3.2    Template Expression Language

The process of data extraction from Web pages which in general are XML documents is not a simple task without specialized tools deployed. There are known solutions of this kind publicly available, e.g., XPath or XQuery. Based on these technologies and in accordance to TDSP needs, a new EL extending TDSP templates' capabilities has been designed and developed.

A valid expression in EL consists of *text* and *command* parts interspersed. There is also a possibility of *commands* recursive nesting. The *text* part is an ordinary string of characters but it cannot contain any *special characters* ('$', '!', '\', '{', '}'). Use of *special characters* in the *text* part has to be preceded by a backslash character ('\'). The *command* part consists of a dollar character, a command name and command arguments — each enclosed in curly brackets.

There are no special commands arguments types defined. Every argument is treated as an ordinary text string. However, each *command* may require the argument text string to be in some specific format, e.g., an XPath expression, a regular expression or a list of indices. Consequently, every *command* returns an ordinary text string.

Due to the application context of CATFC, a set of the most useful predefined EL commands consists of commands related to a Web page data extraction, commands that allow to transform a string of characters and commands responsible for data conversion (hash function, date converter, identifier generator, etc.) [2].

In order to illustrate EL syntax and returned results, the following examples are prepared. Each template instruction parameter is an EL expression. Thus, a string like "`topicID`" is an expression containing only a *text* part passed to the output without any changes. A more complex expression is:

```
http://www.example.com/topic/$u{topic=(t[0-9]+)}{1}
```

It is the result of the concatenation of a *text* part and a *command* part. Like in the case of the first expression, the *text* part remains unchanged while the *command* is being evaluated to "t001". The text string obtained as a result is:

```
http://www.example.com/topic/t001
```

Even more complicated expressions may contain nested commands. An example of such an expression is as follows:

```
http://www.example.com/user/$ue{$x{./div[1]}{Author: (.*)}{1}}
```

Similarly to previous examples, the beginning of this expression remains unchanged. The next part is an `urlencode` *command* with a nested `xpath` *command*. The inner *command* (`xpath`) uses a dot symbol indicating the use of a "base node" set by the `for-node` loop. The loop iterating over XML nodes sets the correct "base node" in every iteration. Therefore, the output is:

```
http://www.example.com/user/Michal%20B
```
or `http://www.example.com/user/Mariusz%20G`
or `http://www.example.com/user/Andrzej%20Sz`

The TDSP parsing process is elaborated in [2].

## 3.3    Implementation

The TDSP is written in Java. The modular architecture of CATFC is based on an OSGi framework. Thus, TDSP similarly to other modules of CATFC is implemented in the form of an OSGi bundle. After being deployed to the Glassfish application server and registered as the OSGi service, TDSP may be immediately used by the *Parser Manager* module.

EL expressions are described in the Section 3.2. Nevertheless, it is worth to explain how *EL Interpreter* module is designed. The *EL Interpreter* is implemented using the *JParsec* library. The library allows the construction of recursive text parsers of grammar defined in the native Java language. A single EL command is defined as a regular Java class method. Definitions of the class and the method must be successively preceded by dedicated annotations: *@CommandDefinitions* for the class and *@Command* for the method. These annotations, used together with an annotation processor, provide the means for the registration of the command definition by the *EL Interpreter*. The use of Java annotations allows adding new commands in an easy way.

While the main tool for selecting the appropriate information in XML documents is XPath, an important issue was to make this process fast. The current implementation of XPath Evaluator is based on a Cached XPath API provided by the Xalan library. The Cached XPath API is used instead of the default Java XPath API implementation because of the caching feature which makes the XPath look-up process much faster than when the commonly used XPath API is used.

Jena framework has been used to store the extracted data as RDF triples. It is an open source solution that provides a versatile API for management of RDF graphs.

## 4    Evaluation

This section presents an evaluation of TDSP to the selected existing solutions mentioned in the Section 2. As TDSP is a data extraction tool — i.e., it does not perform any other related tasks like Web pages fetching — the scope of the comparisons has been strictly limited to data extraction functions. The systems have been compared from two perspectives: qualitative (Section 4.1) and quantitative (Section 4.2). The first one concerns the template formats: the TDSP template format has been compared with the template formats used in case of the existing solutions. The second perspective of the analysis is reflected by processing times measurements.

### 4.1    Qualitative Evaluation

Functions offered by the template format are usually the biggest strength of the particular Web parser. The first of worth noting features is the *availability of a visual template designer*. All of the discussed template formats provides a dedicated tool for this purpose.

Each of the evaluated systems allows to use ontologies while creating RDF documents. This means that all of the template formats under the comparison gives the ability to express the semantics of extracted data.

The next examined feature is *native support for RDF creation*. Only WH does not provide an immediate support for RDF output. In the case of other solutions included in presented comparison the structure of the RDF document can be created in a simple way by using visual generators. For this purpose the TDSP format contains dedicated instructions for RDF triples creation, i.e., *resource, property, literal* [2].

The last aspect of the comparison is the *availability of custom scripting languages* that provide the ability to manipulate extracted data. WPT and OXPath do not have such ability. The TDSP template format has build-in EL of capabilities described in Section 3.2. Similarly to TDSP, WH offers the possibility to use a few scripting languages, i.e., BeanShell, Groovy and Javascript.

### 4.2   Quantitative Evaluation

In this section a quantitative evaluation of TDSP against the WH has been presented. WPT was not included in the set of evaluated systems, as it is a template format, for which a parser implementation is not available. OXPath has been also omitted in the presented comparison due to the fact that the measurement of the duration of the data extraction process would cause an interference with the source code of OXPath.

All the experiments have been conducted on the same machine equipped with an Intel Core i5-430M processor working at 2.27 GHz, with 4 GB of RAM, running Linux (kernel 3.11) and Oracle Java Development Kit 7u45.

**Methodology** The HTML page that was used — to perform the experiments — is available as a file on the WH official website [1]. A TDSP template and a WH configuration file for the selected Web page have been prepared. Both TDSP and WH were processing the data contained on the Web page into an RDF/XML document. Thus, in case of each of the compared systems the output document had the same structure and the same content. For both the tools the measurement of the data extraction task duration was repeated 50 times. Results of the measurements are shown in the Figures 1 and 2.

**Results** Figure 1 presents the results of 50 consecutive measurements of Web page processing tasks duration. The results of the first measurements for TDSP and WH, differ significantly from each other. Such behavior may be explained by a reference to the specification of the Java Virtual Machine (JVM). During the first invocation of tools all required classes are initialized by the JVM [6] what results in an additional time overhead. Other time measurements are much more similar to one another and maintain a decreasing trend. Figure 2 is a kind of approximation of the last 25 measurements from the Figure 1. As it may be seen on the two diagrams, both the approaches result in a comparable processing time, with a little predominance on the favor of WH. A confirmation of this observation may be the average processing times for both tools — 51 miliseconds for WH and 60 miliseconds for TDSP. Based on the results, we claim that the processing time of TDSP does not differ significantly from this of the leading Web parser referenced in the relevant literature, which is WH.

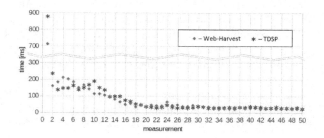

**Fig. 1.** 50 subsequent measurements

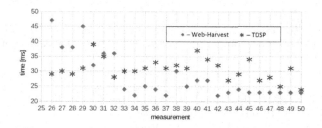

**Fig. 2.** Selected measurements from 26 to 50

## 5    Conclusions

To the best of our knowledge, Template-Driven Semantic Parser is the first parser that is capable to semantically annotate extracted Web data according to a template constructed specifically for ontology-defined RDF output creation and enriched by a specialized Expression Language enabling data manipulation and post-processing.

Moreover, as demonstrated in Section 4.2, the support for RDF as a Web parser's native output format does not necessarily leads to a significant increase of the parsing time. It is important to note that the lack of the native RDF output format support would degrade the usefulness of visually-aided template creation tools because using such tools would require manual RDF triples formatting.

# 6   Future Work

There are at least a few directions of the further development of TDSP template format or EL. TDSP may be transformed into an independent tool operating outside CATFC. However, such a redesign would require introducing an interface providing a remote access to the parser functionality, e.g., in the form of a Web service. Another possible way of increasing the parser capabilities would be to extend the set of the predefined EL commands. A construction of the *EL Interpreter* that is responsible for processing the special expressions makes that new commands could be added in a simple way. One may also consider modifying the module responsible for evaluating expressions of an XPath language.

**Acknowledgments.** This work is supported by research project O ROB 0025 01 financed by The National Center for Research and Development, and by Poznan University of Technology under grants 04/45/DSPB/0122 and 04/45/DSMK/0124.

# References

1. Web-Harvest Version 2.0. http://web-harvest.sourceforge.net/ (accessed June 20, 2013)
2. Blinkiewicz, M., Galler, M., Szwabe, A.: Template-Driven Semantic Parser as a Focused Web Crawler Module. Tech. Rep. 660, Poznan University of Technology (2014)
3. Chan, K., Kwok, L.: Assisting the authoring process of IMS-LD using web parsing technique. In: Luo, X., Spaniol, M., Wang, L., Li, Q., Nejdl, W., Zhang, W. (eds.) ICWL 2010. LNCS, vol. 6483, pp. 21–30. Springer, Heidelberg (2010). http://dx.doi.org/10.1007/978-3-642-17407-0_3
4. Furche, T., Grasso, G., Schallhart, C.: Effective web scraping with OXPath. In: WWW (developer track), pp. 23–26 (2013)
5. Gandhi, C., Singh, M.: Crawlers: A Review. The Journal of Computer Science and Information Technology 4(1), 47–49
6. Gosling, J., Joy, B., Steele, G., Bracha, G., Buckley, A.: The Java Language Specification. http://docs.oracle.com/javase/specs/jls/se7/html/ (accessed December 09, 2013)
7. Hitzler, P., Krotzsch, M., Rudolph, S.: Foundations of semantic web technologies, chap. 2. Chapman and Hall/CRC (2011)
8. Kumar, M.S., Neelima, P.: Article: Design and Implementation of Scalable, Fully Distributed Web Crawler for a Web Search Engine. International Journal of Computer Applications 15(7), 8–13 (2011). Published by Foundation of Computer Science
9. Manukyan, A., Manukyan, A., Mayilyan, A., Sayadyan, A.: Website Parse Templates, April 2008. http://tools.ietf.org/html/draft-manukyan-website-parse-templates-00
10. van Ossenbruggen, J.R., Hardman, H.L., Rutledge, L.: Hypermedia And the Semantic Web: A Research Agenda, vol. 3 (2002)
11. van der Vlist, E.: RELAX NG. O'Reilly Media (2003). http://www.google.pl/books?id=A3a-Tw2PQZsC

# Search and Querying

# A Keyword Exploration
# for Retrieval from Biomimetics Databases

Kouji Kozaki[1(✉)] and Riichiro Mizoguchi[2]

[1] The Institute of Scientific and Industrial Research, Osaka University,
8-1 Mihogaoka, Ibaraki, Osaka 567-0047, Japan
kozaki@ei.sanken.osaka-u.ac.jp
[2] Japan Advanced Institute of Science and Technology, 1-1 Asahidai, Nomi,
Ishikawa 923-1292, Japan
mizo@jaist.ac.jp

**Abstract.** Biomimetics contributes to innovative engineering by imitating the models, systems, and elements of nature. Biomimetics research requires the development of a biomimetics database including widely varied knowledge across different domains. Interoperability of knowledge among those domains is necessary to create such a database. Ontologies clarify concepts that appear in target domains and help to improve interoperability. Furthermore, linked data technologies are very effective for integrating a database with existing biological diversity databases. In this paper, we propose a keyword exploration technique to find appropriate keywords for retrieving meaningful knowledge from various biomimetics databases. Such a technique could support idea creation by users based on a biomimetics ontology. This paper shows a prototype of the biomimetics ontology and keyword exploration tool.

**Keywords:** Biomimetics · Biological diversity · Ontology · Linked data · Keyword exploration for retrieval

## 1    Introduction

Learning from nature aids development of technologies. Awareness of this fact has been increasing, and biomimetics[1] [1, 2], innovative engineering through imitation of the models, systems, and elements of nature, has caught the attention of many people. Well-known examples of biomimetics include paint and cleaning technologies that imitate the water repellency of the lotus, adhesive tapes that imitate the adhesiveness of gecko feet, and high-speed swimsuits that imitate the low resistance of a shark's skin. These results integrate studies on the biological mechanisms of organisms with engineering technologies to develop new materials.

Facilitating such biomimetics-based innovations requires integrating knowledge, data, requirements, and viewpoints across different domains. Researchers and engineers need to develop a biomimetics database to assist them in achieving this goal.

---

[1]  http://www.cbid.gatech.edu/

© Springer International Publishing Switzerland 2015
T. Supnithi et al. (Eds.): JIST 2014, LNCS 8943, pp. 361–377, 2015.
DOI: 10.1007/978-3-319-15615-6_27

Because ontologies clarify concepts that appear in target domains [3], we assume that it is important to develop a biomimetics ontology that contributes to improvement of knowledge interoperability between the biology and engineering domains. Furthermore, linked data technologies are very effective for integrating a database with existing biological diversity databases. On the basis of these observations, we developed a biomimetics ontology and keyword exploration tool based on linked data techniques. The tool allows users to find important keywords for retrieving meaningful knowledge from various biomimetics databases. This paper describes a prototype of our proposed biomimetics ontology and keyword exploration tool.

In Section 2, we outline biomimetics databases and discuss requirements for keyword exploration for retrieving knowledge from them. Section 3 discusses a prototype of a biomimetics ontology we created. Section 4 shows a keyword exploration tool based on our biomimetics database ontology. Section 5 discusses results and problems of our ontology and the proposed system. In section 6, we discuss related work and in section 7, we draw conclusions and discuss future work.

# 2     Biomimetics Database

## 2.1     Biomimetics Database Requirements

We are developing a biomimetics database as a part of the "Innovative material engineering based on biological diversity" project supported by the KAKENHI program of the Japan Society for the Promotion of Science (JSPS). The project aims to create a new academic domain that systematizes a new paradigm of engineering through learning from biological diversity; this domain is called "biomimetics." The biomimetics database is intended to be an open innovation platform bridging various domains, such as natural history, biology, agriculture, materials science, mechanical engineering, information science, environment and policy science, and sociology. Concretely, our research goal is to develop a biomimetics database based on museum-owned biological resource inventories and on information techniques to assist engineers in creating innovative ideas for new technological development through knowledge of biological diversity.

The requirements for such a database are as follows.

**Requirement 1. Variety of Data and Knowledge**
It is not an ordinary database storing information about a particular domain. It is an interdisciplinary database comprising not only databases of papers on biomimetics but also databases of all kinds of biological data, such as inventories, electron microscopy images, and experimental data. Therefore, interoperability of various kinds of data must be considered.

**Requirement 2. Exploratory Search with Idea Creation Support**
Engineers in biomimetics are not familiar with biological databases. Hence, they need substantial assistance in finding useful information about organisms that can be a source of creative ideas for developing innovative engineering products.

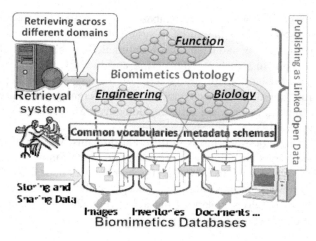

**Fig. 1.** An overview of the biomimetics databases

### Requirement 3. Publishing as an Open Innovation Platform

Because linking knowledge on various domains is expected to facilitate new innovations, it is important to provide mechanisms that enable everyone to use the database and add new knowledge to it. These could contribute to self-growing of the database.

### 2.2    Basic Design of a Biomimetics Database Retrieval System

Fig. 1 shows an overview of biomimetics databases with an accompanying retrieval system. With the aim of satisfying the requirements discussed in the previous section, we develop and use a biomimetics ontology to satisfy requirements 1 and 2. Various databases can be integrated by adding metadata to them based on the biomimetics ontology. In addition, we satisfy requirement 3 by publishing the ontology as Linked Open Data (LOD) to enable it to be linked to a variety of data on the Web.

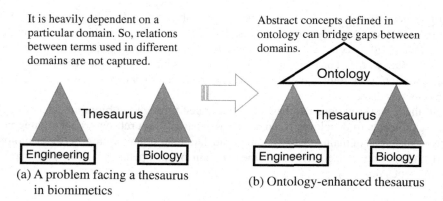

(a) A problem facing a thesaurus in biomimetics

(b) Ontology-enhanced thesaurus

**Fig. 2.** Theasurus vs. ontology-enhanced thesaurs

Users retrieving information from databases need to find appropriate keywords. Some retrieval systems use a thesaurus, a systematic collection of related terms, to help the users with that task. Note here that biomimetics databases include a variety of data in different domains, such as engineering and biology. Because a thesaurus is developed for a particular domain, all the terms in it are necessarily dependent on the domain. Therefore, relations between terms used in different domains are not included in a thesaurus. Such gaps between different domains are a problem for biomimetics thesauruses (Fig. 2(a)).

To bridge the gaps between terms in different domains, we use abstract concepts defined in an ontology. Our biomimetics ontology provides common vocabularies and schema based on systematized knowledge for biomimetics databases. Through mappings between terms in the thesaurus and concepts in the ontology, an ordinary thesaurus becomes an ontology-enhanced thesaurus (Fig. 2(b)). Technically, we offer two-step mediation using our ontology as follows.

**Method 1: Keyword Translation via an Ontology-Enhanced Thesaurus**
When concepts under consideration appear in two domains but are represented by different terminologies in the two domains, keywords used in one domain (e.g., engineering) are translated to keywords used in other domains (e.g., biology) through the biomimetics ontology, using natural language resources [4]. That is, the ontology provides abstract concepts to enhance an exiting thesaurus to bridge the gaps between domains caused by terminological differences. We propose using functional ontology [5, 6] for this purpose because functions are what biomimetics engineers want to realize by learning from the functions realized in nature.

**Method 2: Keyword Exploration Based on Ontology Exploration**
Concepts appearing in a domain (e.g., engineering) do not always appear in others (e.g., biology). We call such a situation a missing link case. Missing links between the domains have to be compensated in order to allow engineers to find more useful information. In such a case, since no thesaurus can help, we apply an ontology-exploration tool [7] to help bridge the gap caused by such missing links.

Furthermore, the biomimetics ontology is published as LOD. This enables users to access other databases easily (e.g., just by clicking links). Large-scale linked data such as DBpedia[2] could be especially useful for acquiring an overview of a selected concept (keyword). Through such information obtained from other databases, users can find more important and appropriate keywords that they can then use to search the biomimetics database.

In these ways, various databases are integrated through the biomimetics ontology. This enables us to develop a biomimetics database with a retrieval system integrated across different domains. The system could facilitate collaboration between biology and engineering and contribute to the creation of an open innovation platform for biomimetics research.

---

[2] http://dbpedia.org/

In this paper, we focus on keyword exploration to find appropriate keyword for retrieving biomimetics databases by compensating for the missing link between biology and engineering.

## 2.3 Keyword Exploration for Retrieving from Biomimetics Databases

To consider how keyword exploration works, let us see a scenario showing these two methods in action.

Imagine an engineer working in a house-construction company. He or she is asked by his or her boss to suggest ecofriendly for floors and walls. The engineer might think of keywords such as "easy to clean" or "stain-resistant," but not the technical term "antifouling" used in databases as an index. At first, the engineer finds "antifouling" from synonyms of "stain-resistant" through a normal thesaurus. Then he or she uses proposed two methods as follows.

Method 1 (keyword translation) can help the engineer find appropriate technical terms that best capture his or her intention. When the engineer types "antifouling," the biomimetics database will return relevant information if it contains information about biological organisms that confer antifouling properties.

However, it is not realistic to expect all useful pieces of information to be indexed by "antifouling." The engineer will never notice the information that is not indexed. Hence, there remains the issue of how to compensate for such missing links, in other words, how to fill the gap between his or her goal and the biological database with respect to "antifouling."

Method 2 (keyword exploration) can help to solve this issue based on a biomimetics ontology. The ontology provides information about concepts related to organisms and their functions with relationships among them, such as functional decomposition, characteristic behaviors or features, and living environment. A keyword explorer helps engineers find a set of appropriate keywords for their purposes by allowing them to explore the keyword space spanned by the ontology. Using the keyword explorer, he or she might find organisms such as lotus, snail, or sandfish some of which are not directly indexed by "antifouling" function but somewhat related to it. Then the engineer will be able to find information related to sandfish as well as lotuses and snails, because a sandfish lives in a desert without being soiled with sand. Note here that, the method 2 provides only keywords for search. After that, the engineer conducts detailed search using them to find related documents and so on from other biomimetics databases.

This scenario shows that it is essential to find appropriate keywords to search for information across different domains. This is why we separate the tasks of finding keywords using the keyword explorer from information retrieval tasks on the biomimetics database. Technically, keyword translation is implemented by mapping between an existing thesaurus and the functional ontology, and the keyword exploration is developed based on our ontology exploration tool [7] using the biomimetics ontology. Though we use existing technologies for keyword exploration, this means conceptually that the existing tool becomes a new tool for keyword exploration.

The important point here is that we divide the keyword finding task into two problems, the problem of different terminology and the problem of missing links at the conceptual level. These are supported by keyword translation and exploration, respectively. The two methods are used not only sequentially but also freely whenever they are needed. For instance, if a user finds an organism as a candidate for the target of a search, then he or she can use method 1 to solve a terminological problem before beginning the search.

### 2.4    Keyword Exploration Reasoning

Keyword exploration for a biomimetics database aims to suggest users appropriate keywords to assist them in finding meaningful information for innovative biodiversity-based engineering. Please note here that the proposed system gives users not the perfect solution for their requirements, but hints to stimulate idea creation. To stimulate idea creation, keyword exploration does not require strict reasoning, because it shows only well-known knowledge that users expect, whereas they want to create unexpected ideas. That is, rough inference is more suitable for idea creation, even though its results could include theoretically useless information. Of course, users must investigate the result of retrieval before developing a product based on it. However, this is not conducted in the first stage of the development, but in a later stage.

Our ontology exploration tool supports such a rough inference and helps users find unexpected relationships among concepts [7]. For the same reason, an ontology must cover broader concepts without much detailed definition when it is used for keyword exploration. In the following sections, we discuss our biomimetics ontology and keyword exploration tool in accordance with these considerations.

## 3    A Biomimetics Ontology

### 3.1    Biomimetics Database Search Requirements

Before we began developing a biomimetics ontology, we conducted interviews with engineers working with biomimetics regarding their requirements for biomimetics database search. When we asked, "What do you want to search for in a biomimetic database?" they said they wanted to search for organisms or organs that perform functions that they were trying to develop in their new products. In fact, most successful examples are imitations of capabilities that organisms possess, such as the water repellency of a lotus and the adhesiveness of a gecko's feet. Therefore, we proposed that it is important to search the biomimetic database for functions or goals that they want to achieve.

On the other hand, someone engaged in cooperative research with engineers and biologists reported that engineers do not have knowledge that is very familiar to biologists. For instance, when an engineer had a question about functions of projections shown in an electron microscopy image of an insect, a biologist (entomologist) suggested that it could have an anti-slip capability, because the insect often clings to slippery surfaces. This suggests that a biomimetic ontology must bridge knowledge gaps between engineers and biologists, as discussed in section 2.3. In this example, we can consider that a key concept in compensating for the missing link is the relationship between the supposed function and the characteristic behavior of the insect. Its living environment is another key concept, because environment is closely related to behavior.

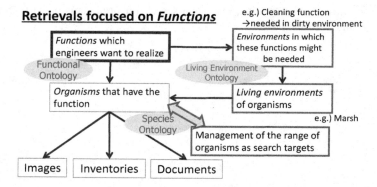

**Fig. 3.** Components of the biomimetics ontology and how they are used to retrieve biomimetics databases

## 3.2    Basic Design of the Biomimetics Ontology

Considering the requirements discussed in the previous section, we set the first requirement for biomimetics ontology as to be able to search for related organisms by the function the user wants to perform. At the same time, we propose that it should support various viewpoints to compensate for missing links among domains. We created the basic design of the biomimetics ontology based on these observations. Fig. 3 shows components of the biomimetics ontology and how they are used to retrieve biomimetics databases.

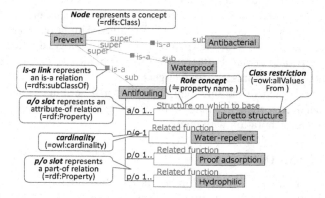

**Fig. 4.** An example ontology representation in Hozo and its correspondence with OWL

As discussed previously, *functions* are main keywords for biomimetics database search. Therefore, a functional ontology is a core component of our biomimetics ontology in representing functions that engineers want to include in their new products. We plan to use existing functional ontologies [5, 6] with mappings to thesauruses for keyword translation. Functions defined in our functional ontology are related to biological concepts, such as *organisms* and their *body parts* that perform those functions. Through these relationships, users can search for organisms or parts that could perform the functions they want.

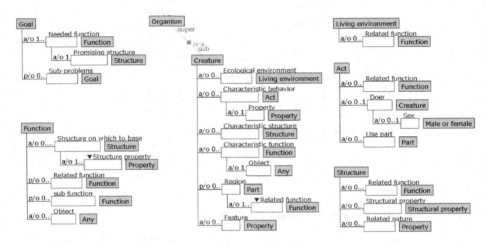

**Fig. 5.** Representative upper concepts defined in the biomimetics ontology

In addition, we build a biological ontology including notions, such as *living environments*, which are closely related to characteristic functions and behaviors of organisms. For example, some fish living in bogs have cleaning functions that prevent being dirtied by a bog. Through relationships among concepts in functional ontology and the *living environment* ontology, users can search for "organisms that have a function" via "a living environment where they might need such function." This means that they can search from different viewpoints related to functions. Similarly, users can perform searches from a wider variety of viewpoints using other biological ontologies, such as *structure*, *property*, and *action*. Furthermore, *is-a* hierarchies of *species* can be used to manage the range of organisms as search targets.

The concepts defined in the ontology are used for describing metadata regarding biomimetics databases such as images, inventories, and documents. The retrieval system is used in collaboration with them.

### 3.3   A Biomimetic Ontology

For verification of how keyword exploration using a biomimetics ontology works, we developed a prototype of the biomimetics ontology using Hozo[3] [8]. We use two

---

[3] http://www.hozo.jp/

kinds of source information for building the ontology. The first is metadata for scanning electron microscopy images stored in the biomimetic database. The metadata are provided by the biologists who took the pictures, and it includes not only inventory information, such as species name, gathering location, and magnification, but also keywords related to the target organism. The second source is introductory documents about typical examples of products used in biomimetics studies.

Fig. 4 shows an example of ontology representation in Hozo. Each concept is defined using two kinds of slots, part-of (denoted by p/o) and attribute-of (denoted by a/o). A slot consists of a role name, class constraint, and cardinality, corresponding roughly to restrictions on properties in a Resource Description Framework Schema (RDF(S)) and Web Ontology Language (OWL)[4]. For instance, Fig. 4, representing bicycle (class), has the restrictions owl:allValuesFrom *Wheel* and owl:cardinality 1 on the *front-wheel* property.

As a result, we built a biomimetics ontology that includes 379 concepts (classes) and 314 relationships (properties), except for the *is-a (sub-class-of)* relation. Fig. 5 shows representative upper concepts defined in the ontology. *Goal* represents a goal that engineers want to achieve in their new products. It has relationships to functions that are needed to achieve that goal. Some goals can be decomposed into *sub-goals*. *Function* can also be decomposed into *sub-functions* and may have relationships to a related *Structure* or *Function*. *Organism* is an important concept in compensating for missing links between engineering and biology. It may have relationships such as *Ecological environment*, *Characteristic behavior*, *Characteristic structure*, *Characteristic function*, and *Region Part*. *Living environment*, *Act*, and *Structure* may also have relationships to *Function*. In these ways, the biomedical ontology defines primarily concepts and relationships that are important for bridging missing links between domains.

# 4    Keyword Exploration Based on Biomimetics Ontology

## 4.1    Divergent Ontology Exploration Tool

We developed the ontology exploration system for biomimetics databases based on ontology exploration techniques proposed in our previous work [7]. Fig. 6 outlines the ontology exploration framework. The framework enables users to freely explore a sea of concepts in the ontology from a variety of perspectives according to their own motives. Exploration stimulates their way of thinking and contributes to deeper understanding of the ontology and hence its target world. As a result, users can discover what interests them. This could include findings that are new to them, because they might find unexpected conceptual chains from the ontology exploration that they would otherwise never have thought of.

---

[4] The details of the OWL representation of the Hozo ontology is discussed in [9].

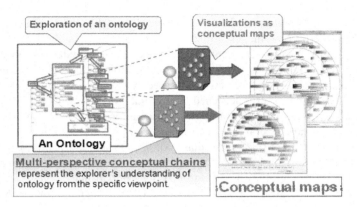

**Fig. 6.** Divergent exploration of ontology

Exploration of an ontology can be performed by choosing arbitrary concepts from which *multi-perspective conceptual chains* can be traced, according to the explorer's intention. We define the viewpoint for exploring an ontology and obtaining multi-perspective conceptual chains as the combination of a *focal point* and *aspects*. A focal point indicates a concept to which the user pays attention as a starting point of the exploration. The aspect is the manner in which the user explores the ontology. In the context of ontology exploration, we can regard relationships among concepts defined in ontologies as directed links to related concepts. Then, the aspect can be represented by a set of methods for extracting concepts according to its links and a direction to follow (upward or downward).

The left side of Fig. 7 shows an example of a conceptual chain obtained by exploring the biomimetics ontology from *Antifouling* as the focal point. It shows the path "*Antifouling → Water-repellent → Lotus*." In this example, the former path "*Antifouling → Water-repellent*" is obtained by following its *Related function* link (relationship) upward, and the latter path "*Water-repellent → Lotus*" is obtained by following its *Characteristic function* link (relationship) downward. Similarly, other conceptual chains are also obtained through other links according to other aspects. A multi-perspective conceptual chain is obtained by combining them and visualizing them in a user-friendly form, i.e., a conceptual map (see right side of Fig. 7).

## 4.2    Architecture of the Keyword Exploration System

Based on the ontology exploration tool, we developed the keyword exploration system for retrieving information from biomimetics databases as a web application to assist the user in using the results easily for searching other databases, while the previously described system was developed as a Java client application. We implemented the ontology exploration tool using HTML5 and Java Script to enable it to work on web browsers on many platforms, including not only PCs but also tablets and smartphones.

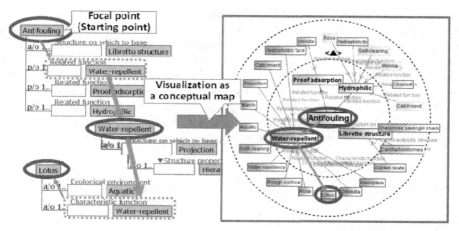

**Fig. 7.** An example of conceptual chain obtained by exploring the biomimetics ontology

The biomimetics ontology was translated from the original Hozo format into the Resource Description Framework (RDF) and stored in an RDF database. We used AllegroGraph[5] as the database for the ontology, but we can also use other RDF databases, because we implemented the exploration methods based on Simple Protocol and RDF Query Language (SPARQL) queries. That is, we developed the keyword exploration system to be applicable to other ontologies and linked data. These are technological differences between the keyword exploration system and the ontology exploration tool.

### 4.3    Keyword Exportation Through SPARQL

#### RDF Translation of the Biomimetics Ontology

To store the biomimetics ontology in an RDF database, we translated the ontology in Hozo format into RDF. As discussed in section 2.4, keyword exploration needs neither strict reasoning nor detailed definitions. Consequently, we took the approach of translating the ontology into a simple RDF model similarly to the way in which we published our disease ontology as Linked Data [10], in contrast to the Hozo approach of having a function to export its ontology in OWL. Fig. 8 shows an example of RDF representation translated from the concept *Antifouling* defined in the biomimetics ontology. In this RDF model, all restrictions of the *Antifouling* concept (class) in OWL are omitted, and properties related to them are represented as properties of the class. The *Antifouling* class also has an *rdfs:type* property to represent a reprehensive upper class[6]. It is used to judge the top-level category of RDF resources explored by SPARQL queries discussed in the next section.

---

[5] http://franz.com/agraph/allegrograph/

[6] We chose seven upper concepts Function, Goal, Behavior, Property, Structure, Organism, and Living environment as top-level categories.

## SPARQL Queries for Keyword Exploration

For keyword exploration, the system searches all combinations of aspects (links) to generate conceptual chains from a concept selected as a starting point to those specified by the user. It easy to obtain such information from the biomimetics ontology stored in the RDF database using simple recursive SPARQL queries, such as "select ?s ?p ?o where {?s ?p ?o}." However, the system must create as many queries as the number of all possible combinations of aspects, if it uses such simple queries. Repeating SPARQL queries multiple times is obviously inefficient. Furthermore, the response time for queries sent across networks could be even longer.

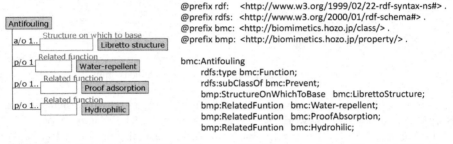

(a) Definition of the concept "Antifouling" in Hozo.

b) RDF representation of the concept "Antifouling" translated from Hozo format.

**Fig. 8.** An example of an RDF translated from the Hozo ontology

Consequently, we consider efficient queries for keyword exploration. Fig. 9 shows an example of a SPARQL query to obtain conceptual chains from the starting point to the end point specified by the user. The system can obtain a conceptual chain of length exactly three from the query. To obtain all conceptual chains whose length is equal to or less than three, the system gathers three queries, such as the query in Fig. 9. More generally, the system can obtain all conceptual chains whose depth is N (where N is a natural number) using N queries. These queries can be formalized recursively in terms of the starting point, the end point, and the depth N.

### 4.4     Examples

Fig. 10 shows one result of ontology exploration using the system. In this example, the user selected *Antifouling-antibacterial coating* as the starting point and obtained conceptual chains to some *Organism* as the end point. The system shows all conceptual chains between the selected concepts as a conceptual map. In the conceptual maps, the nodes represent resources (classes) and the links represent properties among them. The nodes are colored on the map according to their top-level categories. By clicking the nodes on the map, the user can labels the nodes on the selected paths.

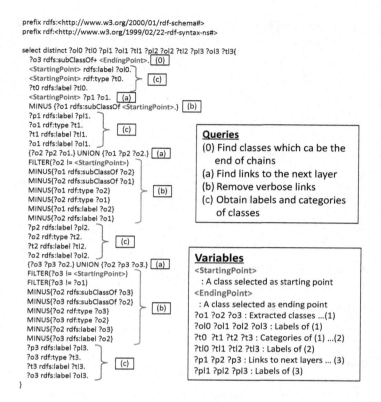

**Fig. 9.** A SPARQL query to obtain conceptual chains whose length is exactly three

Through the map, we can find a variety patterns of paths such as:

1. Goal → Function → **Organism**
   e.g. → *Antifouling*  → *Sunfish*
2. Goal → Function → Living environment →**Organism**
   e.g. → *Antifouling*  → *Marsh* → *Snail*
     → *Antifouling*  → *Water-repellent* → *Catchment* → *Desert* → *Sandfish*
3. Goal → Function → Structure →**Organism**
   e.g. → *Self-cleaning* → *Oil-proof* → *Winkle* → *Morpho butterfly*

Fig.11 shows right click commands that the keyword exploration system provides to the users. The user can see the selected path in another visualization. The user can also explore the next paths if he/she wants to know. Furthermore, the user can use the selected information to search other linked data such as DBpedia and other data-bases (see Fig. 11). Though the current version supports only a few LODs and data-bases, it can be easily extended to others.

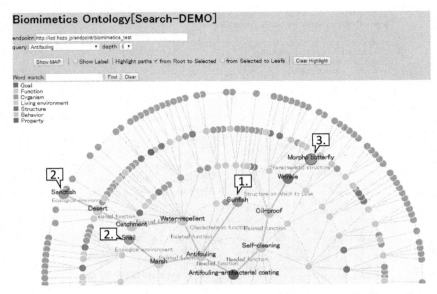

**Fig. 10.** An example of the result of the keyword exploration from "*Antifouling-antibacterial coating*" to "*Creature*"

## 5    Discussion

We developed the biomimetics ontology and published it on the keyword exploration system in order to verify their potential. They were demonstrated at some symposiums on biomimetics and group meetings of the project. We also received comments from project members who used the system.

Although the current ontology is a small prototype, some interesting conceptual chains (paths) can be obtained from it, such as those shown in Fig. 10. One of interesting examples is the path *Antifouling-antibacterial coating* → *Antifouling* → *Water-repellent* → *Catchment* → *Desert* → *Sandfish*. Sandfish (Scincus scincus) is a kind of skink that lives in desert. It was a reasonable but unexpected candidate that might have antifouling function although it is well known to have a low-friction-surface skin since it moves in sands. This is an example of a good result the system could support because it is based on rough inference for abstract concepts that bridge missing links among domains. However, ontologies must be extended to cover a wide range of biodiversity-related knowledge.

The following topics must be considered:

1. Which organisms should be included in the ontology?
2. What kinds of relationships should be defined?
3. How can we construct large-scale biomimetics ontology?
4. How can the system support the users to choose appropriate keywords when a multiple results are obtained at the same time?
5. How should keyword exploration cooperate with keyword translation?

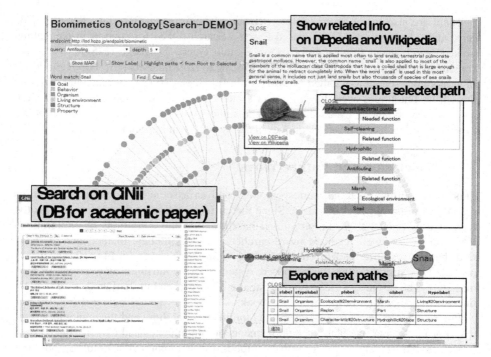

**Fig. 11.** Right click commands of the keyword exploration system

Defining too much knowledge in the ontology can cause the system to show too much information. Therefore, design principles are the most important problem. Building methods for large-scale ontologies are also an important issue. We are investigating methods to support extensions of the biomimetics ontology from several approaches using text mining techniques and exiting LOD.

## 6 Related Work

Biomimetics databases are being developed by other research groups, for example, BIOMIN-GLOBE[7] by TU Bergakademie Freiberg and Ask Nature[8]. There are mainly for providing case studies related to biomimetics while they support a thesaurus or terminology for biomimetics. As discussed in section 2.2, thesauri are not sufficient for use with biomimetics databases because they are domain dependent. Cheong [4] and Stroble [11] proposed methods to translate biological terms to engineering terms via functional terms. They are corresponded to keyword translation introduced in section 2.2. On the other hand, this paper focuses on keyword exploration for retrieving information from biomimetics databases.

---

[7] http://tu-freiberg.de/en/exphys/biomineralogy-and-extreme-biomimetics
[8] http://www.asknature.org/

There are also many approaches to Semantic Search using SPARQL. For example, Ferré and others propose Query-based Faceted Search (QFS) for support in navigating faceted search using Logical Information System Query Language (LISQL) [12] and implement it based on SPARQL endpoints to scale to large datasets [13]. Popov proposes an exploratory search tool called Multi-Pivot [14], which extracts concepts and relationships from ontologies according to a user's interest. These are visualized and used for semantic search among instances (data). The authors took the same approach as Popov. Considering how to use these techniques in our system is an important future work.

# 7     Conclusion and Future Work

This paper outlined a biomimetics ontology and an associated keyword exploration system for a biomimetics database. Since the current version of the system is a prototype, it uses only a small ontology and has limits on the conditions of exploration. However, it was well received by researchers on biomimetics. In fact, one of them said that the resulting path from *Antifouling* to *Sandfish* shown in Fig. 10 was unexpected. This suggests that the proposed system could contribute innovations in biomimetics. The researchers also plan to use the biomimetics ontology and system as an interactive index for a biomimetics textbook.

Future work includes extensions of the biomimetics ontology and the exploration system. For the former, we plan to use documents on biomimetics and existing linked data related to biology and consider some methods for semi-automatic ontology building using them. We also plan to align the biomimetics ontology with existing functional ontologies and thesauri. For later, we are exploring potentially useful patterns through discussion with biomimetics researchers and ontology engineers. We also plan to develop more detailed exploration functions such as refinement of search results, faceted search, etc. Evaluation of the keyword explorer is also an important future work.

The current version of the proposed system is available at the URL http://biomimetics.hozo.jp/ontology_db.html .

**Acknowledgements.** This work was supported by JSPS KAKENHI Grant Number 25280081 and 24120002. The authors are deeply grateful to biologists Shuhei Nomura, Shinohara Gento, Takeshi Yamasaki, and others for providing us with broad biological knowledge. The authors would also like to thank members of National Deliberation Committee for ISO/TC266 Biomimetics, especially, Masatsugu Simomura and Naoyuki Tsunematsu. The authors would like to thank Enago (www.enago.jp) for the English language review.

# References

1.  Shimomura, M.: Engineering Biomimetics: Integration of Biology and Nanotechnology, Design for Innovative Value Towards a Sustainable Society, pp. 905–907 (2012)
2.  Vattam, S., Wiltgen, B., Helms, M., Goel, A., Yen, J.: DANE: Fostering Creativity in and through Biologically Inspired Design. In: Proc. First International Conference on Design Creativity, pp. 115–122, Kobe, Japan, November 2010

3. Gruber, T.: A translation approach to portable ontology specifications. In: Proc. of JKAW'92, pp. 89–108 (1992)
4. Cheong, H., et al.: Biologically Meaningful Keywords for Functional Terms of the Functional Basis. Journal of Mechanical Design, vol. 133 (2011). doi:10.1115/1.4003249
5. Hirtz, J., Stone, R.B., et al.: A Functional Basis for Engineering Design: Reconciling and Evolving Previous Effort, NIST Technical Note 1447 (2002)
6. Kitamura, Y., Segawa, S., Sasajima, M., Tarumi, S., Mizoguchi, R.: Deep Semantic Mapping between Functional Taxonomies for Interoperable Semantic Search. In: Domingue, J., Anutariya, C. (eds.) ASWC 2008. LNCS, vol. 5367, pp. 137–151. Springer, Heidelberg (2008)
7. Kozaki, K., Hirota, T., Mizoguchi, R.: Understanding an Ontology through Divergent Exploration. In: Antoniou, G., Grobelnik, M., Simperl, E., Parsia, B., Plexousakis, D., De Leenheer, P., Pan, J. (eds.) ESWC 2011, Part I. LNCS, vol. 6643, pp. 305–320. Springer, Heidelberg (2011)
8. Mizoguchi, R., Sunagawa, E., Kozaki, K., Kitamura, Y.: : Hozo. J. of Applied Ontology 2(2), 159–179 (2007)
9. Kozaki, K., Sunagawa, E., Kitamura, Y., Mizoguchi, R.: Role Representation Model Using OWL and SWRL. In: Proc. of 2nd Workshop on Roles and Relationships in Object Oriented Programming, Multiagent Systems, and Ontologies, Berlin, pp. 39–46 (2007)
10. Kozaki, K., Yamagata, Y., Imai, T., Ohe, K., Mizoguchi, R.: Publishing a Disease Ontologies as Linked Data. In: Kim, W., Ding, Y., Kim, H.-G. (eds.) JIST 2013. LNCS, vol. 8388, pp. 110–128. Springer, Heidelberg (2014)
11. Stroble, J., Stone, R., McAdams, D.A., Watkins, S.: An Engineering-to-Biology Thesaurus To Promote Better Collaboration, Creativity and Discovery. In: Proceedings of the CIRP DESIGN 2009 International Conference (2009)
12. Ferré, S., Hermann, A.: Reconciling faceted search and query languages for the semantic web. IJMSO 7(1), 37–54 (2012)
13. Guyonvarch, J., Ferré, S.: Scalewelis: a scalable query-based faceted search system on top pf SPARQL endpoints. In: Proceedings of the 3rd Open Challenge on Multilingual Question Answering over Linked Data (QALD-3), Valencia, Spain (2013)
14. Popov, I.O., Schraefel, M.C., Hall, W., Shadbolt, N.: Connecting the Dots: A Multi-pivot Approach to Data Exploration. In: Aroyo, L., Welty, C., Alani, H., Taylor, J., Bernstein, A., Kagal, L., Noy, N., Blomqvist, E. (eds.) ISWC 2011, Part I. LNCS, vol. 7031, pp. 553–568. Springer, Heidelberg (2011)

# Choosing Related Concepts for Intelligent Exploration

Kouji Kozaki[(✉)]

The Institute of Scientific and Industrial Research (ISIR), Osaka University,
8-1 Mihogaoka, Osaka, Ibaraki, Japan
kozaki@ei.sanken.osaka-u.ac.jp

**Abstract.** How to explore semantic data such as linked data, knowledge graph, ontologies etc. and get appropriate information from them are very important techniques for intelligent application systems based on them. In this article, we focus on intelligent exploration of ontologies since ontologies provide systematized knowledge to understand target domain and contribute to deep understanding of semantic data. The authors propose a novel conceptual search method called "Multistep Expansion based Concept Search" to get appropriate concepts from ontologies according to the user's intentions and purpose.

**Keywords:** Ontology · Concept search · Intelligent exploration

## 1    Introduction

Intelligent exploration of semantic data such as linked data, knowledge graph, ontologies and so on, is one of important techniques to make the most use of vast amounts of information being developed from day to day. In particular, we focus on intelligent exploration of ontologies since ontologies provide systematized knowledge to understand target world and contribute to deep understanding of semantic data.

In previous works, the authors proposed divergent exploration of ontology and showed it contribute to the user's understanding of ontology according to their intention and interests [1, 2]. Ontology explorations are done through finding concepts related with a focused concept and trace them repeatedly. When an ontology exploration system finds related concepts, it has to consider not only concepts which are directly connected with the focused concept through some relationships but also concepts which are indirectly related through conceptual structures in the target ontology. For example, when we explore a dish ontology in Fig.1, meat is directly connected with meat dish and indirectly related with beef steak. These related concepts are acquired by concept search techniques.

In order to get good results through ontology exploration, we need a technique of concept search to acquire appropriate concepts requested by the user. Though there are some approaches for concept search in ontologies, there are some rooms to improve it in order to deal with conceptual structure of ontologies more efficiently. It is

© Springer International Publishing Switzerland 2015
T. Supnithi et al. (Eds.): JIST 2014, LNCS 8943, pp. 378–386, 2015.
DOI: 10.1007/978-3-319-15615-6_28

partly because that most concept search
techniques tend to get all concepts which
satisfy search conditions while ontology
explorations needs flexible classification of
search result. To overcome these issues,
the authors propose a novel conceptual
search method called "Multistep Expansion
based Concept Search" which the user can
get appropriate concepts from ontologies
according to the user's intentions and
purpose.

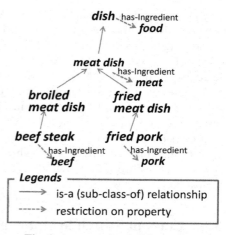

**Fig. 1.** An example of dish ontology

This paper is organized as follows. The
next section overviews related works about
semantic search. Section 3 outlines basic
characteristics of concept search in ontology. In Section 4, we propose Multistep
Expansion based Concept Search and its implementation is discussed in Section 5.
Finally, we present concluding remarks with a discussion of future work.

## 2    Related Works

There are many approaches for Semantic Search. They are classified into instance
search and concept (class) search. Faceted Search is the most major method for instance searches. For example, Ferré proposes Query-based Faceted Search (QFS)
which support to navigate faceted search by Logical Information System Query Language (LISQL) [3]. Simple Protocol and RDF Query Language (SPARQL) is the
most major query language for Semantic Web. Because it is a query language for
RDF, it also can be used for concept searches in ontologies which are written in RDF
such as RDFS and OWL. However, SPARQL is not for concept search since it does
not support semantics of ontology language while SPARQL 1.1 supports some reasoning such as property chain. For example, when we want to get classes which has a
restriction using SPARQL, we have to know RDF graph representation of the restriction.

On the other hands, some systems support concept search more efficiently. Protégé
4 provides DL Query tab for concept search. It enables the user to search concepts
whose definitions match a search condition described by the user in Manchester OWL
Syntax. This is a fundamental method for concept search.

Concept search are used for semantic search systems which use semantic information of ontologies. Dimitrios formalizes queries for concept search and develops a
support system for DL query description using a graphical user interface [4]. This is
an approach to help the user to describe queries. Popov proposes exploratory search
called Multi-Pivot [5]. It extracts concepts and relationships from ontologies according to a user's interest. They are visualized and used for semantic searches. This is a

good example, conceptual structures in ontology are used for semantic searches for instances. Ravish proposes a hybrid search which combines keyword search by natural language and semantic search [6]. PowerAqua is a multi-ontology based question answering system [7]. They are also conceptual structure in ontologies, and concept search is one of key techniques to realize them.

Our approach shares a lot with concept search methods used in these existing semantic search system. However, our main concern is not how to query and get search results but how to classify search results. Many concept search systems show the search results just a list and some others show them with some categorizations in tag style or hierarchy structure. Although *is-a* hierarchies of concepts are used to show such categories, we want to introduce other classification of search results from ontological point of view.

## 3     Concept Search in Ontology

In ontologies, concepts are defined by relationships with other concepts. Therefore, concept search in an ontology is not simple string matching but semantic search using relationships between them. The most essential relationships are *is-a* (sub class of) relationships between sub concepts (classes) and super concept (classes). Because *is-a* relationships have semantics that sub concepts inherit all definition of their super concept, they are used to systematize concepts which have some similar definitions. Other relationships are used to represent definitions of concepts. They are used to describe search conditions for concept search.

For example, when a user searches "dish whose ingredient is meat" in a dish ontology shown in Fig.1, a search system find concepts which has *has-Ingredient* relationship (property) with *meat*. In the case of ontology in OWL, it means to find sub classes of dish which has a restriction on *has-ingredient* property as *meat*.

Here, according to semantics of *is-a* relationships discussed the above, all sub concepts of concepts which satisfy the search condition also satisfy it. However, it depends on the user's intention and purpose which concepts are needed as the search result. In the case of the above example, we can consider some kinds of purposes as follows;

1. When the user wants to get the most general definition of *dish* whose ingredient is *meat* to know its common characteristics, the search result should be only *meat dish*.
2. When the user wants to know *dishes* common to all kind of *meat* (not specific kinds of *meat*), the search result should be *meat dish*, *broiled meat dish* and *fried meat dish* whose *ingredients* are (not specific kinds of) *meat*.
3. When the user wants to know all *dish* whose *ingredient* is *meat*, the search result should be all *meat dishes* shown in Fig.1.

In order to deal with such intentions and purpose for concept search, we have to consider ontological meanings of concept search and classifications of search result. This topic is discussed in the following sections.

# 4    Multistep Expansion Based Concept Search

## 4.1    Basic Idea

In most semantic concept search methods, main search condition is definitions of concepts in which the user is interested. That is, its results are concepts whose definitions satisfy the search condition. Here, all sub concepts (sub classes) of the resulting concept also satisfy the search condition because all sub concepts inherit definition of their upper concepts (super classes). Therefore, it is a useful way to browse the search results one by one according to there *is-a* hierarchy.

In this method, though the search result is considered as a single set of concepts whose definitions satisfy the search condition, conceptual differences among these concepts are unclear. For example, when a user searches "dish whose ingredient is meat" in an ontology shown in Fig.1, it is not clear whether its result should include "dish whose ingredient is beef or pork" or it should be only "dish whose ingredient is (not specific kinds of) meat". It depends on intention of the user. So, in order to represent the user's intention appropriately, the search result should be classified systemically.

It is important to systematize search results according to feature of concepts on which the user focuses so that the user can use conceptual definition in ontology efficiently. On the basis of this observation, the authors propose a novel search method, which is named *Multistep Expansion based Concept Search*, which extracts concepts in ontology according to the user's interests represented as search condition. The proposed method consists of two expansion methods for concept hierarchies. One is *result expansion* which use *is-a* hierarchies of resulting concepts. The other is *condition expansion* which use *is-a* hierarchies of concepts which appear in search conditions.

### 4.1.1    Result Expansion

In concept search in ontologies, its results are concepts whose definition satisfies its search condition. In the followings, we call the results resulting concepts. All sub concepts of a resulting concept also satisfy the search condition since all sub concepts inherit definitions from its super concept according to *is-a* hierarchy of them. Based on this feature of *is-a* hierarchy, we can expand a resulting concept according to its *is-a* hierarchy and get its sub concepts as another resulting concepts. We call such a method result expansion.

For example, when a user searches "dish whose ingredient is meat" and gets *meat dish* as a resulting concept, the user also gets its sub concepts such as *broiled meat dish* and *fried meat dish* as resulting concepts using result expansions.

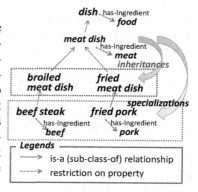

**Fig. 2.** Examples of inheritances and specializations of restrictions

Here, we can classify result expansions into three levels by considering inheritances and specializations of restrictions on properties. By specializations of restrictions we mean restrictions which specializes a restriction inherited from its super concept. In the case of an example in Fig. 2, *meat dish* has a restriction on *has-ingredient* property as *meat*. Its sub concepts such as *broiled meat dish* and *fried meat dish* inherit the restriction on *has-ingredient* as is. Furthermore, in their sub concepts such as *beef steak* and *fried pork*, restrictions on *has-ingredient* are specialized from *meat* to *beef* and *pork*. Note here that all of these sub concepts satisfy definition inherited form their super concept, in this case *meat dish*, whether some restrictions are specialized or not. Therefore, when a user searches "dish whose ingredient is meat", all of them satisfy the search condition. On the basis of the above considerations, we introduce three levels of result expansions as follows;

- **Result Expansion: Level 0** The user does not use result expansion and gets only top (the most upper) concept which satisfy a search condition as resulting concepts.
- **Result Expansion: Level 1** The user gets resulting concepts which inherit the restriction on property specified in a search condition as is without specializations. That is, when the restriction is specialized in sub concepts, they are not expanded.
- **Result Expansion: Level 2** The user gets all resulting concepts inherit the restriction on property specified in a search condition whether it is specialized or not. That is, all sub concepts of the resulting concepts gotten in level 0 are expanded.

Here, we call concepts whose definition have a restriction specified as the search condition *search condition directly match concepts*. And we call concepts whose definition have a specialization of the restriction specified as the search condition is *defined specialized search condition match concepts*.

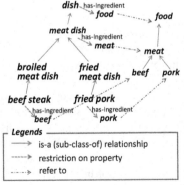

**Fig. 3.** A dish ontology

### 4.1.2    Condition Expansion

When a user searches concepts whose definition refers a concept (denoted by *C*) in a property restriction, concepts whose definition refers a sub concepts of *C* also satisfy the search condition. That is, when a search condition is "concepts whose definitions include a restriction on property *p* as concept *C*", "concepts whose definitions include a restriction on property *p* as sub concepts of *C*" are also its resulting concepts. In this way, we can expand a concept which is referred to in search condition according to its *is-a* hierarchy and get its concepts whose definitions refer to its sub concepts as another resulting concepts. We call such a method condition expansion.

For example, when a user searches "dish whose ingredient is meat" in a dish ontology shown in Fig.3, its search condition is "dish whose definition includes a restriction on *has-ingredient* property as *meat*". In a case that the user does not use condition expansion, he/she gets only *meat dish*, which has a restriction on *has-ingredient* property as *meat*, as the search result. On the other hand, in a case that the user use condition expansions, the search condition is expanded to "dish whose definition includes a restriction on *has-ingredient* property as *beef* or *pork*" according to *is-a* hierarchy of *meat*, and the user gets *beef steak* and *fried pork* as the search result.

When we use condition expansions, we do not consider *is-a* hierarchies of resulting concepts. That is, resulting concepts by condition expansions do not include concepts which inherit the restriction on property specified in a search condition as is. It means that resulting concepts by condition expansions represent boundary where definition specified in a search condition is specialized.

## 4.2    Combination of Result Expansion and Condition Expansion

Before we discuss combination of result expansion and condition expansion, we consider differences of resulting concepts by two expansions. When we use result expansion, all resulting concepts are sub concepts of concepts which directly satisfy an original search condition. On the other hand, when we use condition expansion, search condition is expanded according to *is-a* hierarchy of a concept which is referred to in an original search condition. Therefore, resulting concepts by condition expansion could include concepts other than resulting concepts by result expansion.

For example, we suppose that a user searches "dish whose ingredient is meat" in a dish ontology shown in Fig.4 (its search condition is "dish whose definition includes a restriction on has-Ingredient property as meat"). In this case, the user gets *meat dish* as a search condition directly match concepts. Then, the user gets sub concepts of *meat dish* (e.g. *beef steak*, *fried pork*) as resulting concepts by result expansions. On the other hand, the user can also get *beef salad* and *pork salad* as resulting concept by condition expansions while they are not included in resulting concepts by result expansions.

In this way, resulting concepts by result expansions and condition expansions are different. It suggests us that combinations of two expansions more detailed classification of search result. Because result expansion has three levels and condition expansion has two levels (apply or not), we can consider six patterns of combinations as follows;

(1)    Result expansion: level 0 + Condition Expansion: not applied
(2)    Result expansion: **level 1** + Condition Expansion: not applied
(3)    Result expansion: **level 2** + Condition Expansion: not applied
(4)    Result expansion: level 0 + Condition Expansion: **applied**
(5)    Result expansion: **level 1** + Condition Expansion: **applied**
(6)    Result expansion: **level 2** + Condition Expansion: **applied**

In the followings, we discuss each pattern using the same example shown in Fig.5.

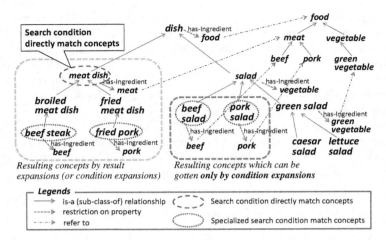

**Fig. 4.** Differences of resulting concepts by result expansions and condition expansions

Pattern (1) is concept search without result expansions nor condition expansion. Its resulting concept are only search condition directly match concepts (e.g. *meat dish*).

Pattern (2) and (3) get sub concepts of search condition directly match concepts, which are result by (1), as resulting concepts. In the case of (2), resulting concepts are concepts which inherit the definition specified in the original search condition as is (e.g. *broiled meat dish*, *fried meat dish*). In the case of (3), resulting concepts include concepts whose definitions are specialized from the definition specified in the original search condition (e.g. *beef steak*, *fried pork*). That is, resulting concepts by pattern (3) are all sub concepts of resulting concept by (1).

Pattern (4) gets not only search condition directly match concepts (e.g. *meat dish*) but also specialized search condition match concepts. Some of them (e.g. *beef salad* and *pork salad*) are not sub concepts of the search condition directly match concepts as discussed the above.

Pattern (5) and (6) also gets sub concepts of result by (4) as resulting concepts. Although pattern (5) expands only sub concepts which inherit the definition specified in expanded search conditions by (5) as is, its resulting concepts are the same with result by (6) because specialized search condition match concepts are already expanded by (4). That is, (5) and (6) can be integrated to one pattern and its resulting concepts include all concepts which satisfy the original search condition. Furthermore, when all specialized search condition match concepts are sub concepts of search condition directly match concepts, resulting concepts by (3), (5) and (6) are the same.

# 5     Implementation

The authors developed a Multistep Expansion based Concept Search System based on the considerations in section 4. It is implemented using Java with Hozo Core which is

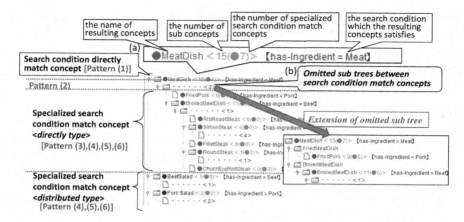

**Fig. 5.** An example of search result in Multistep Expansion based Concept Search System

API for ontology built by Hozo[1]. Fig.5 shows an example of search result when a user searched "dish whose definition includes a restriction on has-Ingredient property as meat" in a dish ontology (It is another ontology than the dish ontology in Fig.4).

The search result is visualized using graphical user interface in a tree. Resulting concepts are represented by tree nodes according to classifications of resulting concepts. A Node with icon shows the name of resulting concepts, the number of its sub concepts, the number of specialized search condition match concepts in its sub concepts, the search condition which the resulting concepts satisfies (see Fig.5(a)). Nodes whose label is " · · · · · " represents sub trees between search condition match concepts is omitted to show. The number with the node represents the number of omitted concepts. When the user clicks the node, the omitted concepts are shown like Fig.5(b).

Through this tree interface, the user can capture and understand results of concept search systematically according to definition he/she has interested in while it is difficult in existing one by one expansions of *is-a* tree.

## 6    Conclusion

In this paper, we proposed Multistep Expansion based Concept Search and a search system based on the method. Its feature is classifications of search results according to ontological differences of resulting concepts. It could help the users to capture and understand the search result according to their interests. As the result, it could be an essential technique for semantic exploration of ontologies and other semantic data based on ontologies. Future works include an integration of the proposed method and other ontology exploration methods such like our previous works [1, 2], application of

---

[1] http://www.hozo.jp

the system for other ontologies, and improvements of the system through feedbacks from them. Evaluation of the proposed system is also an important future work.

**Acknowledgments.** This work was supported by JSPS KAKENHI Grant Numbers 25280081.

# References

1. Kozaki, K., Hirota, T., Mizoguchi, R.: Understanding an ontology through divergent exploration. In: Antoniou, G., Grobelnik, M., Simperl, E., Parsia, B., Plexousakis, D., De Leenheer, P., Pan, J. (eds.) ESWC 2011, Part I. LNCS, vol. 6643, pp. 305–320. Springer, Heidelberg (2011)
2. Kozaki, K., Saito, O., Mizoguchi, R.: A consensus-building support system based on ontology exploration. In: Proc. of IESD (2012)
3. Ferré, S., Hermann, A.: Semantic search: reconciling expressive querying and exploratory search. In: Aroyo, L., Welty, C., Alani, H., Taylor, J., Bernstein, A., Kagal, L., Noy, N., Blomqvist, E. (eds.) ISWC 2011, Part I. LNCS, vol. 7031, pp. 177–192. Springer, Heidelberg (2011)
4. Koutsomitropoulos, D.A., Borillo Domenech, R., Solomou, G.D.: A structured semantic query interface for reasoning-based search and retrieval. In: Antoniou, G., Grobelnik, M., Simperl, E., Parsia, B., Plexousakis, D., De Leenheer, P., Pan, J. (eds.) ESWC 2011, Part I. LNCS, vol. 6643, pp. 17–31. Springer, Heidelberg (2011)
5. Popov, I.O., Schraefel, M.C., Hall, W., Shadbolt, N.: Connecting the dots: a multi-pivot approach to data exploration. In: Aroyo, L., Welty, C., Alani, H., Taylor, J., Bernstein, A., Kagal, L., Noy, N., Blomqvist, E. (eds.) ISWC 2011, Part I. LNCS, vol. 7031, pp. 553–568. Springer, Heidelberg (2011)
6. Bhagdev, R., Chapman, S., Ciravegna, F., Lanfranchi, V., Petrelli, D.: Hybrid search: effectively combining keywords and semantic searches. In: Bechhofer, S., Hauswirth, M., Hoffmann, J., Koubarakis, M. (eds.) ESWC 2008. LNCS, vol. 5021, pp. 554–568. Springer, Heidelberg (2008)
7. Lopez, V., Motta, E., Uren, V.S.: PowerAqua: Fishing the semantic web. In: Sure, Y., Domingue, J. (eds.) ESWC 2006. LNCS, vol. 4011, pp. 393–410. Springer, Heidelberg (2006)

# A Semantic Keyword Search Based on the Bidirectional Fix Root Query Graph Construction Algorithm

Siraya Sitthisarn[✉]

Division of Computer and Information Technology, Faculty of Science,
Thaksin University, Songkhla, Phattalung, Thailand
ssitthisarn@gmail.com

**Abstract.** In the last few years, the large amount of personal information in RDF format is widely deployed. To access the semantic information, it needs a semantic formal query (e.g. SPARQL query). However, this kind of query requires users to know the ontology structure and master its syntax. This paper proposes the X-SKengine, the semantic keyword search engine for specific expert discovery domain. The X-SKengine transforms the user keywords to the SPARQL query using a bidirectional fix root query graph construction algorithm which is able to compute the query graphs without limitation of the directions of relationships in ontologies. The experiment was conducted to compare the capability of SPARQL query construction between X-SKengine and the previous version. The results show that X-SKengine can automatically construct SPARQL queries relevant to meaning of user keywords for various ontology structures.

**Keywords:** Semantic keyword search · Bidirectional fix root · Query graph

## 1    Introduction

Recently, there are attempts to develop a keyword search interface to help end-users, without technical expertise to query the semantically enriched information in RDF format. Many semantic keyword search approaches [1-5] use a query graph construction technique to generate the semantic formal queries. One is SKengine [4], a semantic keyword search for the specific expert finding domain. It is based on the query graph construction algorithm named "fix root node query graph construction". The algorithm allows application developers to define a type of answer for their information requirement (i.e. person). SKengine is originally designed for the star-shaped ontology, a common structure of the personal ontology. However, it was found, SKengine is not compatible with other ontology structures, not fixed to a position of the central node as well as the direction of relationships between nodes.

This paper reports X-SKengine, the refined version of SKengine and the new query graph construction algorithm named "bidirectional fix root query graph construction" is proposed. The algorithm uses D-LKN index for computing query graphs, without limitation of directions of relationships between nodes.

© Springer International Publishing Switzerland 2015
T. Supnithi et al. (Eds.): JIST 2014, LNCS 8943, pp. 387–394, 2015.
DOI: 10.1007/978-3-319-15615-6_29

The paper is structured in the following way. The related work is given in the next section. Section 3 gives details of the overview architecture of X-SKengine. Section 4 describes the construction of D-LKN index. Then full details of bidirectional fix root query graph construction algorithm, including pseudo code are discussed in Section 5. The evaluation study is provided in section 6. The last section discusses the conclusion and recommendations for the future work.

## 2        Related Work

This section discusses semantic keyword search systems based on the query graph construction approach. In SPARK [1], the Kruskal's minimum spanning tree algorithm is used for constructing query graphs. The query graph involves exploring the RDF graph and discovering the appropriate connecting nodes. The query graph links each path of mapped entities together. Elbassuoni and Blanco [5] proposed a retrieval model for retrieving a set of sub-RDF graphs which match the user keywords. In addition, the statistic model is used for the query graph ranking scheme. Tran et al [3] proposed a query graph construction approach used a traversal graph algorithm for traversing an RDF graph. Then the algorithm finds the neighboring entities of each mapped entity within a limited range. As a result, possible sub graphs which contain the mapped entities and link of their connecting nodes are extracted for generating the semantic query.

Unfortunately, the computational cost is the main drawback of the above approaches. This is because these algorithms require the exploration of the entire RDF data graph. While Q2semantic [2] is different. Q2semantic uses an RDF graph clustering technique to infer an ontological structure from the schemaless RDF graph. Q2semantic adopts a single-level search algorithm in Blink [8] to generate top-k query graphs (distinct root) by exploring the extracted ontology structure.

X-SKengine also adapts single-level index in Blink by combining LKN entry and directions of the relationships between nodes together. The new index is named D-LKN index. The X-SKengine is different from Q2semantic because the query graph construction is simplified by restricting the query graphs with the fixed root. In addition D-LKN index supports query graph calculation based on the bidirectional graph traversing. This enables the query graph construction with various ontology structures.

## 3        Overview of X-SKengine Architecture

The X-SKengine architecture is inherited from the SKengine architecture. As depicted in Fig. 1, it consists of two modules (i.e. pre-processing module, and formal query construction module). The *pre-processing* module consists of two types of indexes: entity index and D-LKN index. The entity index is an inverted file of ontology entity labels/comments and literals of instances in the RDF data. This index is used to support mapping keywords to the corresponding entities in the formal query construction module. The D-LKN index captures an ontology structure including distances of paths, relationships and directions of relationships between nodes. This index is for supporting query graph computation (see detail of D-LKN index in section 4).

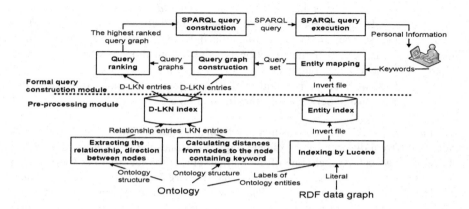

**Fig. 1.** The X-SKengine architecture

The *formal query construction* module is to compute query graph*s*. After a user enters a keyword phrase in the search box, the *entity mapping component* maps the keywords to the entity index and then prepares query sets which obtain initial nodes for the query graph construction component. The *query graph construction component* will then construct query graphs from a set of query sets. This component will produce all the possible query graphs by interpreting the meanings captured by semantics. The *query graph ranking component* will rank and select the most relevant query graph that matches the meaning of user keywords. The selected query graph is forwarded to the *SPARQL construction component* which will translate the query graph into a string conforming to the corresponding SPARQL syntax. The *SPARQL query execution component* will run *SPARQL query* and return the results to the user.

## 4    D-LKN Index

The D-LKN index is adapted from the LKN index in Blink [8]. In general, the LKN index is designed for star-shaped ontology structure. It captures the distance between nodes but it does not take into account on directions of relationships between them. To address the limitation of LKN index, the combination of LKN entries and the direction of relationships between nodes are required. As a result the new type of index named D-LKN index is proposed to support the query graph construction with more complicated ontology structure. The D-LKN structure consists of the LKN entry and attributes about directions of relationships as shown below.

LKN entry

*-dist*: is the shortest distance between a node to the node containing keyword (*knode*)

*-node*: is any node that its connection path can reach to *knode*

*-first*: is the *first* node on the shortest path from *node* to *knode*

*-knode*: is the class that its label contains the keyword

*-property*: is the relationship between *node* and *first*

-If the direction of the relationship between *node* and *first* is "source", then we define, the domain is *node* and range is *first*. However, if the direction is "sink", then domain will capture *first* and range contains *node*.

Direction

*-pnumber*: is the number of RDF statements associated to each relationship between the domain and range.

To construct D-LKN index, paths that link from any *nodes* to *knode* are computed according to different distances. For example, Fig. 2 illustrates the D-LKN list of keyword "Publication". As can be seen, one of the entries reflects the fact that the shortest path from *"class: Article"* to *"class: Publication"* has distance 1 and *first* is *"class: Publication"*. The *"class:Article"* and *"class:Publication"* are connected with object property *"isIncludeIn"*. For the direction, we can see that, *"Class: Article"* is *domain* and *"class: Publication"* is *range*. In addition, all subclasses of *knode* and *node* will also be indexed.

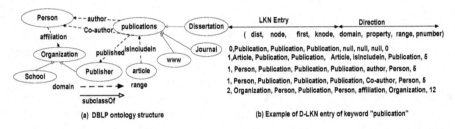

(a) DBLP ontology structure                 (b) Example of D-LKN entry of keyword "publication"

**Fig. 2.** The Structure of DBLP Ontology and the D-LKN (Publication)

## 5    Bidirectional-Fix-Root Query Graph Construction

The basic principle of this algorithm is to convert all query sets into query graphs. Fig. 3 illustrates the example of a query set corresponded to user keywords ("W3C XML website"). To construct query graph, the disjointed initial set of nodes will be connected and traced right to the root node step by step.

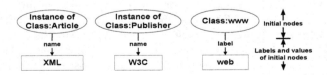

**Fig. 3.** An example of the query sets associated with user keywords "W3C XML website"

The algorithm of bidirectional fix root query graph construction is shown in Fig. 4. Input to the algorithm is query set S= {initial_node$_1$, initial_node$_2$,...,initial_node$_n$} where n = the number of keyword terms; and  r = the root node of interest. The output is a query graph.

```
ConstructGraph(r,initial_node₁,…,        expansion (nodeᵢ)
            initial_nodeₙ)              16{p_nodeListᵢ.add(exploreDLKN(nodeᵢ))
1 {for i  ∈[1,n] do                     /* explore D-LKN index to find
2 {nodeᵢ= new node (initial_nodeᵢ)      possible expanding nodes */
3 nodeᵢ.status = true}                  17 if(p_nodeListᵢ.empty = ture)
4 for i  ∈ [1,n] do                        /* no expanding node*/
5 {checkDisjoint(nodeᵢ)                 18 {expanding_node = null
6  checkRelation(nodeᵢ)}                19  nodeᵢ.graph_flag = false /* it
7 while (checkEnd (∀ i∈[1,n]:           cannot construct query graph*/  }
nodeᵢ)// stopping condition             20  else
8 { for  i  ∈ [1,n] do                  21 S_Li= for-
9  { if nodeᵢ.status  = true {          ward_Expansion(p_nodeListᵢ)
10  node.ᵢ ₙₑw = expansion(nodeᵢ)       22 If (S_Li ==1) return
11  nodeᵢ = node.ᵢ ₙₑw                  p_nodeLListᵢ.get(0).pos_node
12  checkDisjoint(nodeᵢ )               23 else
13  checkRelation(nodeᵢ )} }}           24  expand_node = backward(S_Li)}

14  }}                                  38  Return expand_node}
```

**Fig. 4.** Algorithm of Bidirectional fix root query graph construction

Initially, the construction process starts with placing all the initial nodes as the lowest level nodes of the query graph. The statuses of all initial nodes are *"active"*, (line 3: node.status = true). The query graph construction uses a bottom-up approach based on two basic operations - node merging and node expansion (see Fig.5).

**Node Merging**
For each node (node$_i$), merging verification will be taken with others (node$_j$) in round-robin (line 4-6). Merging between two nodes can take place when one of the following function is true: (i) the function: checkDisjoint is used to check if each pair of nodes has the same class or if one is a super class of the other. The status of the merged node is "non-active"; (ii) the *function: checkRelation* is used to check if the pair of nodes has a relationship with each other. The D-LKN index is used to determine the active node. The node with a shorter distance to the root is active. Unfortunately, if any couple of nodes have the same distance, the domain of the relationship is active.

**Node Expansion**
Each active node will expand to a new connecting node, in the round robin (line 8-10). This is an iterative process with the aim to move up one level at a time towards the

root node. The bidirectional traversal approach is used to determine which node will be chosen.

*Forward Expansion*

Given the possible node list (p_nodeList$_i$), the distance from each possible nodes to the root is considered as a forward expansion condition. If the distance is null then that node will not be chosen because it never reaches the root.

*Backward Expansion*

The remaining possible nodes from forward expansion ($S_{L1}$) are further determined by using backward expansion condition; (i) if there is the only one possible node in $S_{L1}$ and it can be merged with other active nodes, then it will be the node in the query tree constructed so far. (ii) In the case that there are more than one possible nodes (i.e. $S_{L1}$.lenght > 1), the selected node is the one which has the most frequency of merging with other nodes. (iii) Unfortunately, if the previous condition is inadequate to determine, a node with the shortest length to the pre-defined root will be chosen. Finally the expanding node is returned with the notice of the direction of relation between the expanding node and the old node.

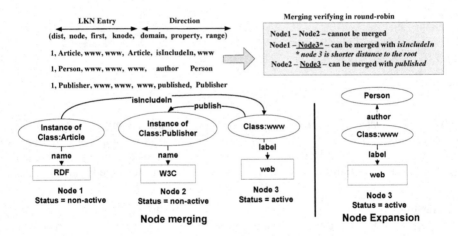

**Fig. 5.** Two basic operations - node merging and node expansion

*Stopping*

The loop of node merging and node expansion is repeated until stopping condition which is checked by *function:CheckEnd* (line7). The algorithm stops whenever, all current nodes are non-active because the node all converge to the root node. In this case the algorithm returns a query graph. Otherwise, the graph status is false and no result is returned.

The query graph construction process may result more than one possible query graphs with different meanings. Therefore, ranking will be taken to compute the most relevant query graph to the user's information need. X-SKengine still uses the same ranking schemes as SKengine.

# 6     Evaluation

## 6.1     Experiment Setup

This experiment was to compare the capability of X-SKengine with the SKengine (i.e. the previous version) as regards the semantic query construction with different ontology structures. The test set of ontologies includes OntoLife [7], RDFResume [9], DBLP [10], FOAF [6] and Researcher (i.e. the designed ontology for describing researcher information). OntoLife is the star-shaped ontology for comparing with the others. Five sets of simulated RDF information associated with each ontology were constructed. In addition, five colleagues of the author were asked to provide 20 keyword queries along with the description in natural language. An example of a keyword query for DBLP ontology is "WC3 XML website" and the related description is "Who is author of title XML in W3C web site?"

For assessing the effectiveness of the generated semantic queries and rankings, Mean Reciprocal Rank (MRR) [11] was adopted with regard to the system created a candidate list of SPARQL queries in order by 'probability of acceptability'. The acceptable SPARQL query referred to the query that was equivalent to the manually generated SPARQL query. MRR $= \frac{1}{|N_{KP}|}\sum_{i=1}^{N_{KP}}\frac{1}{rank_i}$ where |NKP| is the number of keyword phrases in the test set; $rank_i$ = order of the acceptable SPARQL query of each keyword phrase. For the experiment process, each ontology, RDF data graph and the test set of user keywords were input to X-SKengine and SKengine. After being run, the rank of acceptable SPARQL query of each keyword phrase was recorded, Finally the MRR value was calculated.

## 6.2     Result and Analysis

The results are summarized in Table 1. As can be seen, SKengine can  provide the high values of MRR for the star-shaped ontology as Ontolife. Unfortunately for other ontology structures, it returned significantly lower values of MRR. This points that SKengine is only specific for star-shaped ontology. While X-SKengine provides the high value of MRR for  all tested ontology structures. That means the X-SKengine has capability to generate and rank the acceptable semantic query relevant to the meaning of user keywords without limitation of ontology structures.

**Table 1.** Result of the experiment

| Ontologies | Size of RDF data (MB) | MRR | |
|---|---|---|---|
| | | **SKengine** | **X-SKengine** |
| RDF Resume | 10.6 | 0.15 | 0.75 |
| Ontolife | 9.0 | 0.78 | 0.78 |
| DBLP | 11.1 | 0.10 | 0.72 |
| FOAF | 12.0 | 0.28 | 0.75 |
| Researcher | 12.5 | 0.30 | 0.81 |

# 7    Conclusion

This paper proposed X-SKengine with the novel bidirectional fix root query graph construction algorithm. This is for solving limitation of query graph construction in the previous work. The experiment shows that, X-SKengine has capability to generate and rank the acceptable semantic query relevant to the meaning of the user keyword without limitation of ontology structure. An important next step of X-SKengine involves the investigation of machine learning techniques for improving the ranking capability.

**Acknowledgement.** The author is grateful to the National Science and Technology Development Agency (NSTDA) for the financial support under grant agreement no. SCH-NR2012-511 (The Extended Semantic Keyword Interface Engine).

# References

1. Zhou, Q., Wang, C., Xiong, M., Wang, H., Yu, Y.: SPARK: Adapting keyword query to semantic search. In: Aberer, K., et al. (eds.) ASWC 2007 and ISWC 2007. LNCS, vol. 4825, pp. 694–707. Springer, Heidelberg (2007)
2. Wang, H., Zhang, K., Liu, Q., Tran, T., Yu, Y.: Q2Semantic: A lightweight keyword interface to semantic search. In: Bechhofer, S., Hauswirth, M., Hoffmann, J., Koubarakis, M. (eds.) ESWC 2008. LNCS, vol. 5021, pp. 584–598. Springer, Heidelberg (2008)
3. Tran, T., Cimiano, P., Rudolph, S., Studer, R.: Ontology-based interpretation of keywords for semantic search. In: Aberer, K., et al. (eds.) ASWC 2007 and ISWC 2007. LNCS, vol. 4825, pp. 523–536. Springer, Heidelberg (2007)
4. Sitthisarn, S., Lau, L.M.S., Dew, P.M.: Semantic keyword search for expert witness discovery. In: 2011 International Conference on Semantic Technology and Information Retrieval (STAIR), Kuala Lumpur, Malaysia, pp. 18–25. IEEE (2011)
5. Elbassuoni, S., Blanco, R.: Keyword search over RDF graphs. In: Proceedings of the 20th ACM International Conference on Information and Knowledge Management (CIKM 2011) (2011)
6. FOAF Vocabulary Specification 0.91. http://xmlns.com.foaf.spec/
7. Kargioti, E., Kontopoulos, E., Bassiliades, N.: OntoLife: An ontology for semantically managing personal information. In: Iliadis, L., Vlahavas, I., Bramer, M. (eds.) Artificial Intelligence Applications and Innovations III. IFIP, vol. 296, pp. 127–133. Springer, Boston (2009)
8. He, H., Wang, H., Yang, J., Yu, P.S.: BLINKS: ranked keyword searches on graphs. In: SIGMOD Conference (2007)
9. Bojars, U., Breslin, J.G.: ResumeRDF: Expressing skill information on the semantic web. In: The 1st International ExpertFinder Workshop, Berlin, Germany (2007)
10. DBLP. http://datahub.io/dataset/l3s-dblp
11. Voorhees, E.M.: The TREC-8 Question Answering Track Report, NIST (2000)

# Applications of Semantic Technology

# An Ontology-Based Intelligent Speed Adaptation System for Autonomous Cars

Lihua Zhao[1]([✉]), Ryutaro Ichise[2], Seiichi Mita[1], and Yutaka Sasaki[1]

[1] Toyota Technological Institute, Nagoya, Japan
{lihua,smita,yutaka.sasaki}@toyota-ti.ac.jp
[2] National Institute of Informatics, Tokyo, Japan
ichise@nii.ac.jp

**Abstract.** *Intelligent Speed Adaptation* (ISA) is one of the key technologies for *Advanced Driver Assistance Systems* (ADAS), which aims to reduce car accidents by supporting drivers to comply with the speed limit. Context awareness is indispensable for autonomous cars to perceive driving environment, where the information should be represented in a machine-understandable format. Ontologies can represent knowledge in a format that machines can understand and perform human-like reasoning. In this paper, we present an ontology-based ISA system that can detect overspeed situations by accessing to the ontology-based *Knowledge Base* (KB). We conducted experiments on a car simulator as well as on real-world data collected with an intelligent car. Sensor data are converted into RDF stream data and we construct SPARQL queries and a C-SPARQL query to access to the Knowledge Base. Experimental results show that the ISA system can promptly detect overspeed situations by accessing to the ontology-based Knowledge Base.

**Keywords:** Ontology · *Intelligent Speed Adaptation* (ISA) · Autonomous car · Knowledge Base · SPARQL · RDF stream

## 1 Introduction

*Advanced Driver Assistance Systems* (ADAS) are designed to improve car safety by perceiving a driving environment and making decisions for safe driving. Overspeed is one of the main causes of car accidents, which should be automatically detected to warn the drivers or the autonomous cars. *Intelligent Speed Adaptation* (ISA) technology is one of the most cost-efficient way to improve roadway safety by gaining the speed limits from digital maps or vehicle-to-roadside wireless communication [7]. Advanced digital maps include not only road topological information, but also speed limits, nearest *Points-Of-Interests* (POI) such as schools and restaurants.

Current intelligent autonomous cars sense and perceive surrounding environment with sensors such as a camera, lidar, and GPS. These sensors and digital map information should be represented in the format that autonomous cars can understand. Therefore, we use ontologies to represent knowledge, which is

© Springer International Publishing Switzerland 2015
T. Supnithi et al.(Eds.): JIST 2014, LNCS 8943, pp. 397–413, 2015.
DOI: 10.1007/978-3-319-15615-6_30

defined as an explicit specification of a conceptualization [9]. Ontologies are the structural frameworks for organizing information and are used in artificial intelligence, Semantic Web, biomedical informatics, and information architecture as a form of knowledge representation about the world or some part of it.

*Resource Description Framework* (RDF) is designed for conceptual description that provides a clear specification for modeling data [3]. An instance is described by a collection of RDF triples in the form of <subject, property, object>, where property is also called predicate [20]. The *Web Ontology Language* (OWL) is a semantic markup language developed as a vocabulary extension of the RDF with more vocabularies for describing classes and properties [6]. To represent the sensor stream data from the sensors equipped on the autonomous cars, we use timestamp-based temporal RDF representation to construct RDF Stream data [10]. *SPARQL Protocol and RDF Query Language* (SPARQL) is a powerful RDF query language that enables Semantic Web users to access to static RDF data [11]. C-SPARQL is an extension of SPARQL designed to express continuous queries such as RDF stream data [5].

In this paper, we introduce an ontology-based Knowledge Base and an ISA system that detects overspeed at real-time. The ISA system receives RDF stream data from sensors and query on the ontology-based Knowledge Base to gain speed limits of different roads and school zones to send overspeed warning.

The remainder of this paper is organized as follows: In Section 2, we introduce some of the related research papers that utilize ontologies for autonomous cars. We introduce the flowchart of the ISA system and the ontology-based Knowledge Base in Section 3. Section 4 shows experimental results with car simulator and real-world sensor data collected with an intelligent car. In Section 5, we discuss the advantages of our approach and some issues occurred in the experiments. At last, we conclude and state future work in Section 6.

## 2   Related Work

Many researchers have utilized ontologies for ADAS or autonomous vehicle controlling. A simple ontology that includes context concepts such as Mobile Entity (Pedestrian and Vehicle), Static Entity (Road Infrastructure and Road Intersection), and context parameters (isClose, isFollowing, and isToReach) is modeled to enable the vehicle to understand the context information when it approaches road intersections [4]. By applying 14 rules written in *Semantic Web Rule Language* (SWRL), the ontology is able to process human-like reasoning on global road contexts.

Two ontology models about autonomy levels and situation assessment for *Intelligent Transportation Systems* (ITS) are introduced for co-driving [14]. One ontology defines the relationship between the automation levels and the algorithmic needs. The other ontology is related to the situation assessment level including the concepts of driver, ego-vehicle, communication, free zone, moving obstacles, and environment. Inference rules are defined to link the situation

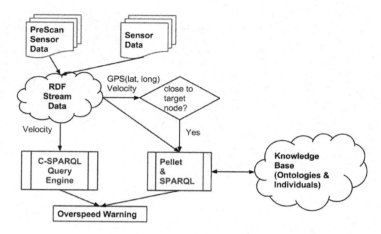

**Fig. 1.** ISA System Flowchart

assessment ontology to the automation level ontology, which contains five control levels of a car: longitudinal control, lateral control, local planning, parking, and global planning.

A complex intersection ontology, which contains concepts of car, crossing, road connection (lane and road), and sign at crossing (traffic light and traffic sign) is introduced to form a lean ontology to facilitate fast reasoning that is close to real-time [12]. Another ontology-based traffic model that can represent typical traffic scenarios such as intersections, multi-lane roads, opposing traffic, and bi-directional lanes is introduced in [15]. Relations such as opposing, conflicting, and neighboring are introduced to represent the semantic context of the traffic scenarios for decision making.

In contrast to the above research, our approach aims for constructing an ontology-based Knowledge Base, which is based on map, control, and car ontologies. Ontology reasoning is conducted only once at the beginning of the experiment and process sensor RDF stream data by executing SPARQL and C-SPARQL queries. We make the most of Semantic Web technology to detect overspeed, which is one of the most effective way to improve driving safety.

## 3    Approach

We propose an ISA system that enables the autonomous cars to perceive driving environment for overspeed detection. An ontology-based Knowledge Base is the essential component of the ISA system that contains machine-understandable knowledge of environment such as speed limit information. In this section, we will explain the flowchart of the ISA system and describe the ontology-based Knowledge Base in detail. We will also give some examples of SWRL rules, SPARQL, C-SPARQL queries that are used for the ISA system.

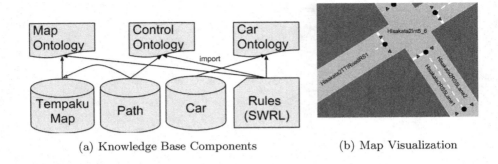

(a) Knowledge Base Components          (b) Map Visualization

**Fig. 2.** Knwledge Base

### 3.1 System Flowchart

Figure 1 shows the flowchart of ISA system. There are two resources of sensor data: one is from PreScan simulator and the other is from the sensors installed on an intelligent car. These sensor data are converted into RDF stream data for further process.

C-SPARQL and SPARQL queries are performed on the RDF stream data to detect overspeed. With C-SPARQL query, we check if a car's velocity exceeds its own maximum speed limit. We execute SPARQL queries on the inferred Knowledge Base to retrieve current lane, speed limit, and update the next target node according to the received real-time sensor data. SPARQL queries are only performed when the car approaches to the closest position from the current target node. By comparing a car's velocity with the speed limit retrieved from the Knowledge Base, we can send overspeed warning to the autonomous car.

### 3.2 Knowledge Base

*Knowledge Base* (KB) is indispensable for autonomous cars to perceive driving environment and to make safe driving decisions. To construct a machine-understandable Knowledge Base, we construct several ontologies, instances, and rules for autonomous cars.

In the following, we introduce the ontologies used for ISA system to detect overspeed situations at real-time. Figure 2 shows the Knowledge Base, where Fig. 2a shows the ontology-based Knowledge Base designed for the ISA system and Fig. 2b shows the visualization of the map representation. The Knowledge Base consists of three ontologies: map ontology, control ontology, car ontology; three instance files of the map (Tempaku ward near the campus of Toyota Technological Institute), path, car; and rules based on the ontologies.

#### 3.2.1 Ontologies

For practical usage of ontologies, they should support large-scale interoperability, to be well-founded and axiomatized to be generally understood [18]. Ontologies

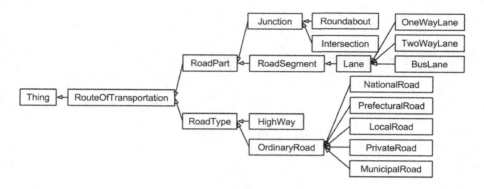

**Fig. 3.** Map Ontology

are expressed in an ontology language, which usually deal with the following kinds of entities [8]:

- **Classes** are interpreted as a set of instances in the domain defined by owl:Class. Classes are also called concepts that are the main entities of an ontology.
- **Properties** are also called predicates or relations, mainly categorized into owl:ObjectProperty and owl:DatatypeProperty.
- **Instances** are interpreted as particular individuals of a domain, which are defined by owl:Thing.
- **Rules** are statements in the form of an if-then sentence that describe the logical inferences.

Protégé ontology editor is used for developing ontologies, which is an open development environment for Semantic Web applications [13]. In the following, we describe the main parts of three ontologies that are used for the ISA system.

1. **Map Ontology**
   Digital map is one of the resources that an autonomous car can retrieve knowledge about road and surrounding environment information. Therefore, we constructed a map ontology that can describe the road, intersection, lane, and speed limit.
   Figure 3 shows the main classes of the map ontology, where the concepts are linked with rdfs:subClassOf relation. A road consists of connected road segments and junctions, where a road segment contains arbitrary number of lanes. Each road has speed limit, which is described using a DatatypeProperty map:speedMax[1]. The DatatypeProperty map:boundPOS (circle marks in Fig. 2b) is used to describe boundary GPS point of junctions and road segments. The ObjectProperty map:isLaneOf is defined to assert the relation between lanes and road segments. A lane is described with map:enterPos and map:exitPos (triangle marks in Fig. 2b), which indicates the enter and exit

---

[1] PREFIX map: http://www.toyota-ti.ac.jp/Lab/Denshi/COIN/Map#

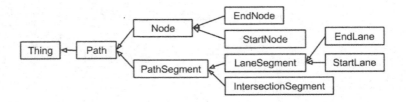

**Fig. 4.** Control Ontology

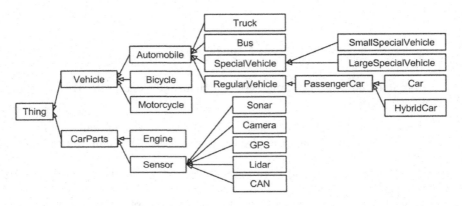

**Fig. 5.** Car Ontology

GPS points, respectively. The DatatypeProperty map:osm_ref is designed to link the road in our Knowledge Base with OpenStreetMap road resources.

2. **Control Ontology**

   The classes shown in Fig. 4 are used to represent paths of autonomous cars. Instead of collecting all the GPS points of a trajectory, we use instances of control:pathSegment[2], which can be either an intersection or a lane. The Node class contains startNode and endNode, which are the start and end GPS positions. We assigned a DatatypeProperty control:pathSegmentID to index path segments and use ObjectProperty control:nextPathSegment to link connected path segments.

3. **Car Ontology**

   The car ontology contains concepts of different types of vehicles and devices which are installed on a car such as sensors and engines as shown in Fig. 5. The ObjectProperty car:usedSensor[3] is designed to relate sensors with a car, and the DatatypeProperty car:distance_front is used to measure the distance from a sensor to the front of a car. We also added other DatatypeProperties such as car_ID, car_length, and velocity to describe a car.

---

[2] PREFIX control: http://www.toyota-ti.ac.jp/Lab/Denshi/COIN/Control#

[3] PREFIX car: http://www.toyota-ti.ac.jp/Lab/Denshi/COIN/Car#

**Table 1.** An example of map ontology based instances

| Subject | Property | Object |
|---|---|---|
| tempaku:Hisakata2Road | rdf:type | map:LocalRoad |
| tempaku:Hisakata2Road | map:speedMax | "40"^^kmh |
| tempaku:Hisakata2Road | map:osm_ref | osm_way:49559442 |
| tempaku:Hisakata2Road | map:hasIntersection | tempaku:Hisakata2Int5_6 |
| tempaku:Hisakata2Road | map:hasRoadSegment | tempaku:Hisakata2TTIRoadRS1 |
| tempaku:Hisakata2Int5_6 | rdf:type | map:Intersection |
| tempaku:Hisakata2Int5_6 | map:boundPos | 35.107663, 136.983845 |
| tempaku:Hisakata2Int5_6 | map:boundPos | 35.107846, 136.983889 |
| tempaku:Hisakata2Int5_6 | map:boundPos | 35.107860, 136.983683 |
| tempaku:Hisakata2Int5_6 | map:boundPos | 35.107708, 136.983604 |
| tempaku:Hisakata2Int5_6 | map:isConnectedTo | tempaku:Hisakata2RS5 |
| tempaku:Hisakata2Int5_6 | map:isConnectedTo | tempaku:Hisakata2TTIRoadRS1 |
| tempaku:Hisakata2RS5 | rdf:type | map:RoadSegment |
| tempaku:Hisakata2RS5 | map:boundPos | 35.107663, 136.983845 |
| tempaku:Hisakata2RS5 | map:boundPos | 35.107357, 136.984067 |
| tempaku:Hisakata2RS5 | map:isConnectedTo | tempaku:Hisakata2Int5_6 |
| tempaku:Hisakata2RS5Lane1 | rdf:type | map:OneWayLane |
| tempaku:Hisakata2RS5Lane1 | map:enterPos | 35.107353, 136.984054 |
| tempaku:Hisakata2RS5Lane1 | map:exitPos | 35.107657, 136.983832 |
| tempaku:Hisakata2RS5Lane1 | map:isLaneOf | tempaku:Hisakata2RS5 |
| tempaku:Hisakata2RS5Lane2 | rdf:type | map:OneWayLane |
| tempaku:Hisakata2RS5Lane2 | map:enterPos | 35.107667, 136.983856 |
| tempaku:Hisakata2RS5Lane2 | map:exitPos | 35.107362, 136.984081 |
| tempaku:Hisakata2RS5Lane2 | map:isLaneOf | tempaku:Hisakata2RS5 |

### 3.2.2  Instances

Instances are also known as individuals that model abstract or concrete objects based on the ontologies. With the three ontologies, we model instances such as maps, paths, and cars. In the following, we give some examples of instances in the Knowledge Base.

1. **Tempaku Map Instance**
   Map instances include roads, road segments, intersections, lanes, schools, etc. Table 1 shows instances of a road tempaku:Hisakata2Road, an intersection tempaku:Hisakata2Int5_6[4], a road segment tempaku:Hisakata2RS5 connected to the intersection, and two lanes tempaku:Hisakata2RS5Lane1 and tempaku:Hisakata2RS5Lane2 on tempaku:Hisakata2RS5 road segment as shown in Fig. 2b. The instances are based on the map ontology, which uses map:hasIntersection or map:hasRoadSegment to link a road with an intersection or a road segment. We use the ObjectProperty map:isConnectedTo to link intersections with road segments and use map:isLaneOf to relate lanes with road segments. The tempaku:Hisakata2Road relates to the OpenStreetMap instance osm_way:49559442[5], which has speed limit of 40km/h.
   By accessing to this ontology-based map, an intelligent car can retrieve the current lane, the speed limit of current road, and the position to update the target node, which is either the position of a car entering from an intersection to a lane or from a lane to an intersection. Here, the target node is the map:enterPos of the next lane or the map:exitPos of the current lane.

---

[4] PREFIX tempaku: http://www.toyota-ti.ac.jp/Lab/Denshi/COIN/TempakuMap#
[5] PREFIX osm_way: http://www.openstreetmap.org/way/

**Table 2.** An example of a path instance

| Subject | Property | Object |
|---------|----------|--------|
| path:Path1 | rdf:type | control:Path |
| path:Path1 | control:startLane | tempaku:TTICampRS1Lane1 |
| path:Path1 | control:endLane | tempaku:TTICampRS1Lane2 |
| tempaku:TTICampRS1Lane1 | rdf:type | control:StartLane |
| tempaku:TTICampRS1Lane1 | control:pathSegmentID | 0 |
| tempaku:TTICampRS1Lane1 | control:nextPathSegment | tempaku:Hisakata2TTIRoadInt2_3 |
| tempaku:Hisakata2TTIRoadInt2_3 | control:pathSegmentID | 1 |
| tempaku:Hisakata2TTIRoadInt2_3 | control:pathSegmentID | 99 |
| tempaku:Hisakata2TTIRoadInt2_3 | control:nextPathSegment | tempaku:Hisakata2TTIRoadRS2Lane2 |
| tempaku:Hisakata2TTIRoadInt2_3 | control:nextPathSegment | tempaku:TTICampRS1Lane2 |

**Table 3.** An example of a TOYOTA car with a GPS-IMU sensor

| Subject | Property | Object |
|---------|----------|--------|
| ex:ToyotaCar | rdf:type | car:Car |
| ex:ToyotaCar | car:carID | Z00000001 |
| ex:ToyotaCar | car:usedSensor | ex:GPSSensor |
| ex:GPSSensor | rdf:type | car:GPS |
| ex:GPSSensor | car:distance_from | "2500"^^mm |

2. **Path Instance**

   A path instance is constructed based on the Tempaku map and control ontology as shown in Fig. 2a. It is predefined for an autonomous car so that it can retrieve the next path segment. A path segment is either a lane or an intersection. Table 2 shows a path instance path:Path1[6], which contains a start lane and an end lane. Each path segment has control:pathSegmentID and next path segment linked with control:nextPathSegment.

3. **Car Instance**

   Intelligent cars are equipped with sensors such as cameras, CAN, and GPS. Car instances are constructed based on the car ontology. Table 3 is an example of a TOYOTA car instance ex:ToyotaCar[7] with a sensor instance ex:GPSSensor installed on the car. The car's ID is Z00000001, and the distance from the position of GPS sensor to the front of the car is 2500mm.

### 3.2.3  Rules

The *Semantic Web Rule Language* (SWRL) is a proposed language for the Semantic Web that can be used to express rules as well as logic, combining OWL DL or OWL Lite with a subset of the Rule Markup Language [1].

   Pellet reasoner provides standard and cutting-edge reasoning services for OWL ontologies, which can be used for inferring rules [17]. It implements a tableau-based decision procedure for general TBoxes (subsumption, satisfiability, classification) and ABoxes (retrieval, conjunctive query answering). TBox is

---

[6] PREFIX path: http://www.toyota-ti.ac.jp/Lab/Denshi/COIN/Path#
[7] PREFIX ex: http://www.toyota-ti.ac.jp/Lab/Denshi/COIN/Example#

**Table 4.** Query the next path segment

| |
|---|
| SELECT DISTINCT ?next |
| WHERE { |
| tempaku:currentPathSegment control:nextPathSegment ?next. |
| tempaku:currentPathSegment control:pathSegmentID  "currentID"^^xsd:int. |
| ?next                      control:pathSegmentID  ?nextID. |
| Filter ( ?nextID = (currentID + 1) ) } |

a Terminological Component and ABox is an Assertion Component, both make up a Knowledge Base.

The following two rules indicate that if a car is running on a road near a kindergarten, the speed limit is 30km/h even the default speed limit is higher than 30km/h.

Intersection(?rs), Kindergarten(?place), nearTo(?place, ?rs) -> SpeedLimit30(?rs)
Kindergarten(?place), RoadSegment(?rs), nearTo(?place, ?rs) -> SpeedLimit 30(?rs)

In the above rules, the ObjectProperty map:nearTo describes that an intersection or a road segment is near to a kindergarten. For example, an RDF triple <tempaku:Takasaka_Kindergarten, map:nearTo, tempaku:Hisakata2RS2> means that the tempaku:Takasaka_Kindergarten is located near to the road segment tempaku:Hisakata2RS2. Therefore, if a car is running on any lane of the road segment tempaku:Hisakata2RS2, it should run at maximum 30km/h.

## 3.3  Query

SPARQL and C-SPARQL queries are applied to access to the ontology-based Knowledge Base and RDF Stream data, respectively. In the following, we will give some query examples used for the ISA system.

### 3.3.1  SPARQL Query

SPARQL is a powerful RDF query language that enables Semantic Web users to access to the ontology-based Knowledge Base. Three main SPARQL queries are designed for the ISA system.

Table 4 shows a SPARQL query to retrieve the next path segment with current path segment tempaku:currentPathSegment and current pathSegmentID "currentID"^^xsd:int. The first pathSegmentID is 0 and increments by 1. By assigning the pathSegmentID, we can easily identify the next path segment even the current path segment has more than one pathSegmentID. For example, the intersection tempaku:Hisakata2TTIRoadInt2_3 has two pathSegmentIDs 1 and 99 as shown in Table 2. With the SPARQL query in Table 4, we can retrieve the next path segment tempaku:Hisakata2TTIRoadRS2Lane2, which has the pathSegmentID as 2.

**Table 5.** Query the speed limit of currentPathSegment

| | | |
|---|---|---|
| SELECT ?max | | |
| WHERE { | | |
| { tempaku:currentPathSegment | map:isLaneOf | ?roadsegment. |
| ?roadsegment | map:isRoadSegmentOf | ?road. |
| ?road | map:speedMax | ?max. |
| } UNION { | | |
| ?road | map:hasIntersection | map:currentPathSegment. |
| ?road | map:speedMax | ?max. } |
| } | | |

**Table 6.** C-SPARQL query for checking if a car overspeeds its own speed limit

```
REGISTER QUERY OverSpeedCheck AS
SELECT ?car
FROM STREAM ⟨http://www.toyota-ti.ac.jp/Lab/Denshi/COIN/stream⟩
   [RANGE 500ms STEP 50ms]
WHERE { ?car car:velocity ?speed . }
GROUP BY ?speed
HAVING (AVG(?speed) >= maxSpeed )
```

Table 5 shows a SPARQL query that retrieves the maximum allowed speed of current path segment, which can be either a road segment or an intersection. SPARQL query "ASK { map:currentPathSegment rdf:type map:SpeedLimit30}" is executed to check if the currentPathSegment is near to a kindergarten, in other words, check if the current path segment has speed limit of 30km/h. The speed limit is inferred from the SWRL rules introduced above.

Another SPARQL query is to update a target node GPS position. A target node is defined as the map:exitPos of a lane if the car is currently running on a lane, or map:enterPos of the next lane if the car is currently running on an intersection. The distance of the car and a target node is calculated at real-time according to the Haversine Formula [16]. When the car approaches to the target node, we retrieve the next target node using SPARQL query and update it.

### 3.3.2 C-SPARQL Query
RDF stream data are represented in the format of RDFQuadruple <Subject, Property, Object, Timestamp>. For example, the RDFQuadruple <ex:Toyota Car, car:velocity, "11.11"^^xsd:float, 1406360324543> represents the velocity of a ToyotaCar at time 1406360324543, where the timestamp uses the time in milliseconds. We use geo:lat[8] and geo:long to represent the latitude and longitude information from the GPS sensor.

We register the OverSpeedCheck C-SPARQL query as shown in Table 6, to query on the RDF stream data. The C-SPARQL query checks if a car's average

---

[8] PREFIX geo: http://www.w3.org/2003/01/geo/wgs84_pos#

(a) PreScan Simulator Car                    (b) Intelligent Car

**Fig. 6.** PreScan simulator car and intelligent car

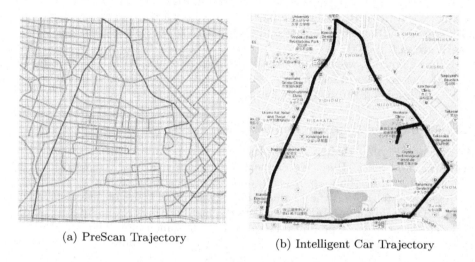

(a) PreScan Trajectory                    (b) Intelligent Car Trajectory

**Fig. 7.** Trajectory for the experiment

velocity in the past 500ms exceeds its own allowed maximum speed (maxSpeed, i.e. 120km/h). The RANGE is the duration to receive sensor stream data for analysis and the STEP size is the frequency of a sensor receiver.

## 4 Experiment

In this section, we introduce the Knowledge Base used for our experiments, and discuss experimental results of our *Intelligent Speed Adaptation* (ISA) system. The experiments are conducted in two ways. First, we test with PreScan simulator by setting up the path for a simulator car as shown in Fig. 6a. Then, we test with the real-world data collected using an intelligent car (Fig. 6b) run on the same path.

### 4.1   Experiment Settings

The computer specification for both experiments are as follows

| Experiment | Operating System | CPU (64 bits) | RAM |
|---|---|---|---|
| PreScan Simulator | Windows 7 Professional | i7-4770 CPU @3.40GHz | 32GB |
| Real-World Data | Ubuntu 12.04 LTS | i7-4770 CPU @3.40GHz | 16GB |

The trajectory for the experiment is shown in Fig. 7, which is around Toyota Technological Institute (TTI) campus in Tempaku ward in Nagoya, Japan. The path contains 101 path segments and each path segment is assigned a pathSegmentID from 0 to 100.

In the following, we will show the experimental results with PreScan simulator and with real-world sensor data collected with the intelligent car. The sensor data is transmitted through User Datagram Protocol (UDP) [2] at real-time.

### 4.2   Knowledge Base for Experiments

As shown in Fig. 2a, the Knowledge Base for an experiment contains ontologies, instances, and rules. Here, we explain the instances in our Knowledge Base, which are based on the map, control, and car ontologies.

As shown in Table 7, the first column lists different types of instances in the Knowledge Base, the second column lists the number of corresponding instances in it. The last column represents the speed limit of some roads. The PrivateRoad is the road inside TTI campus that allows 25km/h as maximum speed. There are one MunicipalRoad that allows maximum 50km/h, and five LocalRoads where three of them allows 40km/h and two of them allows 50km/h.

The Knowledge Base is inferred with Pellet reasoner at the beginning of the experiment. Then the ISA system accesses to the inferred Knowledge Base to retrieve knowledge of the environment. Pellet API[9] with OWL API[10] are

**Table 7.** Instances in the Knowledge Base

| Types of Instances | Number of Instances | Speed Limit |
|---|---|---|
| Intersection | 54 | |
| RoadSegment | 59 | |
| Kindergarten | 1 | |
| BusLane | 5 | |
| OneWayLane | 162 | |
| PrivateRoad | 1 | 25km/h |
| MunicipalRoad | 1 | 50km/h |
| LocalRoad | 5 | 40km/h (3), 50km/h (2) |
| Path | 1 | |

[9] http://clarkparsia.com/pellet/
[10] http://owlapi.sourceforge.net/

applied for reasoning and Jena API[11] is used for executing SPARQL queries on the Knowledge Base.

The Pellet reasoner infers that three road segments and two intersections near the Takasaka kindergarten are instances of SpeedLimit30 according to the SWRL rules. Therefore, even though the road allows maximum 40km/h on the road, the car should run under 30km/h on these SpeedLimit30 area.

### 4.3 PreScan Simulator Experiment

PreScan is specifically aimed at designing and evaluating ADAS and *Intelligent Vehicle* (IV) systems that are based on sensor technologies such as radar, laser, camera, ultrsonic, GPS and C2C/C2I communications [19]. PreScan can communicate with MATLAB (2011b) [12] and Simulink [13] during simulations. We collect the sensor data in the simulator with a C++ program on simulink. The OpenStreetMap near TTI campus is imported to create the simulation environment as shown in Fig. 7a. We used a car actor with C2C/C2I Receiver & Transmitter Antenna for receiving its own velocity and GPS position with 50ms frequency. Instrument Control Toolbox[14] in Matlab is applied to set up a UDP server for the simulation to send the sensor data at real-time.

For the simulation test, we manually set up the trajectory as shown in Fig. 7a. A trajectory in PreScan consists of a speed profile and a path. We set up the speed between 8m/s to 18m/s by performing smooth acceleration, smooth deceleration, and constant speed in the middle of the trajectory. For the start and end of the path, we set up smooth acceleration and smooth deceleration to finish the trajectory.

Figure 8a shows the experimental result of overspeed detection, where the cross marks are the positions that the car exceeded the speed limit. Cross marks in the circle represent the overspeed positions, which are close to the Takasaka kindergarten. The reasoning time with Pellet reasoner took 242ms and the three SPARQL queries took 11ms on average to detect if the car overspeeds or not. The maximum query time is 23ms and minimum 2ms for executing three SPARQL queries.

The average distance from a sensor to the front of a car at the point of updating the next target node is 3.2113m, with maximum 7.3440m and min 1.861m. However, the optimal average distance should be approximately 2.5m. This difference on PreScan simulator is caused by the GPS position shifts in PreScan simulator. The number of lanes are not included in some parts of the roads and there are also some error connections when importing the OpenStreetMap into PreScan. After manually fixing the imported environment, there occurs some GPS shifts and mismatches to the real positions. However, by matching to the ontology-based map and path, we can correctly detect the speed limit on the running road and detect overspeed situations.

---

[11] http://jena.apache.org/documentation/ontology/
[12] http://www.mathworks.com/products/matlab/
[13] http://www.mathworks.com/products/simulink/
[14] http://www.mathworks.com/products/instrument/

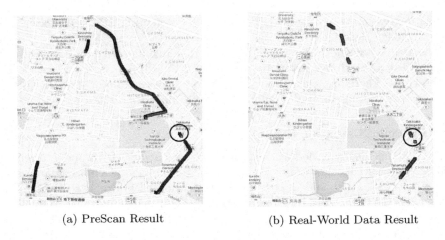

(a) PreScan Result                    (b) Real-World Data Result

**Fig. 8.** Experiment result of overspeed detection

### 4.4    Real-World Data Experiment

The real-world data is collected with an intelligent car, which is a hybrid car equipped with many sensors such as GPS-IMU sensor. The GPS-IMU sensor sends sensor data with the frequency of 5ms. We collected sensor data by running on the same path as shown in Fig. 7b, which took about 13 minutes. The ISA system is tested on the collected sensor data to detect overspeed situations.

Figure 8b shows the experimental result with the real-world data. The cross marks represent overspeed positions and the cross marks in the circle represent overspeed positions near the Takasaka kindergarten. The experimental result shows that even an experienced driver may drive faster than the speed limit, which is normally shown with a road sign on the road.

The running time for reasoning with Pellet reasoner is 177ms and the execution time for three SPARQL queries is 11ms. The maximum query time is 23ms and minimum 3ms. The distance from the GPS-IMU sensor to the front of the intelligent car is 2.5m. Therefore, the position where we decide to update the target node is where the closest distance to the target node is approximately 2.5m. According to the experimental result, the average distance for updating the next target node is 2.514m, maximum 2.554m, and minimum 2.134m. The minimum distance appeared at two positions, where the car had sudden acceleration that caused a big movement from previous position.

With the ontology-based Knowledge Base and precise GPS positions from the GPS-IMU sensor, we can promptly detect the current running lane or intersection and retrieve the speed limit information. The accuracy of the overspeed detection may be affected by the delay of data transmission from the GPS-IMU sensor or by the execution time of three SPARQL queries. Therefore, the ideal execution time for overspeed detection should be less than the frequency of the GPS-IMU sensor, which is 5ms in our experiment.

# 5    Discussion

We proposed an Intelligent Speed Adaptation system that can effectively detect overspeed situations by accessing to the ontology-based Knowledge Base. The system works on both PreScan simulator and real-world data collected with an intelligent car. The advantage of our approach is that we can represent the sensor data into machine-understandable RDF stream data and query with SPARQL and C-SPARQL to retrieve knowledge from the Knowledge Base. The Knowledge Base contains map, control, and car ontologies, instances of map, path, and car based on these three ontologies, and rules written in SWRL. By accessing to the Knowledge Base, an autonomous car can easily perceive the driving environment by gaining machine-understandable knowledge. However, we face some problems using the PreScan simulator and GPS-IMU sensor for experiments.

Since the simulator cannot perfectly simulate the same environment as real-world environment, there exist shifts of GPS positions and missing number of lanes on some roads. In order to deal with these mismatches, we have to test several times to find out the maximum shift distance to set up allowed shift threshold for finding out the position for updating target nodes. Because of the mismatches, the car running on the left lane sometimes appears as running on the right lane in the PreScan simulator.

The frequency of sending sensor data for GPS-IMU is 5ms. However, we found there are delays of data transmission in the collected GPS sensor data. The average frequency is 5.16ms, which has maximum frequency of 61ms and minimum frequency of 5ms. Therefore, if we assume that the velocity is 50km/h, the ISA system cannot receive the data during it advances approximately 85cm in the worst case.

# 6    Conclusion and Future Work

The main purpose of our research is to apply Semantic Web technology to the *Intelligent Speed Adaptation* (ISA) system, which is one of the most cost-efficient way to improve roadway safety. In this paper, we proposed an ontology-based Knowledge Base and an ISA system that accesses to the Knowledge Base. The ISA system utilized Semantic Web technology such as reasoning, SPARQL, C-SPARQL, and SWRL rules to process sensor RDF stream data. Experimental results show that the ISA system can effectively detect overspeed situations on PreScan simulator and on the real-world data collected by an intelligent car.

This system can be further extended by adding more knowledge such as traffic light data and traffic regulations to improve driving safety. In future work, we will add more instances and traffic regulations to the Knowledge Base so that the autonomous car could perceive the environment with sensors and make decisions by obeying the traffic regulations. Furthermore, we will add links to other data resources and apply ontology integration method as introduced in [21] to discover hidden knowledge from linked instances for autonomous cars.

**Acknowledgments.** We would like to express our gratitude to Dr. Yoneda and Dr. Huy for their assistance and technical support. This research is partly supported by NII Collaborative Research Grant.

# References

1. SWRL: A Semantic Web Rule Language Combining OWL and RuleML. http://www.w3.org/Submission/SWRL/
2. UDP - User Datagram Protocol. http://ipv6.com/articles/general/User-Datagram-Protocol.htm
3. Allemang, D., Hendler, J.A.: Semantic Web for the Working Ontologist - Effective Modeling in RDFS and OWL, 2nd edn. Morgan Kaufmann (2011)
4. Armand, A., Filliat, D., Ibañez-Guzman, J.: Ontology-based context awareness for driving assistance systems. In: IEEE Intelligent Vehicles Symposium, pp. 227–233 (2014)
5. Barbieri, D.F., Braga, D., Ceri, S., Grossniklaus, M.: An execution environment for C-SPARQL queries. In: 13th International Conference on Extending Database Technology, pp. 441–452. ACM (2010)
6. Bechhofer, S., van Harmelen, F., Hendler, J., Horrocks, I., McGuinness, D.L., Patel-Schneider, P.F., Stein, L.A.: OWL Web Ontology Language Reference. W3C Recommendation (2004). http://www.w3.org/TR/owl-ref/
7. Eskandarian, A.: Handbook of Intelligent Vehicles. Springer (2012)
8. Euzenat, J., Shvaiko, P.: Ontology Matching. Springer, Heidelberg (2007)
9. Gruber, T.R.: A Translation Approach to Portable Ontology Specifications. Knowledge Acquisition **5**(2), 199–220 (1993)
10. Gutierrez, C., Hurtado, C.A., Vaisman, A.: Introducing Time into RDF. IEEE Transactions on Knowledge and Data Engineering **19**(2), 207–218 (2007)
11. Heath, T., Bizer, C.: Linked Data: Evolving the Web into a Global Data Space. Morgan & Claypool (2011)
12. Hülsen, M., Zöllner, J.M., Weiss, C.: Traffic intersection situation description ontology for advanced driver assistance. In: IEEE Intelligent Vehicles Symposium, pp. 993–999. IEEE (2011)
13. Knublauch, H., Fergerson, R.W., Noy, N.F., Musen, M.A.: The Protégé OWL plugin: an open development environment for semantic web applications. In: McIlraith, S.A., Plexousakis, D., van Harmelen, F. (eds.) ISWC 2004. LNCS, vol. 3298, pp. 229–243. Springer, Heidelberg (2004)
14. Pollard, E., Morignot, P., Nashashibi, F.: An ontology-based model to determine the automation level of an automated vehicle for co-driving. In: 16th International Conference on Information Fusion, pp. 596–603 (2013)
15. Regele, R.: Using ontology-based traffic models for more efficient decision making of autonomous vehicles. In: 4th International Conference on Autonomic and Autonomous Systems, pp. 94–99. IEEE Computer Society (2008)
16. Robusto, C.C.: The Cosine-Haversine Formula. The American Mathematical Monthly **64**(1), 38–40 (1957)
17. Sirin, E., Parsia, B., Grau, B.C., Kalyanpur, A., Katz, Y.: Pellet: A Practical OWL-DL Reasoner. Journal of Web Semantics: Science, Services and Agents on the World Wide Web **5**(2) (2007)
18. Staab, S., Studer, R.: Handbook on Ontologies, 2nd edn. Springer, Heidelberg (2009)

19. Tideman, M., Janssen, S.J.: A simulation environment for developing intelligent headlight systems. In: IEEE Intelligent Vehicles Symposium, pp. 225–231 (2010)

20. Zhao, L., Ichise, R.: Instance-based ontological knowledge acquisition. In: Cimiano, P., Corcho, O., Presutti, V., Hollink, L., Rudolph, S. (eds.) ESWC 2013. LNCS, vol. 7882, pp. 155–169. Springer, Heidelberg (2013)

21. Zhao, L., Ichise, R.: Ontology Integration for Linked Data. Journal on Data Semantics 3(4), 237–254 (2014). doi:10.1007/s13740-014-0041-9. ISSN: 1861-2032

# Ontology Based Suggestion Distribution System

Tong Lee Chung(✉), Bin Xu, Xuanyu Yao, Qi Li, and Bozhi Yuan

Department of Computer Science and Technology,
Tsinghua University, Beijing, China
{tongleechung86,yaoxuanyu1993}@gmail.com, xubin@tsinghua.edu.cn,
{zhongguoliqi,lawby1229}@163.com

**Abstract.** The digitization of modern cities has brought cities to a new level. There are still many new areas yet to be discovered in this new ecosystem. Today, there is an urgent need for smarter cities to support the growing population. One particular problem is citizens do not know which city department to give their suggestions to. This paper presents a system for distributing suggestions from citizens to the right city officials based on ontology knowledge base. We use data from official websites to construct our ontology and do experiments with actual suggestions from citizens. The experiments show some promising results.

**Keywords:** Suggestion distribution · Ontology application · City management

## 1 Introduction

City digitization has been going on in China for over a decade and it is believed to be the right time to begin focusing on smarter cities [1]. This is an exciting time of the century where new intelligent systems is expected to appear in cities. But, with great opportunities comes great challenges. Today, city management has become more sophisticated then ever before. Two main reasons can be concluded, one is the growing population and the other is the increasing complexity of city departments. One particular problem is for city officials to hear the voices to citizens and for citizens to give suggestions to the right officials. Today, the main strategy for dealing with is this is set up different portal for different departments or distributing suggestions manually. There are many practical issues in these two types of approaches. Citizens do not know the actual duties and roles of these departments, and it is not feasible to understand all duties and roles just Lto give a suggestion. With the huge city population, it is unreliable for humans to handle the tedious job.

Many researchs have been carried out for smart city development, including using semantic technology and ontologies.There have been many research with constructing onotology for city management and e-governance [2] [3] [4] and in the task of data integration using ontologies [5] [6]. There have also been a number of reseach on boosting the . But using these ontology is a challenging task. This paper investigates the possiblity of using ontology in city managment and discuss some advantage of using ontology.

© Springer International Publishing Switzerland 2015
T. Supnithi et al.(Eds.): JIST 2014, LNCS 8943, pp. 414–421, 2015.
DOI: 10.1007/978-3-319-15615-6_31

We envision that computers will be able to assist humans in doing many repetitive jobs in city management. We also believe that ontology technology will play an important role in future smart cities. In this paper, we present a suggestion distribution system based on ontology knowledge base. We design a system takes a new suggestion as an imput and gives the department that is most likely in charge of relying to the suggestion as an output. Our system uses an ontology to search for the corresponding department. Our method of searching for the corresponding department differs from many modern classification methods where classification model is trained using another set of suggestions. Our approach is based on an ontology that describes the roles and duties of each department. Our system is flexible and robust to change. Our experiment shows that our system has satisfying results when compared to some baseline methods. We crawl data from official websites and implement an demonstration of the system.

The remaining of this paper is structured as following. Section 2 will give a formal definition of our problem and challenges will also be discussed.In section 3, the details of implementation of the system will be presented. Section 4 will look into data for the experiment and experiment results. We will finally conclude our paper and point out directions for future research.

## 2 Implementation of Suggestion Distribution System

The process of inferring a department from a suggestion can be divided into two different parts. The first part is matching an entity that is most related to the . The second part is inferring the corresponding data from the ontology. Figure 1 shows the inferring process in detail. When a suggestion arrives, it is first matched to the most related concept in the ontology. This step is for pruning down the search space so that trivial search is not necessary. The entities of the most related concept acts as the candidate entities that will be used for matching in the next step. In the second part, reasoning in the ontology is done to infer the most corresponding department.

### 2.1 Concept Matching

Given a new suggestion, we would like to first know which concept it most probably belong to. In an ontology, there are many concept, and each concept has a number of entities. Using this information, We propose an unsupervised approach to match the suggestions.Our idea is to capture the similarity using TF-IDF [7]. We need to first create a feature vector that is used to calculate the similarity. Using the idea of term frequency reverse document frequency from information retrieval, we treat each entities in a concept as a document. To deal with robustness of the entities, we also include all datatype label and it value as part of its content. For example, datatype property of personnel concept include name, position, and age. These can be very useful when matching suggestion, for example when a suggestion contain a position, it is more likely a personnel

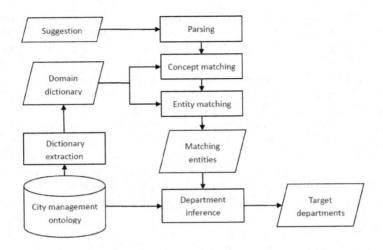

**Fig. 1.** Department inferring process of department from suggestion

concept. The matching weight of a word in a suggestion to a concept is calculated as:

$$\wp\left(w\right) = \log\left(tf_w + 1\right) \times log\left(idf_w\right) \tag{1}$$

where term frequency for a word $w$ in the suggestion is given by $f_w = f_w/\left|W\right|$, where $f_w$ is the frequency of the word $w$, and $\left|W\right|$ is the total number of words in the suggestion. Inverse document frequency $idf_w = \left|N\right|/n_w$, where $N$ is the total number of entities in a concept and $N_w$ is the number of documents that contain the word $w$. So given the matching weight of every word in the suggestion, the matching weight of a whole suggestion is given as the summation of all its words matching weight:

$$\mathcal{R}\left(sugg\right) = \frac{1}{count\left(w\right)} \times \sum_{w \in sugg} \wp\left(w\right) \tag{2}$$

As every word has a non-negative weight that has a maximum of 1, the summation of every word will not exceed one. The tfidf weight is consider high when the word appears a lot in suggestion and in multiple documents. During the matching phrase, we choose the concept that has the highest matching weight, but in reality, a suggestion may belong to multiple concepts. In the selecting stage, we set a treatment for this by first selecting the concept with highest match and then choosing concepts that satisfies $weight > weight_{max}$.

## 2.2   Entity Matching

After getting the matching concepts, we consider matching only entities that belongs to the matched concept. To deal with robustness problem, we use the same treatment as above and use entity label, datatype property, and value of datatype property as entity feature. We use cosine similarity to calculate the matching of the suggestion and the concept. Cosine similarity is calculated using:

$$cSim\,(sugg, concept) = \frac{sugg \times concept}{|sugg| \times |concept|} \tag{3}$$

where $|sugg|$ is a count of each word in the suggestion and $|concept|$ is the count of each word in the concept. $sugg \times concept$ is

### 2.3 Department Reasoning

After finding the most corresponding entity, we need search for a related department. We assume that an entity is related to all its department when one of its office has a certain role and this entity has a direct or in-direct upward link to this role. By up-ward link, we consider the basic ontology, because of the structure of the ontology, a department has some office, office has some roles, these roles will include some processes, processes are made up of activity and these activity can include some personnel, objects, documents. This is the down flow of city management. By upward link, we mean that the links go from lower layer entities to department entity. To customize our own rule set in OWLIM-Lite rule set, we need to modify one of the pre-defined rule set and add our own rules.

And finally, we have to return a weight that determines the relatedness of a suggestion and a department. The first feature we consider is similarity weight of suggestion and matched entity. Another feature we consider is the length of the evidence, the longer the evidence the less relatedness of the department. The final weight of a department to a suggestion if the summation of all evidence that lead to the department. The relatedness is calculated as below:

$$rel_{dep}\,(sugg) = \frac{1}{R} \sum_r cSim\,(sugg, concept) \times lenght\,(r) \tag{4}$$

Where $r$ is a route for a concept in matched concept set to its related department, $R$ is the total number of routes to a department $dep$. $lenght\,(r)$ is the lenght of the route from concept to suggestion.

### 2.4 UI

As it is shown in the Figure 2, our ontology based suggestion distribution system provides a easy and convenient platform for users to access and deal with suggestion data. A list of suggestions will take place on the left of the screen, and detailed information, including title, content, user's advice, the suggested category provided by the system, etc., will be shown when one of the suggestions in the list is once clicked. If the user wish to let the suggestion distributed to a specific bureau, he can choose among the radio boxs and press "Confirm" button. The system can automatically apply different themes to suggestions, with the undistributed in red, the correctly distributed in green and the wrongly distributed in yellow, which provides convenience distinguish suggestions of different status.

**Fig. 2.** User Interface

# 3  Experiment

## 3.1  Ontology Construction

Constructing the city management ontology was a challenging task for the experiment. We first design a ontology schema for city management, Figure 3 shows the ontology schema for city management. We define eight classes in city management, namely, Department, Office, Role, Process, Activity, Document, Person and Object.

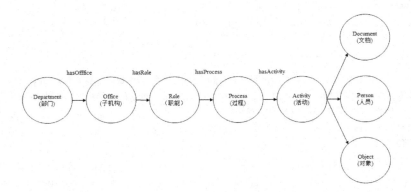

**Fig. 3.** Ontology schema for city management

We use web data to populate the city management ontology. We crawl city management guide from official websites and use it as source for constructing the city management ontology. We consider that the department and office are the corresponding department website and office. Then we populate the our ontology by using the management guide as process. Then by using document structure, we can further populate our ontology. By using the document hierarchical structure, we capture management process knowledge of a certain department. Altogether, we pull 105 processes to construct the management ontology for these five departments.

## 3.2   Testing Data

To get testing data to evaluate our system, we crawl the official website for suggestions. We consider the corresponding department as the target of the system. All of the suggestions that are posted on the web have a response. We clean our data using the response by removing the suggestion of which the response says 'go to some other department'. After data cleaning, we choose 30 suggestions from each department to evaluate our system.

## 3.3   Results

We test our system using the test suggestion described in the previous section. We use two meausres to evaluate our system. Top one match is when the department with the highest match is the target department. Since our system requires no training data, we use all data for testing. Another measure is top two match, where we consider that the system is correct when the target is in the two highest match. Table 1 shows the result of our system. Our system have a top one match of over 70 percent. From the results, we find that in a certain department the accurarcy is very low. After looking at the suggestions that we crawled, we find that the suggestion isn't relevant to any process in our ontology. The only way we can think of to fix this problem is to include related knowledge into our database.

**Table 1.** Experiment result for matching suggestions to department

|  | Correct | Total | Accurarcy |
|---|---|---|---|
| Overall | 66 | 90 | 73% |
| Traffic | 16 | 30 | 53% |
| Civil Affairs | 24 | 30 | 80% |
| Tourism | 26 | 30 | 86% |

We compare our method with some machine learning model with using knowledge from ontology. We choose Naive Bayes [8] classification method and Logistic Model [9] as our baseline method. We preform the a 5 fold cross-validation experiment using WEKA [10]. Table 2 shows that using ontology gives better performance compared to Naive Bayes and close to the performance of Logistic Model. Another important fact about our ontology based system is that it doesn't depend on training data. Machine learning models will out-perform our system when training data large enough. Using ontology, on the other hand will have better preformance if more knowledge is used.

**Table 2.** Experiment result for matching suggestions to department

|  | Accurarcy |
|---|---|
| Ontology System | 73% |
| NaiveBayes | 66% |
| Logistic | 75% |

We believe that there is more potentials in our system, we can use better matching model in the system. But this paper shows that ontology system can performs just as well as machine learning models. The advantange of using ontology is that knowledge is highly portable and can be used in many different applications, whereas new data has to be required in most machine learning models. The ontology models the behaviour of city management and can be used solved related problem. This high reusability of data cannot be compared by most learning models.

## 4    Conclusion

With the rapid development of computer science technology, smart city applications can reach a certain level of intelligences. Semantic web technology is a set of tool that is especially suited for AI applications. This paper aims to discover some possible domains where semantic web technology can enhance city management. We look into the task of distributing suggestions of citizens to a corresponding city department.We propose a ontology based suggestion distribution system. Our system uses a city management ontology to refer entities that is most relevant to the suggestion and infers the department using reasoning tools. The main contribution of this paper is to prove the possiblity of using ontology to assist humans in city management. We design a framework for implementing semantic technology in city management systems. Finally, we test our system using real life data from the web. The experiments show that our system have some promising results. The advantage of ontology is that the knowledge is portable and can be used in other systems and applications, using the same framework, we can implement other similar systems.

For our future work, we are looking into two different directions. One is to come up with more ideas that can enhance city management. The other direction is methods to populate the city management ontology using web data.

**Acknowledgements..** This work is supported by the China National High-Tech Project (863) under grant No SS2013AA010307

## References

1. Su, K., Li, J., Fu, H.: Smart city and the applications. In: 2011 International Conference on Electronics, Communications and Control (ICECC), pp. 1028–1031, September 2011

2. Goudos, S.K., Loutas, N., Peristeras, V., Tarabanis, K.: Public administration domain ontology for a semantic web services e-government framework. In: IEEE International Conference on Services Computing, 2007. SCC 2007, pp. 270–277, July 2007

3. Anthopoulos, L.G., Vakali, A.: Urban planning and smart cities: interrelations and reciprocities. In: Álvarez, F., Cleary, F., Daras, P., Domingue, J., Galis, A., Garcia, A., Gavras, A., Karnourskos, S., Krco, S., Li, M.-S., Lotz, V., Müller, H., Salvadori, E., Sassen, A.-M., Schaffers, H., Stiller, B., Tselentis, G., Turkama, P., Zahariadis, T. (eds.) FIA 2012. LNCS, vol. 7281, pp. 178–189. Springer, Heidelberg (2012)

4. Fraser, J., Adams, N., Macintosh, A., McKay-Hubbard, A., Lobo, T.P., Pardo, P.F., Martnez, R.C., Vallecillo, J.S.: Knowledge management applied to e-government services: the use of an ontology. In: Wimmer, Maria A. (ed.) KMGov 2003. LNCS (LNAI), vol. 2645, pp. 116–126. Springer, Heidelberg (2003)

5. Zhai, J., Jiang, J., Yu, Y., Li, J.: Ontology-based integrated information platform for digital city. In: 4th International Conference on Wireless Communications, Networking and Mobile Computing, WiCOM 2008, pp. 1–4, October 2008

6. Anand, N., Yang, M., van Duin, J.H.R., Tavasszy, L.: Genclon: An ontology for city logistics. Expert Systems with Applications **39**(15), 11944–11960 (2012)

7. Roelleke, T., Wang, J.: Tf-idf uncovered: A study of theories and probabilities. In: Proceedings of the 31st Annual International ACM SIGIR Conference on Research and Development in Information Retrieval, SIGIR 2008, pp. 435–442. ACM, New York (2008)

8. John, G.H., Langley, P.: Estimating continuous distributions in bayesian classifiers. In: Proceedings of the Eleventh Conference on Uncertainty in Artificial Intelligence, UAI 1995, pp. 338–345. Morgan Kaufmann Publishers Inc., San Francisco (1995)

9. Sumner, M., Frank, E., Hall, M.: Speeding up logistic model tree induction. In: Jorge, A.M., Torgo, L., Brazdil, P.B., Camacho, R., Gama, J. (eds.) PKDD 2005. LNCS (LNAI), vol. 3721, pp. 675–683. Springer, Heidelberg (2005)

10. Hall, M., Frank, E., Holmes, G., Pfahringer, B., Reutemann, P., Witten, I.H.: The weka data mining software: An update. SIGKDD Explor. Newsl. **11**(1), 10–18 (2009)

# Road Traffic Question Answering System Using Ontology

Napong Wanichayapong[1（✉）], Wasan Pattara-Atikom[1], and Ratchata Peachavanish[2]

[1] National Electronics and Computer Technology Center,
112 Phahonyothin Road, Khlong Nueng, Khlong Luang 12120,
Pathum Thani, Thailand
{napong.wanichayapong,wasan}@nectec.or.th
[2] Department of Computer Science,
Thammasat University, Phra Nakhon Si
Ayutthaya, Thailand
rp@cs.tu.ac.th

**Abstract.** Many people use social media to report and receive road traffic information, e.g., car accidents and congestions. We have implemented a Twitter-based traffic-related information reposting (retweeting) system, which users usually referred to as @traffy. To improve on our works, we propose an ontology-based Thai-language question answering system that gathers real-time traffic data from Twitter. The data collected are converted into traffic incident knowledge of *what* is happening and *where* it is happening. The system can then infer which points of interest (POIs) are affected by the incidents. Users can use natural (Thai) language to query the system against the ontology to receive traffic-related information. The system is currently deployed for demonstration on the web and developers can utilize it via REST API.

**Keywords:** Intelligent Transport System · Traffic information · Question answering system · Ontology · Twitter

## 1    Introduction

Social media are used by many people to both report and receive real-time traffic information, e.g., accidents and congestions. We have implemented a Twitter-based traffic reposting system [1] that collects tweets from well-known twitter users and keyword searches. To improve on our system, we propose to use ontology to incorporate semantics into our system that allows for question answering.

Ontologies can be applied for many purposes in Intelligent Transport System (ITS). Some applied ontology to geospatial application [2][3], others applied to traffic information and integration [4]-[6] and traffic management [7][9].

In our propose traffic question answering system, we use ontology to semantically identify elements of traffic report tweets and infer points of interest (POIs) that are affected by the reported incidents.

© Springer International Publishing Switzerland 2015
T. Supnithi et al. (Eds.): JIST 2014, LNCS 8943, pp. 422–427, 2015.
DOI: 10.1007/978-3-319-15615-6_32

# 2     Ontology Design for Traffic Question Answering System

The ontology was an improved version of what we previously proposed [9] and is shown in Fig. 1.

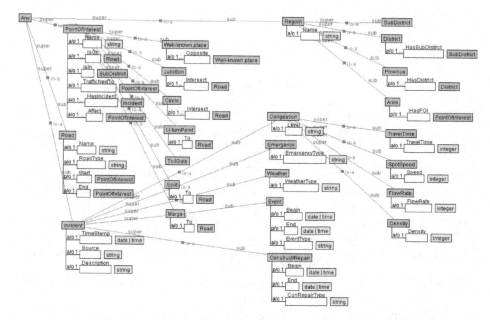

**Fig. 1.** Ontology for traffic question answering system

## 2.1     Definitions

Some important classes of the ontology are described as follows. A **Road** is a pathway for vehicles. An alley is also a Road. A frontage road running parallel to the main road is considered a separate entity from the main road because they do not share the same traffic. A single two-way road is also considered to be two separate entities for the same reason. A **Point of Interest (POI)** is a location that people refer to by a name. A **Region** is an officially (governmental) designated geographic area within a regional hierarchy. An **Area** is a Region that encompasses a set of POIs that people named, usually unofficially.

An **Incident** affects road users and/or road condition. It has at least one POI, Road, or Region, and at least one of any incident cue word. Five types of incidents are:

- **Event** - affecting road users usually before and after the event takes place, e.g., a concert.
- **Construction/Repair/Malfunction** - affecting road users usually during the activity, e.g., light rail construction.

- **Emergency** - needing immediate response, e.g., car accident, building fire.
- **Weather** - natural phenomena affecting wide-area region, e.g., rain, fog.
- **Congestion** - regular traffic congestion classified into 3 levels of severity, i.e., high, medium, and low.

## 2.2    Inference Rules

The ontology has inference rules shown in Table 1.

**Table 1.** Inference Rules

| IsOn | If a Road is mentioned, POIs that have IsOn relation with this Road must be mentioned. |
|---|---|
| IsIn | If a SubDistrict is mentioned, POIs that have IsIn relation with this SubDistrict must be mentioned. |
| Affect | If a POI is mentioned directly (not by Affect relation inference), other POIs that this POI has Affect relation with must be mentioned. |
| HasSubDistrict | If a District is mentioned, SubDistricts that have HasSubDistrict relation with this District must be mentioned. |
| HasDistrict | If a Province is mentioned, Districts that have HasDistrict relation with this Province must be mentioned. |
| HasPOI | If an Area is mentioned, POIs that have HasPOI relation with this Area must be mentioned. |
| TrafficNextTo | If a part of a Road is mention as Start POI and End POI (of Incident), POIs that can use TrafficNextTo relation from the Start POI recursively to the End POI must be mentioned. |
| Opposite | If a Well-known place is mentioned, other Well-known places that this Well-known place is Opposite to must be mentioned. |

A POI that has **TrafficNextTo** relation with another POI implies that driver can drive from the first POI to the second POI as both are on the same road. A Well-known place that has **Opposite** relation with another Well-known place implies that by standing in front of the first one, user can see the second one located on the opposite side of the road.

# 3    Process

The system uses the ontology as the knowledge base for interpreting text input strings from tweets. A tweet can either be an incident report, or a question asking for traffic information.

**Fig. 2.** This traffic question answering system process

The first step is **Tweet Acquisition.** We use multiple methods to obtain relevant traffic-related tweets. First, we use predefined keywords as query terms against Twitter's search API to obtain tweets. Second, we use Twitter's *user_timeline* API to obtain tweets from well-known traffic reporters, e.g., traffic radio station Twitter accounts. Lastly, we use Twitter's *mention_timeline* API to get tweets that mention @traffy, which are very likely to be relevant to us. Next, we perform **Preprocessing** as Thai language writing does not have spacing between individual words. We use Lexto[10] to tokenized text strings into individual words, followed by word filtering to remove stop words and other irrelevant words. Finally, we convert synonyms and abbreviations into full official words.

Tweet interpretation starts with **Tagging** where we use all words in ontology and cue word dictionaries to identify the type of each word by string matching, i.e., road, start POI, and whether the tweet is a question. Words that can be identified as more than one type will be tagged all those types. The process is followed by **POI Inferencing** where we use ontology inference rules to infer which POIs are affected by the tweet and tag them accordingly. **Incident Class Classification** is then performed to determine the type of incident.

If the tweet is an incident report, **Information Converting** is performed. The system generates combinations of POI tags and Incident tags. For example, consider an incident report tweet:

*"Car crash in front of Paragon. It causes long traffic jam from Pathumwan junction to Chaloem Phao junction"*

The process tokenizes the string and converts informal word *Paragon* into *Siam Paragon Shopping Center*. Official names are identified – *Pathumwan Junction* and *Chaloem Phao Junction*. Cue words are identified, i.e., *long traffic jam, from, to, cause, car crash, in front*. The POI Inferencing identifies the two junctions as the Start POI and End POI, which makes up a part of a road. The TrafficNextTo inference rule determines that the Siam SkyTrain Station and the Siam Paragon Shopping Center are both affected by the incident. The Incident Class is identified from the cue words and the system, in the **Store Information** step, stores all combinations of POI tags and Incident tags into the database. In this case, the resulting tags are *Pathumwan Junction-Congested, Siam SkyTrain Station-Congested, Siam Paragon Shopping Center-Congested, Chaloem Phao Junction-Congested, Pathumwan Junction-Car Crash, Siam SkyTrain Station-Car Crash, Siam Paragon Shopping Center-Car Crash, Chaloem Phao junction-Car Crash*.

The same Tagging, POI Inferencing, and Incident Class Classification processes are performed if the tweet is a question. The question interpretation in **Has Question Tag** step is the method that we proposed in An Ontology Design for Traffic Incident

Q&A System [9]. Then, **Information Acquisition** process is performed by generating database queries using tags and related POIs, retrieving all related incidents that has been tagged in the same POIs as the tweet and happened in past hour (time frame is adjustable). The **Summarization** step summarizes incidents for the POIs relevant to the question. For traffic congestion, we use a weighting technique to estimate the level of congestion as there may be conflicting reports due to rapidly changing traffic condition. We use majority voting to determine the congestion level and we give more weight to the more recent reports.

## 4      Evaluation

The system uses incident classification method that we proposed in Social-based Traffic Information Extraction and Classification [1] which has 76.85% accuracy for one-incident-one-POI items and 93.23% accuracy for one-incident-one-part-of-road items.

In this paper, we re-evaluate question interpretation method that we proposed in An Ontology Design for Traffic Incident Q&A System [9] then evaluate the process of information acquisition and summarization. We used 300 mentions @traffy tweets obtain during a week in February 2014 as input. Our question interpretation method can detect question with accuracy of 85%, precision of 67.18% and recall of 97.78%. For Information Acquisition and Summarization, we used 88 tweets that question interpretation method and we identified as question tweets. At the time we evaluate method, there are 11,174 records in knowledge base and 8,847 of them are implied. The system can acquire information from the knowledge base that store incident that happen in past hour and correctly summarize the answer with accuracy of 76.92%, precision of 79.49% and recall of 75.61%.

## 5      Discussion and Future Work

There are many further improvements that can be done to the system. More data sources can be used, i.e. Facebook, to obtain more coverage. Duplicated information should be detected and managed. For improving the summarization process, the system should detect incident that will happen in the future or has happened in the past. Summarization process accuracy can be improve by analyzing incidents that cover more than one road.

## 6      Conclusions

We propose a traffic question answering system that uses ontology to infer POIs that affect by incidents. This system collects tweets from Twitter and answer user by tweeting automatically. Users can try our service via web demo and Twitter. In addition, developers can user our service via REST API.

# References

1. Wanichayapong, N., Pruthipunyaskul, W., Pattara-Atikom, W., Chaovalit, P.: Social-based traffic information extraction and classification. In: 2011 11th International Conference on ITS Telecommunications (ITST), pp. 107–112. IEEE (2011)
2. Bishr, Y., Kuhn, W.: Ontology-based modelling of geospatial information. In: 3rd AGILE Conference on Geographic Information Science, pp. 25–27. Citeseer, May 2000
3. Wiegand, N., García, C.: A Task-Based Ontology Approach to Automate Geospatial Data Retrieval. Transactions in GIS. 11, 355–376 (2007)
4. Zhai, J., Chen, Y., Yu, Y., Liang, Y., Jiang, J.: Fuzzy semantic retrieval for traffic information based on fuzzy ontology and rdf on the semantic web. Journal of Software 4, 758–765 (2009)
5. Yang, W., Zhu, Q.: Ontology-based semantic fusion of traffic information. In: 2012 International Conference on Computer Science & Service System (CSSS), pp. 769–772. IEEE (2012)
6. Samper, J.J., Tomás, V.R., Martinez, J.J., van den Berg, L.: An ontological infrastructure for traveller information systems. In: Intelligent Transportation Systems Conference, ITSC 2006, pp. 1197–1202. IEEE (2006)
7. Dongli, Y., Suihua, W., Ailing, Z.: Traffic accidents knowledge management based on ontology. In: Sixth International Conference on Fuzzy Systems and Knowledge Discovery, FSKD 2009, pp. 447–449. IEEE (2009)
8. Lécué, F., Schumann, A., Sbodio, M.L.: Applying Semantic Web Technologies for Diagnosing Road Traffic Congestions. In: Cudré-Mauroux, P., et al. (eds.) ISWC 2012, Part II. LNCS, vol. 7650, pp. 114–130. Springer, Heidelberg (2012)
9. Wanichayapong, N., Peachavanish, R., Pattara-Atikom, W.: An ontology design for traffic incident Q&A system. In: 20th ITS World Congress 2013 (2013)
10. Sansan Thai Lexeme Tokenizer by NECTEC. http://www.sansarn.com/

# Author Index

Printed in the United States
By Bookmasters